Thermoluminescence Dating

STUDIES IN ARCHAEOLOGICAL SCIENCE

Founding Editor
Geoffrey W. Dimbleby
Emeritus Professor
Institute of Archaeology
University of London
London, England

Consulting Editor
Donald K. Grayson
Burke Memorial Museum
University of Washington
Seattle, Washington

Chaplin: The Study of Animal Bones from Archaeological Sites
Reed: Ancient Skins, Parchments and Leathers
Tite: Methods of Physical Examination in Archaeology
Evans: Land Snails in Archaeology
Limbrey: Soil Science and Archaeology
Casteel: Fish Remains in Archaeology
Harris: Principles of Archaeological Stratigraphy
Baker/Brothwell: Animal Diseases in Archaeology
Shepherd: Prehistoric Mining and Allied Industries
Dickson: Australian Stone Hatchets
Frank: Glass Archaeology
Grayson: Quantitative Zooarchaeology
Dimbleby: The Palynology of Archaeological Sites
Aitken: Thermoluminescence Dating

Thermoluminescence Dating

M. J. AITKEN

*Research Laboratory for Archaeology
and the History of Art
Oxford University
England*

1985

ACADEMIC PRESS

Harcourt Brace Jovanovich, Publishers
London   Orlando   San Diego   New York
Austin   Montreal   Sydney   Tokyo   Toronto

COPYRIGHT © 1985 BY ACADEMIC PRESS INC. (LONDON) LTD.
ALL RIGHTS RESERVED.
NO PART OF THIS PUBLICATION MAY BE REPRODUCED OR
TRANSMITTED IN ANY FORM OR BY ANY MEANS, ELECTRONIC
OR MECHANICAL, INCLUDING PHOTOCOPY, RECORDING, OR
ANY INFORMATION STORAGE AND RETRIEVAL SYSTEM, WITHOUT
PERMISSION IN WRITING FROM THE PUBLISHER.

ACADEMIC PRESS INC. (LONDON) LTD.
24–28 Oval Road
LONDON NW1 7DX

United States Edition published by
ACADEMIC PRESS, INC.
Orlando, Florida 32887

BRITISH LIBRARY CATALOGUING IN PUBLICATION DATA
Aitken, M. J.
    Thermoluminescence dating.—(Studies in
archaeological science)
    1. Archaeological dating   2. Thermoluminescence
dating
    I. Title   II. Series
    930.1′028′5   CC78.8

LIBRARY OF CONGRESS CATALOGING-IN-PUBLICATION DATA
Aitken, M. J. (Martin Jim)
    Thermoluminescence dating.

    (Studies in archaeological science)
    Bibliography: p.
    Includes index.
    1. Thermoluminescence dating.   I. Title.   II. Series.
CC78.8.A37   1985   930.1′028′5   84-24326
ISBN 0-12-046380-6 (alk. paper)
ISBN 0-12-046381-4 (paperback)

PRINTED IN THE UNITED STATES OF AMERICA

85 86 87 88   9 8 7 6 5 4 3 2 1

CONTENTS

Preface   ix
Units and Conversion Factors   xi

1. Introduction   1

    1.1 Starting Point   1
    1.2 Historical Background; Other Materials   2
    1.3 Measurement of Thermoluminescence   4
    1.4 Stability of the Thermoluminescence Record; the Plateau Test; Evaluation of the Paleodose   7
    1.5 Radioactivity and Annual Dose   10

2. Basic Pottery Dating   17

    2.1 The Quartz Inclusion Technique   17
    2.2 The Fine-Grain Technique   24
    2.3 Subtraction Dating   29
    2.4 Accuracy and Error Limits   30
    2.5 Authenticity Testing of Ceramic Art Objects   32

3. Thermoluminescence   41

    3.1 The Thermoluminescence Process   41
    3.2 Other Types of Thermoluminescence; Prompt Luminescence; Spurious Thermoluminescence   45
    3.3 Stability of the Thermoluminescence Record   47
    3.4 Anomalous Fading   54

4. Natural Radioactivity: The Annual Dose   61

    4.1 Natural Radioactivity   62
    4.2 Annual Dose   66
    4.3 Measurement   82
    4.4 Summing Up   109

5. Artificial Irradiation   113

   5.1 Gamma Irradiation   115
   5.2 Beta Irradiation   118
   5.3 Alpha Irradiation   128
   5.4 Supralinearity, Sensitization, and Saturation   135
   5.5 Dependence on Dose-Rate   140
   5.6 Dependence on Temperature of Irradiation   141
   5.7 Operator Protection   142

6. Special Methods   153

   6.1 Pre-Dose Dating   153
   6.2 Phototransfer   168
   6.3 Zircon Dating   172
   6.4 Kinetic Methods   178
   6.5 Feldspar Dating   180
   6.6 Subtraction Dating   184
   6.7 Gamma-Thermoluminescence   186

7. Beyond Pottery   191

   7.1 Burnt Flint   191
   7.2 Burnt Stones   196
   7.3 Volcanic Eruptions   197
   7.4 Meteorites and Craters, Metallurgical Slag, and Glass   200
   7.5 Stalagmitic Calcite (Unburnt)   202
   7.6 Bone and Shell; Use of Electron Spin Resonance   210
   7.7 Thermoluminescence Dating of the Lake Mungo and
       Laschamp Geomagnetic Excursions   213

8. Sediment Dating—Solar Resetting   219

   8.1 Introduction   219
   8.2 Methods for Evaluation of Paleodose   221
   8.3 Laboratory Procedures   227
   8.4 Stability; Age Range   235
   8.5 Annual Dose   236
   8.6 Application   236

APPENDIX A: The Age Equation   239
APPENDIX B: Age Evaluation and Assessment of Error Limits   241
APPENDIX C: Attenuation of Annual Dose within Grains   252
APPENDIX D: Sample Collection Instructions   264
APPENDIX E: Kinetic Studies and Evaluation of Trapped-
            Electron Lifetimes   269

APPENDIX F:	Anomalous Fading	274
APPENDIX G:	Annual Dose Evaluation: Summary of Radioactivity Data	282
APPENDIX H:	Gamma and Beta Gradients	289
APPENDIX I:	Cosmic Ray Dose	297
APPENDIX J:	Alpha Counting: Derivation of Formulae; Ranges; Standards	299
APPENDIX K:	The a-Value System of Assessing the Alpha Particle Contribution	308
APPENDIX L:	The Portable 4-Channel Gamma Spectrometer	318
APPENDIX M:	Miscellaneous Phosphor Notes	325

References 331
Index 353

PREFACE

When I put pen to paper in the summer of 1982, uppermost in my mind was the need to provide new practitioners with an introduction to the theory and practice of thermoluminescence (TL) dating, particularly those without a strong background in physics. Assimilating the corpus of specialist papers which at present define the theory and practice is a daunting task—in strong contrast with the easy initial step of grasping the basic notions. But I also had in mind those archaeologists, anthropologists, Quaternary geologists and physical geographers who, utilizing TL dates in their research, deem it prudent to have some insight into the method's difficulties and limitations. These readers will have already found help in Stuart Fleming's book, *Scientific Techniques in Archaeology,* but in the years that have elapsed since its publication there have been substantial advances, particularly with respect to materials other than pottery, such as burnt flint and unburnt sediment.

In the course of writing, it soon became evident that there was no easy line to be drawn which separated off the realm of the specialist researcher. Increasingly it seemed to me that it was wrong to draw a line at all; ultimately the TL practitioner should be fully cognizant of all that is relevant to the particular application being made. Much more than in other dating techniques, an important factor in reliability is the extent to which specialist expertise can be brought to bear on the actual measurement process. For this reason I have included numerous technical notes and substantial appendices, while keeping the main text fairly straightforward so as not to compromise the interests of the raw recruit. It was impossible to write without making development in some of the topics concerned, and so I hope that specialist researchers too will find some of the appendices of interest. I must leave it to readers to judge how successful I have been in weaving the various strands together into a coherent fabric; I am only too conscious of deficiencies.

To some extent I am a scribe writing on behalf of all those who have worked to bring TL dating to its present position. Inevitably, I am biassed towards research in which I have been involved at Oxford, and I apologize to those whose work has not been given adequate treatment: this is not because I undervalue it, but because I thought it more useful to expand in areas in which I have better competence. I have drawn heavily on the unpublished theses of the research students whom I have had the good fortune to supervise over the years; what they have taught me is impressive. I owe them a great debt, together with all of the other TL workers at Oxford: Michael Tite, Jeanette Reid (née Waine), John Alldred, Stuart Fleming, Joan Zimmerman (née Thompson), David Zimmerman, Joan Huxtable, Ann Wintle, Doreen Stoneham, Elisabeth Whittle (née Sampson), Sheridan Bowman, Vyoma Desai, Ian Bailiff, Sheelagh Mobbs, Andrew Murray, Gill Spencer (née Bussell), Nick Debenham, Richard Templer, Iain Watson, Chris Gaffney, Peter Clark, Vicky Griffiths, Adrian Allsop and Scott Wheeler.

I am particularly grateful to several of the above for reading and discussing sections of the text, and also for help in these and other ways to Mohan Francis, Ed Haskell, Leif Løvborg, Vagn Mejdahl, Ross Munro, Helen Rendell and Rafael Visocekas. The task of translating my handwriting, and alterations, into impeccable typescript was undertaken largely by Sarah Loughman and Elena Tiffert, with substantial help also by Heather Holloway and Kay Wood. I am highly appreciative of their expertise and of their tolerance, too, as well as of the promptitude and competence of Judith Takacs in drawing the diagrams.

Finally, with respect to research carried out at Oxford, I would like to acknowledge the debt we all owe to the technical staff of the Laboratory, in particular to Dave Seeley for his excellent and patient construction of much of the apparatus involved.

Martin Aitken

UNITS AND CONVERSION FACTORS

Becquerel: 1 Bq = 1 decay/sec
Curie: 1 Ci = 37 × 10^9 Bq
Gray: 1 Gy = 1 J/kg
Joule: 1 J = 10^7 erg
Rad: 100 rad = 1 Gy
Roentgen: in air exposed to 1 R of X-rays the absorbed dose is approximately 0.87 rad
Sievert: 1 Sv = 100 rem; for radiation having a biological effectiveness of unity, 1 Sv = 1 Gy
Electron-volt: 1 eV = 1.602 × 10^{-19} J
Year: The letter 'a', from the French *an,* is used when specifying annual dose, following practice in most thermoluminescence laboratories. However, for radioactive half-lives, 'yr' has been retained, following practice in many journals.

Avogadro's number = 6.02 × 10^{23} (equal to the number of atoms in *A* grams of an element, where *A* is the atomic weight)

Boltzmann's constant: k = 1.38 × 10^{-23} J/°K (at 17°C, kT = 0.025 eV, where T is the absolute temperature in degrees Kelvin)

Lifetime = (Half-life)/0.693

Wavelength (nanometre) × Quantum energy (eV) = 1240

CHAPTER 1

INTRODUCTION

1.1 Starting Point

If a ground-up sample of ancient pottery, brick, tile, or terracotta figure is heated rapidly to 500°C, there is a weak but measurable emission of light—see Figure 1.1, curve (a). For a second heating of the same sample, the emission—curve (b)—consists only of thermal radiation (i.e., 'red-hot glow' or incandescence; the term 'black-body' is also used). The extra light emitted in the first heating is thermoluminescence that comes from constituent minerals in the pottery and is due to the effect on them of prolonged exposure to the weak flux of nuclear radiation emitted by radioactive impurities in the pottery and the surrounding burial soil: potassium-40, thorium, and uranium, at concentrations of a few parts per million. These radioisotopes have long half-lives (10^9 years or more), and thus the radiation flux is constant; hence the amount of thermoluminescence is proportional[1] to the time that has elapsed since the pottery was fired by ancient man. The act of firing 'drains' the minerals of thermoluminescence acquired during geological times, thereby setting the clock to zero. Hence by measuring both the sensitivity of the sample to acquiring thermoluminescence (by exposure to a calibrated radioisotope source) and the radioactive content of the clay and surrounding soil it is possible to calculate the age, at least in principle. In its basic form the age relation is given by Equation 1.1.

$$\text{age} = \frac{\text{(archaeologically acquired thermoluminescence)}}{\binom{\text{thermoluminescence per}}{\text{unit dose of radiation}} \times \binom{\text{annual dose from}}{\text{radioactive impurities}}}. \quad (1.1)$$

In practice there are many complications. To start with these complications are ignored: this first chapter introduces the main features of thermoluminescence dating, and the second chapter deals as straightforwardly as possible with the two basic techniques that have been developed for baked clay. The technological framework of thermoluminescence dating was developed for this material and is best understood in terms of it even though the reader's primary interest may be in one of the other materials to which the method has now been extended. These other materials are dealt with in later chapters after discussions of some details of the thermoluminescence process (Chapter 3), natural radioactivity and evaluation of annual dose (Chapter 4), use of radioisotope sources for measurement of thermoluminescence sensitivity (Chapter 5), and some special methods used in pottery dating (Chapter 6).

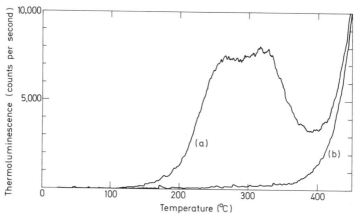

Figure 1.1 Thermoluminescence glow-curve observed from small sample taken from a terracotta statuette (measurement by D. Stoneham). Curve (a) shows the light emission observed during the first heating (at a rate of 20°C/sec) and curve (b) the light observed during a second heating. The latter is the red-hot glow, or incandescence, that occurs whenever a sample is heated, but during the first heating there is in addition substantial emission of thermoluminescence—resulting from exposure during antiquity to the weak flux of ionizing radiation emitted by natural radioactivity in the clay and soil.

1.2 Historical Background; Other Materials

Luminescence phenomena have been studied by scientists for several centuries. In a paper read before the Royal Society on 28 October 1663,

1.2 Historical Background

Robert Boyle gave an account of some observations he had made the previous night 'About a Diamond that Shines in the Dark'. Among other pertinent observations he reported 'I also brought it to some kind of Glimmering Light, by taking into Bed with me, and holding it a good while upon a warm part of my Naked Body'. Elsewhere he comments that it was an extremely sensitive diamond because 'as you know, my constitution is not of the hottest'. This was one of the earliest observations of thermoluminescence, which is a phenomenon exhibited to varying degrees by many minerals. Essentially it is the emission of light when a substance is heated, this light being in addition to ordinary red-hot glow and usually occurring at a less elevated temperature—hence the name 'cold light'. It represents the release of energy which has been stored in the crystal lattice of the mineral, this energy being in the form of electrons that have been trapped at defects in the lattice; the electrons are available for trapping in the first place as a result of exposure to ionizing radiation (though, as we shall see, there are other unwanted agents that can produce the same effect).

Until the advent in the 1940s of the photomultiplier as a very sensitive detector of light, the only use of thermoluminescence was as a geological tool in mineral identification, though it was, of course, a phenomenon studied by physicists and chemists. Then, beginning in the early 1950s its use for measuring exposure to nuclear radiation was developed, stemming from the work of Farrington Daniels at the University of Wisconsin, where its use for geological and archaeological age determination was also suggested (Daniels, Boyd and Saunders, 1953). The thermoluminescence from ancient pottery was first detected at the Universities of Bern (Grogler, Houtermans and Stauffer, 1960) and California (Kennedy and Knopff, 1960). Subsequently, during the 1960s thermoluminescence was developed for archaeological dating at Oxford (Aitken, Tite and Reid, 1964; Aitken, Zimmerman and Fleming, 1968), Kyoto (Ichikawa, 1965), Wisconsin (Mazess and Zimmerman, 1966), Philadelphia (Ralph and Han, 1966), and Denmark (Mejdahl, 1969). Now, around the world there are upwards of 40 laboratories involved in the application of thermoluminescence dating to archaeology and geology or in using it for testing the authenticity of art ceramics–a very powerful role indeed. There is a newsletter, *Ancient TL*, and a specialist seminar of thermoluminescence dating practitioners is held every two or three years.[2]

An early extension of thermoluminescence dating to material other than baked clay was made at the University of Birmingham by Göksu *et al.* (1974); this extension was to burnt flint and potentially it took the method into the largely uncharted paleolithic ages well beyond the range of radiocarbon; the same is true for the later extension to burnt stone and stalagmitic calcite. Volcanic lava was an obvious candidate material with respect to geology, though it has proved unexpectedly difficult in practice (Wintle 1973; Guérin and Valladas, 1980). Extension to various types of sediment, beginning (e.g., Morozov, 1968) with work in the U.S.S.R., was an important development for geology as well as for archaeology, which is currently gaining momentum; windblown and some waterlain sediments can be dated, and it appears that the thermoluminescence clock is set to zero by the 'bleaching' effect of sunlight (Wintle and Huntley, 1979, 1980). In all of these applications the upper dating limit is likely to remain within the last million years; this limit is much less for some types of material and is dependent on the characteristics of the sample itself. Thermoluminescence is also used in elucidating the history of meteorites and lunar material.

It may be mentioned here that an alternative way of detecting trapped electrons is by electron spin resonance (ESR), and dating by this technique is now being developed. This is discussed further in Section 7.6; one advantage of ESR over thermoluminescence is that heating is not an essential part of the measurement, so that samples that decompose on heating, such as bone and shell, can be tackled. The trapped electron population can also be measured by thermally stimulated electron emission[3] (TSEE) and by photostimulated luminescence (PSL); as mentioned in Section 8.2.6, optical dating, based on the latter, has strong advantages, particularly for sediments.

1.3 Measurement of Thermoluminescence

1.3.1 THE PHOTOMULTIPLIER AND THE GLOW-OVEN

There are some geological minerals from which the thermoluminescence is bright enough to be seen with the naked eye, but in dating we are dealing with light levels lower by many orders of magnitude. As indicated above, it was the development of the photomultiplier that made thermoluminescence dating a practical possibility. When light is incident upon the photocathode (see Figure 1.2), there is emission of electrons by the photoelectric material (e.g., of bialkali type such as potassium–caesium) with which the photocathode is coated. These electrons are attracted by the positive voltage on the first dynode, and because of the coating on the dynode (e.g., antimony–caesium) an average of between two and three electrons is emitted for each electron striking it. These electrons are attracted to the second dynode, where further multiplication occurs, and so on. For a photomultiplier having, say, 10 dynodes, several million electrons reach the anode for each electron leaving the photocathode.

1.3 Measurement of Thermoluminescence

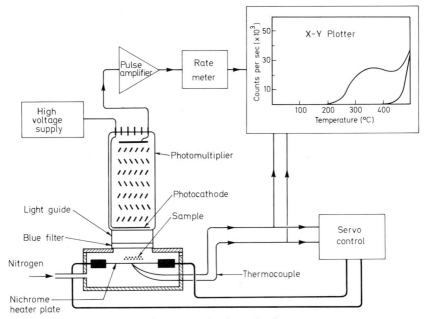

Figure 1.2 Diagram of apparatus for thermoluminescence measurement.

Thus at the anode there is a succession of electrical pulses, each corresponding to the ejection of a photoelectron from the photocathode due to the arrival of a photon of light. However, not every photon gives rise to a photoelectron; the success rate, or quantum efficiency, is between 0 and 25% depending on the wavelength. The anode pulses are amplified and fed to a rate meter, the output of which is proportional to the thermoluminescence intensity—the rate of arrival of photons at the photocathode. This output is fed to the Y axis of an X–Y recorder. Of course, with the advent of microprocessors much more sophisticated systems[4] are now available that permit convenience in data storage and manipulation. However, the photomultiplier remains the vital component in the measurement of thermoluminescence, and no amount of expenditure on silicon chips obviates the need for a high-quality device.

In the basic equipment used for many years in the author's laboratory, the sample is heated electrically by a $\frac{1}{2}$-mm strip of nichrome (or tantalum); it is usually in powder form and either spread directly on the nichrome (henceforth referred to as 'plate' rather than strip—typically it is $2\frac{1}{2}$ cm wide and 4 cm long) or carried on a $\frac{1}{2}$-mm thick, 1-cm diameter aluminium disc which is placed on the plate; discs of stainless steel and silver are also used, as well as planchettes instead of discs. The temperature is measured by a thermocouple, usually chromel–alumel, which is spot-welded onto the underside of the plate. This thermocouple feeds the X-axis of the recorder, thereby enabling the glow-curve, that is, the plot of thermoluminescence versus temperature, to be obtained directly as the heating proceeds. The thermocouple output also goes to a servo system which controls the heater current through the plate.

A rapid heating rate is usual because the intensity of the thermoluminescence (photons per second) is proportional to it, whereas the thermal signal, the red-hot glow, is independent of it. Thus for dim samples there is a strong advantage in having rapid heating, up to 20°C/sec being practical.

It is vital that the actual rate of rise in the temperature of the sample be reproducible from run to run, and this is difficult to achieve for fast rates because of poor thermal contact between sample and plate, particularly if the sample is being carried on a disc or in a planchette. Recently, techniques for direct heating of the sample by means of a short pulse of infrared laser beam have been developed, giving very high heating rates of the order of 10^5 °C/sec (Braunlich *et al.*, 1982).

Statistical fluctuations in the photon rate are a serious limitation for dim samples, and so the solid angle of light collection should be large—π steradians is practical, that is, if the thermoluminescence photons were emitted equally in all directions over a hemisphere, half of them would be collected. To avoid risk that the hot plate causes warming of the photocathode, it is usual to interpose a quartz light guide. Use of an infrared rejecting filter (e.g., Chance Pilkington HA3) is important in reduction of the thermal signal, and for the same reason the photocathode should be of the bialkali type, making it insensitive to wavelengths beyond 600 nm. Further reduction of the thermal signal is achieved by means of a colour filter, typically the Corning 7-59, which transmits only violet light, or the Corning 5-60, which transmits blue/violet; the colour of the thermoluminescence being measured is obviously an important consideration here. The need to reject the thermal signal, which has a significant green component by the time the temperature reaches 500°C, does mean that only samples with thermoluminescence emission reaching to blue and shorter wavelengths can be dated.

1.3.2 SPURIOUS THERMOLUMINESCENCE

As is discussed further in Section 3.2, nuclear radiation is not the only agent that produces latent thermoluminescence. To suppress non-radiation-induced thermoluminescence ('spurious' thermoluminescence), it is essential that the oven be filled with a dry inert gas such as nitrogen or argon, with an impurity content of oxygen and water of less than a few parts per million. Traces of air in the oven would give rise to a strong signal from the surface of the sample grains that would be likely to dominate the 'true', radiation-induced, thermoluminescence. Thus before filling the oven with ultra-high-purity inert gas, it should be evacuated to less than 0.1 Torr (pressure as low as 0.02 Torr, i.e., 2.7 Pa, is necessary with many samples), and it is necessary to have in it a dessicant such as phosphorus pentoxide.[5]

The need to achieve suppression of spurious thermoluminescence cannot be too highly stressed. Otherwise, the age may be substantially overestimated and a fake art ceramic may even be mistaken for genuine. Fortunately the plateau test described in the next section is a good indicator as to whether adequate suppression has been achieved.

Because purity of the gas in the oven is so important, it might be thought that taking the glow-curve with the sample under vacuum might be the solution. Unfortunately this is not practical if the sample is carried on a disc or in a planchette because it likely is heated predominantly by gas conduction; in vacuum there is likely to be a large, non-reproducible thermal lag between the plate and the sample.

1.4 Stability of the Thermoluminescence Record; The Plateau Test; Evaluation of the Paleodose

Although the glow-curve shown in Figure 1.1 is a smooth continuum, it is really composed of a number of overlapping peaks. For a given type of electron trap, the glow-curve is a single peak about 50°C in width; the temperature region in which the peak occurs depends on how securely the electrons are held—for 'deep traps' the degree of thermal vibration of the crystal lattice required to eject electrons is greater than for 'shallow traps'. Similarly, during burial the lifetime of electrons in deep traps is greater than that of electrons in shallow traps, as is discussed in Section 3.3. This explains the absence of thermoluminescence below 200°C in the glow-curve of Figure 1.1; traps for which the associated glow-peak is below 200°C are *usually* so shallow that they have suffered serious loss of electrons during the centuries of burial. For dating purposes it is only traps that have accumulated electrons without leakage that are of interest; this *usually* means traps for which the thermoluminescence glow-peak occurs at 300°C or higher.

It is implicit in Equation (1.1) that it is only the thermoluminescence in the 'stable' region of the glow-curve that is being considered, and so the question immediately arises as to how this stable region is recognized and confirmed. This is done by means of the all-important *plateau test*, in which the shape of the *natural* glow-curve (i.e., the glow-curve observed from a sample which has not received any artificial irradiation in the laboratory) is compared with the *artificial* glow-curve observed as a result of irradiation by means of a radioisotope source. This is illustrated in Figure 1.3. One notes first of all that in the artificial glow-curve there is substantial thermoluminescence appearing below 200°C; secondly, in the plot shown in the lower part of the figure we see that the ratio of natural thermoluminescence to artificial thermoluminescence rises from zero below 200°C to a plateau which begins soon after 300°C as the associated traps become deep enough for negligible leakage during antiquity. The stability corresponding to a given position in the glow-curve increases very sharply with temperature so that, whereas 200°C may correspond to a trapped-electron lifetime of only a few years, 300°C corresponds to a lifetime upwards of several thousand years, and 400°C to 10 million years. Thus a constant ratio between natural and artificial glow-curves gives an indication that, throughout this plateau region, there has been negligible leakage of electrons over the centuries that have elapsed since all traps were emptied in the course of the firing by ancient man.

The actual value of the ratio in the plateau region leads to a convenient way of expressing the level of the natural thermoluminescence; in the

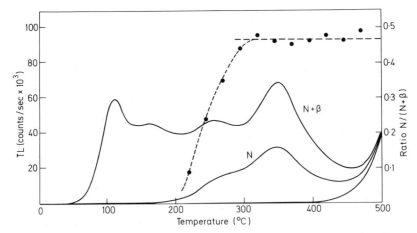

Figure 1.3 Curve N is the 'natural' glow-curve from one portion of the sample and curve N + β is the 'natural + artificial' glow-curve obtained from another portion to which an artificial dose of beta radiation has been administered; the thermal signal is also shown. The dashed line represents the ratio of the two glow-curves. The plateau level of 0.47 indicates that the dose equivalent to the natural thermoluminescence is 0.47/(1 − 0.47), i.e. 0.88, times the artificial dose used (10 Gy in this case).

1.4 Stability of the Thermoluminescence Record

example shown in Figure 1.3, the thermoluminescence due to the artificial dose[6] of 10 Gy is 12% greater than the natural thermoluminescence; hence we may say that the natural thermoluminescence is equivalent to 8.9 Gy. This is not necessarily the correct value for the dose that the sample received during antiquity, the *paleodose,* because of the *supralinearity* effect discussed in Sections 2.1 and 5.4. In terms of paleodose the basic age equation may be rewritten as

$$\text{age} = \frac{\text{paleodose}}{\text{annual dose}}. \qquad (1.2)$$

The symbol P will be used for paleodose; alternative names are *accrued dose, accumulated dose, archaeodose,* and *total dose.* It is usual to reserve the term 'equivalent dose', *ED,* for an evaluation that does not include correction for supralinearity. Using I to represent this correction, it is shown in Appendix A that $P = Q + I$, where, as elsewhere in this book, Q rather than *ED* has been used for equivalent dose.

Since lifetimes have been quoted for different temperatures of the glow-curve, it might be thought that the plateau test is unnecessary and that it would be adequate to utilize the level of thermoluminescence in the 400° region of the glow-curve. This is emphatically not the case for several reasons.

1. The traps of some minerals are afflicted by the malign phenomenon of *anomalous fading,* discussed in Section 3.4; such fading is anomalous in the sense that the observed stability is much less than predicted from the *kinetic* considerations discussed in Section 3.3 on which the lifetimes just quoted are based. Different traps are likely to be affected to different degrees, and so a sample subject to anomalous fading is not likely to show a good plateau. There are exceptions and hence storage tests are also necessary; passing the plateau test is a minimum requirement in this context.

2. If the sample were poorly fired by ancient man, the thermoluminescence clock would not have been properly set to zero, that is, the deeper traps would still contain electrons accumulated during geological times. This would cause the ratio to increase above the plateau value at higher temperatures, although it is more likely that there would be no plateau at all. If the sample were grossly underfired, or not fired at all, then the geological thermoluminescence would be intact. Except for minerals that had suffered comparatively recent geological heating, there would be no plateau at all, or at any rate a delayed plateau that would not start until an abnormally elevated temperature.

3. If the natural glow-curve were to contain a component of spurious thermoluminescence, this would destroy the plateau because it would be unlikely to be present to the same degree in the artificial glow-curve. The relative importance of spurious thermoluminescence usually increases with increasing glow-curve temperature so that the ratio rises, as in the case of poorly fired samples. Spurious thermoluminescence tends to vary from portion to portion; this helps in identification, but it also means that any idea of correcting for it by subtraction must be firmly resisted.

4. Finally, there is the simple possibility that the sample has been contaminated in the course of preparation. This could be with a thermoluminescence-sensitive mineral (such as calcium fluoride, present in the laboratory for thermoluminescence dosimetry purposes), with a detergent or preservative used by the archaeologist in treating his pottery, or, in the case of authenticity testing, with soil or encrustation that had adhered to the surface of the object being tested. The plateau would be destroyed because it is highly unlikely that the contaminant would have the same glow-curve shape as the true sample thermoluminescence. Interference can also arise from an unclean disc or from residual grains on the plate itself.

There are also two non-disqualifying reasons why the plateau may be poor. These reasons are that the supralinearity correction (see Sections 2.1 and 5.4) and the effectiveness of alpha particles relative to beta radiation (see Sections 2.2 and 5.3) may be different for different glow-curve temperatures. However, the misleading indications of the plateau in these circumstances can be avoided by plotting the *age plateau;* for this plot the age is calculated at, say, 25° intervals of the glow-curve, making appropriate corrections for the two effects interval by interval. If the age plateau is poor, then one of the previously mentioned causes must be suspected and the sample discarded.

The plateau evidence regarding reliability is a strong advantage that thermoluminescence dating has over, say, radiocarbon dating. In the latter the measurement produces a number representing the radiocarbon activity of the sample. Unlike a glow-curve, a number has no structure and there is no intrinsic evidence of, say, whether or not the sample has been contaminated with extraneous carbon of a different age.

1.5 Radioactivity and Annual Dose

For most samples the latent thermoluminescence is produced in roughly equal proportions by the nuclear radiations from potassium, tho-

1.5 Radioactivity and Annual Dose

rium, and uranium; the radiation from rubidium and cosmic radiation contributes only a small percentage of the total thermoluminescence. The isotope of potassium that is radioactive is potassium-40, which is present in natural potassium with an atomic abundance of about 0.01%. It emits beta radiation, consisting of particles (electrons) having ranges of up to a few millimetres in pottery; it also emits gamma radiation, which is more lightly ionizing and which penetrates soil to a distance of about 30 cm. The thermoluminescence contribution from beta particles arises from radioisotopes within the sample (except for the surface layer, in which the beta thermoluminescence is transitional between that corresponding to sample radioactivity and that corresponding to soil radioactivity), whereas the thermoluminescence contribution from gamma radiation is almost entirely from the soil (except for samples exceeding 1 or 2 cm in thickness). The thermoluminescence is proportional to the amount of ionization and, in turn, this is proportional to the energy absorbed from the radiation. The thermoluminescence produced for a given amount of absorbed energy, that is, per gray, is the same for beta radiation as for gamma radiation.

There are alpha particles from thorium and uranium in addition. These particles are a heavily ionizing type of radiation with ranges in pottery of between 0.01 and 0.05 mm. The ionization density produced is so great that the thermoluminescence traps lying in the central core of the tracks of the particles get saturated and a much greater proportion of the ionized electrons goes to waste (as far as thermoluminescence is concerned) than in the cases of beta and gamma radiation. As a result the thermoluminescence per gray is substantially less for alpha radiation than for beta and gamma radiation, being lower by a factor of between 0.05 and 0.5, depending on the substance; this factor is known as the k-value or a-value. The age equation (1.1) should really be written

$$\text{age} = \frac{\text{natural thermoluminescence}}{\chi_\alpha D_\alpha + \chi_l(D_\beta + D_\gamma + D_c)}, \qquad (1.3)$$

where χ and D are the thermoluminescence sensitivities and dose-rates of the radiations indicated, the suffix l denotes lightly ionizing radiation, and the suffix c denotes cosmic radiation. As shown in Appendix A, in terms of the paleodose P this can be written as

$$\text{age} = \frac{P}{D'_\alpha + D_\beta + D_\gamma + D_c}, \qquad (1.4)$$

where D'_α is the effective alpha dose-rate, equal to kD_α. In Table 1.1 a value of 0.15 has been assumed for k, as well as 'typical' values for the radioisotope concentrations. Of course, actual samples and soils vary

TABLE 1.1
Annual Radiation doses for 'Typical' Pottery and Soil[a]

	Alpha	Effective alpha[b]	Beta	Gamma	Effective totals
Potassium	—	—	0.83	0.24	1.07
Rubidium	—	—	0.02	—	0.02
Thorium	7.39	1.11	0.29	0.51	1.91
Uranium	8.34	1.25	0.44	0.34	2.03
Cosmic	—	—	—	0.15	0.15
	15.73	2.36	1.58	1.24	5.18

[a] The values are quoted in gray per 1000 years. To obtain values in rad per year, divide by 10. The values given correspond to pottery and soil having 1% potassium, 0.005% rubidium, 10 ppm natural thorium, and 3 ppm natural uranium. These two latter correspond to equal activities such as would give a combined alpha count-rate of 10 per ksec for a thick sample on a scintillator of diameter 42 mm and an electronic threshold factor of 0.835.

[b] The effective alpha contribution assumes a value of 0.15 for k, the alpha effectiveness in inducing thermoluminescence relative to the effectiveness of the beta and gamma radiation.

widely from these typical values; they are given now as a basis for discussion and a yardstick of comparison.

1.5.1 POTTERY INHOMOGENEITY AND ANTI-CORRELATION

Initially attempts were made (e.g., Tite, 1966) to obtain thermoluminescence dates for pottery without making allowance for the reduced thermoluminescence effectiveness of alpha particles (which was then unrealized) and without taking into account the heterogeneous nature of pottery fabric; on an absolute basis the ages obtained were low by a factor of five. While the major part of this discrepancy probably arose due to the former effect, undoubtedly the latter contributed substantially: the thermoluminescence was measured using a powdered sample without mineral separation and without grain-size selection, a naive approach in retrospect. As pointed out by Fremlin and Srirath (1964), because pottery is heterogeneous in its thermoluminescence properties as well as in its radioactivity, it is to be expected that on an absolute basis thermoluminescence dates will be systematically too recent. In the baked clay matrix there are crystalline inclusions, sometimes ranging up to a millimetre in diameter. The thermoluminescence sensitivity of these inclusions, which were either present in the raw clay or added by the potter in order to improve the refractory properties, is at least an order of magnitude higher than that of

1.5 Radioactivity and Annual Dose

the clay matrix in which they were embedded. However, in general the radioactivity is carried only in the clay matrix. Since the alpha particles have an average range of only about 25 μm in pottery, the core of an inclusion is shielded from them, receiving only beta and gamma radiation. Thus part of the highly sensitive thermoluminescent material in pottery does not receive the full dose-rate, and consequently the calculated age is too low. One solution to this problem is the quartz inclusion technique developed by Fleming (1966, 1970) following on from work by Ichikawa (1965). In this technique the quartz grains are extracted and, after etching away the outer skin, the thermoluminescence measurements are made on the residual cores of the grains so that in calculating the age only the beta and gamma dose-rates are taken into account. The other solution is the fine-grain technique developed by Zimmerman (1967, 1971a), in which grains in the size range 1–8 μm are separated out for the thermoluminescence measurements. These grains are small enough for attenuation of the alpha dose to be almost negligible, and so it is appropriate to calculate the age using the full effective dose-rate given in Table 1.1. This dose-rate is termed effective in the sense that allowance has been made for the reduced thermoluminescence efficiency of alpha particles arising from high ionization density. High ionization density is an intrinsic characteristic of this type of radiation, and it should be emphasized that it is quite different from the grain-size effect.

1.5.2 RADIOACTIVE INCLUSION DATING

Although quartz is low in radioactivity, there are other minerals, such as zircon, apatite, and potassium feldspar, for which the radioactivity is high. In large grains of such minerals the thermoluminescence is due predominantly to the internal dose *within the grain*. This has strong advantages for dating because, as discussed in Chapter 6, it means that dependence is reduced not only on the soil radioactivity, making dating of already-excavated samples more feasible, but also on the moisture content of both the sample and soil. As we shall discuss in Chapter 4, the need to estimate the average moisture content over the period of burial seriously limits the dating accuracy attainable with the fine-grain and quartz inclusion techniques.

1.5.3 ABSOLUTE VERSUS RELATIVE DATING

In the early development of thermoluminescence dating it was proposed from time to time that a lot of effort could be avoided if we were to be content with relative dates. In particular, this was seen as a way of avoiding the need for separating out quartz inclusions and fine grains from the sample, the thermoluminescence measurements being made on a powder consisting of all grains. The factor by which the thermoluminescence ages so obtained were low would be obtained by a calibration programme using samples of known age. The fallacy in this procedure is, of course, that the factor differs from sample to sample, depending on the proportion of crystalline grains present. The same objection also applies with respect to sample-to-sample variations due to other effects. This is unlike the situation in radiocarbon dating, where the purpose of the tree-ring calibration is to make allowance for the synchronous worldwide variation of the atmospheric carbon-14 concentration.

Because of sample-to-samples variations, accurate relative dating would require just as detailed allowance for the various effects and mechanisms involved as is required for absolute dating; and, obviously, the latter is preferable.

Technical Notes

[1] In discussions of thermoluminescence dating it is implicit that we are primarily concerned with the trapping of electrons at defects that already exist in the crystal lattice, and hence, until these defects are saturated, to a first approximation there is proportionality to radiation dose. The creation of new defects by the ionizing radiation is unimportant until comparatively heavy dose levels are reached, even in the case of alpha particles.

[2] *Ancient TL* is presently obtainable from I. K. Bailiff, Dept. of Archaeology, Fulling Mill, The Banks, Durham DH1 3EB, U.K. Details of thermoluminescence seminars are announced in the journal *Archaeometry,* obtainable from the Archaeometry Manager, 6 Keble Road, Oxford, OX1 3QJ, U.K. The proceedings of the 1978 and 1980 seminars have been published as volumes 2, 3, and 6 of *PACT*, the journal of the Council of Europe Study Group on physical, chemical, and mathematical techniques applied to archaeology, obtainable from T. Hackens, 28a Avenue Leopold, B-1330 Rixensart, Belgium. The proceedings of the 1982 seminar have been published as volume 9 of *PACT*, obtainable from V. Mejdahl, Nordic Laboratory for TL Dating, Postbox 49, DK-4000 Roskilde, Denmark, and those of the 1984 Seminar in Volume 10 of *Nuclear Tracks and Radiation Measurements*, published by Pergamon Press.

[3] See, for example, the Proceedings of the Seventh International Symposium on Exoelectron Emission, March 1983, Strasbourg.

[4] Commercial manufacturers of equipment designed specifically for thermoluminescence dating include (1) Daybreak Nuclear and Medical Systems, Inc., 50 Denison Drive, Guilford, Connecticut 06437; (2) Littlemore Scientific Engineering Co., Railway Lane, Littlemore, Oxford OX4 4PZ, United Kingdom; (3) Risø National Laboratory, Postbox 49, DK-4000 Roskilde, Denmark.

Technical Notes

⁵ Evacuation is usually accomplished by means of a two-stage rotary pump having a pumping speed of around 50 l/min. It seems that extended flushing with inert gas does not acheive the same measure of suppression as evacuation prior to gas admission; also, it is advantageous to keep the oven under vacuum when not in use—presumably this avoids absorption of oxygen and water vapour on surfaces within the oven. The gas cylinder should be connected to the oven with copper tubing because plastic or rubber tubing allows significant diffusion of air unless the run is short. If gas of high purity is not available, then some form of cleaner will need to be installed (see Sutton and Zimmerman, 1977; Debenham, 1978; Berger, Brown, Huntley and Wintle, 1982).

⁶ The *gray*, abbreviated Gy, is the unit of absorbed dose and is defined as 1 joule per kilogram. Formerly the unit used was the *rad*, and 100 rads = 1 Gy. The *roentgen* was the unit of radiation exposure; in air exposed to 1 roentgen of X-rays the absorbed dose is approximately 0.87 rad. The *sievert* (Sv) and the *rem* are units used in radiation protection which take into account the biological effectiveness of the radiation being measured; for an effectiveness of unity, 1 sievert = 1 gray and 1 rem = 1 rad.

CHAPTER 2

BASIC POTTERY DATING

In this chapter the two basic techniques used for pottery, bricks, and tiles during the first decade of thermoluminescence dating are outlined; namely, the quartz inclusion technique and the fine-grain technique, both already introduced in the first chapter. In addition, indication is given of the way in which the latter is adapted for testing the authenticity of art ceramics. The main text remains simple and should be regarded as a prelude to reading the succeeding three chapters; these chapters deal with the basic framework of thermoluminescence dating. For more details about sample preparation the reader should consult Zimmerman (1978) and, with respect to the fine-grain technique, Huxtable (1978); for a fuller treatment reference should be made to Fleming (1979). Because it was developed more recently, *feldspar dating* is dealt with in Chapter 6.

2.1 The Quartz Inclusion Technique
    (Fleming, 1970)

As we have seen in Chapter 1, the basic notion of the quartz inclusion technique is that thermoluminescence measurements are made on quartz grains from which the outer layer has been etched away, thereby leaving cores into which alpha particles from the clay matrix have not been able to penetrate because of their short range (which does not exceed 0.05 mm). The quartz itself is relatively free of radioactivity so that to a first approximation the dosage received by the cores is from beta, gamma, and cosmic radiation only. By selecting grains of around 0.1 mm diameter before etching there is only small attenuation of the beta dosage. Making a small (10%) correction for this, the age equation becomes

$$\text{age} = \frac{\text{paleodose}}{0.90 D_\beta + D_\gamma + D_c}, \qquad (2.1)$$

where D is the annual dose from the radiations indicated and the subscript c denotes cosmic radiation. The value used for the beta correction is discussed in Appendix C.

2.1.1 SAMPLE PREPARATION

After a 2-mm layer from each surface has been removed by sawing with a diamond-impregnated wheel, the pottery fragment is crushed by squeezing in a vice. The outer layer is discarded because (i) the beta dosage in it is transitional between that corresponding to the pottery radioactivity and that corresponding to the soil radioactivity, (ii) there may also be a reduced level of thermoluminescence in the outer surface because of the effect of sunlight, and (iii) soil contamination must be rigorously avoided because of its high level of geological thermoluminescence. The rubble produced by the 'vicing' is then gently crushed in an agate pestle and mortar, with care taken to avoid degrading large quartz grains into smaller ones because the objective is to obtain grains that were in the size range of 90 to 120 μm while embedded in the pottery matrix; fragments of larger grains must be avoided because of the attenuation of the beta dosage that will have occurred in them. Grains in the desired size range are selected by sieving and then, by means of the well-established mineralogical technique of magnetic separation,[1] crystalline grains (non-magnetic) are separated from the clay matrix (slightly magnetic). Calcite grains are removed from the crystalline fraction with dilute hydrochloric acid and feldspar grains with concentrated hydrofluoric acid,[2] due precaution being taken in the handling of this unpleasant acid. Besides removing the feldspar component, immersion in hydrofluoric acid for about an hour allows etching of the outer layer of the quartz grains to a depth sufficient for the cores remaining to have a negligible component of alpha particle dosage (but see Appendix C). As in all sample preparation for thermoluminescence dating, the procedures have to be carried out in subdued red light to avoid 'bleaching' effects.

2.1.2 EVALUATION OF THE PALEODOSE

The simplest approach to the evaluation of paleodose is by the straightforward procedure of measuring the natural thermoluminescence from a portion of quartz grains and comparing it with the artificial thermoluminescence from that same portion of grains after exposure to a known dosage of radiation from a radioisotope source (e.g., beta particles from strontium-90 or gamma radiation from caesium-137 or cobalt-60). However, this simple procedure usually gives only an approximate value because of the tendency of thermoluminescence sensitivity to be changed during the heating suffered during the first glow. This change in sensitivity occurs in nearly all thermoluminescent minerals and necessitates use of the *additive dose method* (outlined below). One cause for the change is simply that the transparency is affected by the heating; another is the *predose effect* discussed in Section 5.4.

In the additive dose method, measurements are made on a number of weighed portions of the quartz grains, usually in the region of 5 mg each. Several of these portions are used for measurement of the natural thermoluminescence, others for measurement of natural plus artificial thermoluminescence. After normalization[3] according to weight, the thermoluminescence intensities are plotted as in Figure 2.1 and the equivalent dose Q is evaluated. As so obtained Q is not necessarily equal to the paleodose;

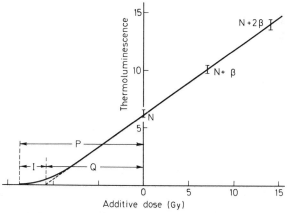

Figure 2.1 The additive method for evaluation of the equivalent dose Q (often referred to as 'ED'). At least two levels of additive dose (which can be administered either with beta or with gamma radiation) should be used in order to check for linearity of response above the level of the natural thermoluminescence. The paleodose P that the sample has received during antiquity is usually greater than Q because of initial supralinearity of response. Evaluation of the correction I for this is as shown in Figure 2.2.

this is because for low doses the growth of thermoluminescence with dose is supralinear as in Figure 2.2, which shows the *second-glow growth characteristic* obtained by measurement of the thermoluminescence from portions which have been irradiated after drainage in the course of the first glow. Using the intercept I as indicated, the paleodose is taken to be $Q + I$. The validity of doing this rests on the assumption that the value of I has not been changed by the first-glow heating; this is discussed in Section 5.4. Plots such as Figures 2.1 and 2.2 are made for various glow-curve regions, usually averaging each 25° interval; hence paleodose values are obtained for each region, and the plot of these values against glow-curve temperature is used for the plateau test. This paleodose plateau test is equivalent to an age plateau test in the case of inclusion dating (but not for fine-grain dating, because the alpha effectiveness may be different in different parts of the glow-curve).

The glow-curve observed from quartz usually consists of a single peak, either at about 325°C or at about 375°C, or of a combination of these two peaks. The 325°C peak has been termed 'malign' because of its tendency to change markedly in sensitivity between first glow and second glow (i.e., the slope of the line in Figure 2.2 is different than that in Figure 2.1), whereas the 375°C peak is benign in this respect. Clearly the assumption that there is no change in I between first glow and second glow is more justifiable with respect to the benign peak. Kinetic studies (see Appendix

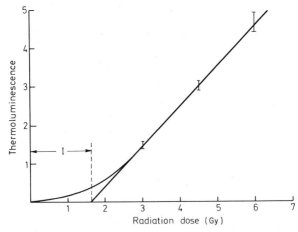

Figure 2.2 Second-glow growth characteristic for evaluation of the supralinearity correction I. These data are obtained with portions from which the natural thermoluminescence has been drained by heating in the course of the first-glow measurement. To avoid interference by pre-dose effects, a fresh portion is used for each measurement. As with the first-glow measurements of Figure 2.1, each data point is the average of several measurements.

2.1 The Quartz Inclusion Technique

E) indicate mean lives, for storage at 15°C, of 100 million years for both peaks, amply long enough for the earliest pottery. Tests made on quartz give no indication that it is prone to anomalous fading. Notwithstanding these assurances with respect to stability, it is important that the plateau test be made, for the reasons given in Section 1.4.

2.1.3 ANNUAL DOSE

As discussed in Chapter 4, a variety of techniques are in use for evaluation of sample radioactivity and annual dose, and there are many complications. Here we shall outline only the simplest, which is by means of thermoluminescence itself. There are some phosphors, such as natural fluorite (calcium fluoride) and artificial calcium sulphate (doped with dysprosium), which are so highly sensitive that a few weeks exposure to the sample induces an accurately measurable level of thermoluminescence. One arrangement, developed by Bailiff (1982), for measuring the annual beta dose is shown in Figure 2.3; others are discussed in Section 4.3.3. About 1 g of powdered pottery is put in a perpex container; in the bottom of this container is a thin window of plastic (thickness 0.18 mm) which allows beta particles to emerge but not alpha particles. A thermoluminescence dosimeter is placed immediately underneath the window; this dosimeter consists of a copper tray (diameter 10 mm) containing the thermoluminescent phosphor set in silicone resin, and the thermoluminescence acquired in several weeks of storage is measured using the standard thermoluminescence oven, the tray being placed directly on the heater plate. Of course, the dose measured by the phosphor is less than that within the pottery because particles reach the phosphor from one side only and because of some absorption in the window. By means of standard samples having known contents of potassium, thorium, and uranium, the ratio

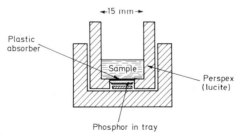

Figure 2.3 Beta thermoluminescence dosimetry unit. The thermoluminescence acquired by the phosphor is measured by placing the copper tray directly on the heater plate of the standard thermoluminescence oven. The plastic absorber stops alpha particles from reaching the phosphor (from Bailiff, 1982).

between the two can be evaluated;[4] it was found to be 3.55 for the arrangement in Figure 2.3. During storage the unit is placed within a lead container having 4-cm thick lead walls to give shielding from cosmic rays and external gamma rays (e.g., from the brick and stone of the building).

The dose-rate within a sample is affected by its moisture content because of absorption of part of the dose by water. It is usual, for convenience, to measure the beta dose-rate with the sample dry and then apply a correction (see Section 4.2.3) to allow for the estimated average moisture content during burial. Uncertainty about this moisture content sets a limit to the accuracy that can be obtained with thermoluminescence dating; however, even when the degree of wetness during burial cannot be estimated, an upper limit to the size of the correction factor is given by the saturation moisture content of sherd, measured before crushing.

The range of beta particles in pottery is only a few millimetres, and consequently a sample diameter of a centimetre or more is effectively infinite; increasing the diameter gives no increase in the dose-rate received by the phosphor. With gamma rays the situation is quite different because of their greater range, of up to 30 cm. This has the important consequence that the gamma dose-rate experienced by the quartz grains

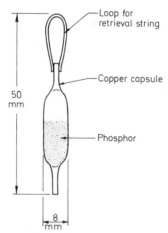

Figure 2.4 Gamma thermoluminescence dosimetry. The capsule is inserted into the soil by means of a 30-cm deep auger hole so that the soil surrounding it is effectively infinite so far as gamma rays are concerned. Alpha and beta particles are prevented from reaching the phosphor by the 0.7-mm copper wall of the capsule. The ends of the capsule are crimped and sealed with hard solder. Before insertion the capsule is heated to at least 300°C for a few minutes to set the thermoluminescence of the phosphor to zero; this can be done in a gas flame, temperature-sensitive paint being a useful indicator. The amount of phosphor should be not less than 0.1 g.

2.1 The Quartz Inclusion Technique

in a pottery fragment not thicker than a few centimetres is almost entirely provided by the surrounding soil. As the weight of soil within a sphere of radius 30 cm is over a quarter of a ton, it might be asked why the gamma dose does not swamp the beta dose and make it unimportant. The reason is that implicit in the greater range is a correspondingly lower coefficient for absorption of energy from the gamma-ray flux.

For measurement of the gamma dose-rate, it is easier to take the thermoluminescent phosphor to the archaeological site, rather than make a laboratory measurement on an amount of soil which for practical reasons has to be less than a quarter of a ton! For on-site burial the thermoluminescent phosphor is contained within a suitable capsule, such as the copper one shown in Figure 2.4. This capsule is inserted, with a string attached for retrieval, into a 30-cm deep auger hole and left for several months or more. It is then taken back to the laboratory for measurement and evaluation of the accumulated dose. This dose also includes the minor contribution made by cosmic rays.

2.1.4 APPLICATION TO A ROMANO–BRITISH POTTERY FRAGMENT

As an example of how the method works, measurements are quoted now for one of the samples kindly supplied during the early testing of the method by Professor Graham Webster from his excavation of the fort on Waddon Hill in Dorset, England. This was a small Roman camp known to have been in use from A.D. 50 to A.D. 60 on the basis of coin evidence and other archaeological considerations. The fragment concerned was from a cooking pot and hence was likely to have been made not more than a few years before its breakage.

The fragment weighed about 5 g and from it 60 mg of etched quartz grains were obtained. Thermoluminescence measurements gave a good plateau, the ratio being within ±5% of the average over the glow-curve region upwards of 320°C. From these measurements a value of 5.3 Gy was obtained for the equivalent dose Q and a value of 0.4 Gy was obtained for the supralinearity intercept I. Hence the paleodose is given by

$$P = Q + I = 5.7 \text{ Gy}.$$

For evaluation of the annual dose from beta radiation the thermoluminescence accumulated by the phosphor dosimeter during 25 days of storage was measured; after subtraction of gamma-ray and cosmic-ray background and inclusion of the calibration factor, the dose-rate was evaluated as 5.33 μGy/day, that is, 1.94 Gy per 1000 years. The measured water content of the fragment as excavated was 15% of the dry weight; from considerations discussed in Section 4.2.3, this would cause the actual beta dose-rate during burial to be less by a factor of 1.19, giving 1.63 Gy per 1000 years as the value for fine grains. The average value for the etched cores of the 100-μm quartz grains is less by 10%, that is, 1.47 Gy/kyear.

For gamma plus cosmic radiation the accumulated dose evaluated for the capsule after burial for 345 days was 1.39 mGy, leading to an annual dose of 1.40 Gy/kyear after a 5% allowance for the smaller attenuation of the gamma dose by the 0.7-mm thick copper walls of the capsule compared to that in the sherd (see Section 4.3.3). Hence the age for the fragment is

$$\frac{5.7}{(1.47 + 1.40) \times 10^{-3}} = 1990 \text{ yr.}$$

Limits of error are discussed in Appendix B, as is the question of variation of the fragment's water content from its 'as dug' value. Taking into account both systematic errors and measurement errors, an uncertainty of about 10% of the age is to be expected, that is ±200 years for the fragment in question, comfortably encompassing the known age. By dating half-a-dozen fragments from the same level and taking the average, the uncertainty arising from random measurement errors is reduced, but not that due to systematic errors. In practice an overall error limit of about ±7% is typical, that is, about ±130 years in the present example. Even so, in the context of Romano–British archaeology these limits are rather wide, and this gives warning that periods for which the chronological framework is accurately established are not going to get much benefit from thermoluminescence dating. The capability of the method should not be overestimated; false claims for its usefulness only lead to disillusion and detract from the importance of the technique in less well-dated periods and regions.

An aspect of dating that is important for good reliability is sample collection; appropriate procedures are indicated in Appendix D.

2.2 The Fine-Grain Technique
   (Zimmerman, 1971)

The fine-grain technique meets two requirements. First, the grains utilized are small enough for there to be full penetration by alpha particles and so, whether or not they are deficient in radioactivity, they get the dosage corresponding to the radioactivity of the clay matrix. It is, of course, essential that the fine-grains utilized are 'true' fine-grains rather than degraded ones produced in the crushing process. Second, the grains

2.2 The Fine-Grain Technique

are obtained in the form of a thin layer suitable for measurement of alpha-particle effectiveness. A basic advantage is that inclusion of the alpha-particle contribution to the thermoluminescence diminishes the importance of the gamma radiation from the soil. Another reason for needing an alternative technique is the paucity of quartz in some pottery fabrics. On the other hand, the fine-grains used in this technique are a mixture of unknown minerals and there is the risk that some among them may have adverse thermoluminescence characteristics, such as anomalous fading. This puts additional emphasis on the stringency of checks and precautions.

2.2.1 SAMPLE PREPARATION

Sample preparation begins as for the quartz-inclusion technique by squeezing the fragment in a vice. By washing the products of this operation in acetone, a suspension of fine-grains is obtained, an ultrasonic bath being used to disperse coagulation. Making use of the fact that the settling time is determined by diameter, grains in the size range of 1 to 8 μm are separated; these sizes correspond to settling times of 20 min and 2 min, respectively, for a 60-mm column. The separated grains are then resuspended in acetone and allowed to deposit on aluminum discs[5] (usually 10 mm in diameter and $\frac{1}{2}$ mm thick) in a thin layer of a few microns thickness; for deposition the discs are placed at the bottom of individual flat-bottomed glass tubes, the acetone being lost by evaporation. Usually sixteen such discs are prepared; each disc carries a few milligrams of sample, and the disc-to-disc scatter in thermoluminescence reproducibility should be not more than ±5%. Such discs are a convenient way of handling the sample for measurement (the disc as a whole being placed on the heater plate using tweezers) and are remarkably robust: although the powder can be removed by wiping, quite often there is no loss if the disc is dropped.

Grains less than 1 μm in diameter are avoided because their greater surface-to-volume ratio results in an enhanced likelihood of a significant level of spurious thermoluminescence.

2.2.2 ALPHA PARTICLE EFFECTIVENESS: k-VALUE AND a-VALUE

The prime consideration in the above form of sample presentation is that the thin layer permits measurement of the thermoluminescence effectiveness of alpha particles. As already mentioned, for a given dosage (in terms of deposited energy) alpha radiation is less effective than beta radiation and gamma radiation, the ratio between the two being termed the k-value for the sample. The alpha sensitivity is measured using a calibrated alpha source such as americium-241, employing the same additive dose method as illustrated in Figure 2.1; this yields the alpha dose Q_α that would induce a level of thermoluminescence equal to the natural thermoluminescence—often termed the equivalent alpha dose. Since Q_α and Q, the equivalent beta dose, are inversely proportional to the respective sensitivities, it follows that

$$k = Q/Q_\alpha. \qquad (2.2)$$

Q is measured using a calibrated beta source, as in the quartz inclusion technique. Since discs are used rather than weighed portions, it is usual to rely on adequate disc-to-disc reproducibility, averaging several discs to even out variations.

The thermoluminescence effectiveness of an alpha particle depends on its energy, and hence the value used for Q_α in Equation (2.2) must have been corrected for the difference between the energy spectrum of the alpha source and that of the natural dosage from thorium and uranium. According to Zimmerman (1971) if 3.7-MeV alpha particles are used for evaluation of the equivalent alpha dose, then the directly obtained value needs to be divided by 0.85 before insertion in Equation (2.2). An important practical consideration with respect to energy is that the layer of fine-grains must be thin enough for there to be full penetration by the particles; otherwise the value obtained for k will be erroneously low.

The k-value system has now been superceded by the a-value system which, although more difficult in concept, has strong practical advantages. This and other aspects of alpha irradiation are discussed in Section 5.3.

2.2.3 ALPHA COUNTING

Although in principle the alpha dose-rate could be determined by thermoluminescence dosimetry, it is not easy in practice owing to the short range of alpha particles. Instead, the technique of 'thick-source' alpha counting is used; in this technique a powdered layer of sample, thick compared with the 50-μm range of the most energetic alpha particles from thorium and uranium, is placed on top of a scintillation screen which is viewed from beneath by a photomultiplier (see Figure 2.5). Each alpha particle striking the screen produces a scintillation, and this produces a burst of photoelectrons from the photocathode which, after amplification, become an electrical pulse at the anode of the photomultiplier. After amplification the pulse is registered on a suitable counting device. The scintillation screen is made by sprinkling zinc sulphide onto sellotape

2.2 The Fine-Grain Technique

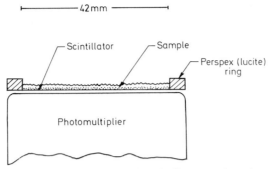

Figure 2.5 Thick-source alpha counting. The scintillator consists of a monograin layer of zinc sulphide on the upper surface of a sellotape screen which is carried by the perpex ring. So far as alpha particles are concerned, a sample depth of only 0.1 mm is more than thick enough to be effectively infinite, and so in spreading the sample it is a matter of using enough sample to ensure complete coverage; typically of the order of a gram is used. Shielding from beta particles and gamma rays is not required, because the scintillations produced are much smaller than those produced by alpha particles; on the other hand, scrupulous care is necessary to exclude the possibility of contaminating alpha activity from the zinc sulphide, the sellotape, and the perspex ring. A less simple form of sample holder is shown in Figure 4.7.

(tapping off any surplus as the coverage must not be too thick; on the other hand, it needs to be complete); the powdered sample is placed in direct contact with this. Alternatively, scintillation screens can be purchased ready-made as zinc sulphide incorporated into plastic. The great advantage of zinc sulphide is that the pulses corresponding to the scintillations produced by alpha particles are much larger than those produced by beta particles and gamma radiation, thus enabling easy rejection of the latter in the electronics.

For a screen diameter of 42 mm, counts are received from the lower 0.2 g of sample, about a quarter of the alpha particles originating in this amount actually reaching the screen. For typical pottery or soil there are about 1000 counts per day, a hundred times greater than the background of counts from the radioactive impurities in the screen itself. It can be calculated (see Table 1.1) that a count-rate of 10 counts/ksecond corresponds to an alpha dose-rate of 15.7 Gy per thousand years for equal activities of thorium and uranium, and that even if the activity is all thorium or all uranium the dose-rate value is within 5% of that quoted.

The beta and gamma dose-rates can also be derived from the alpha count-rate, but there is greater dependence on the thorium/uranium ratio; a sample which is all thorium or all uranium departs by around 20% from the conversion factor for equal activities. However, although extreme variations in the activity ratio are not common, it is preferable to use the thermoluminescence dosimetry methods for the beta and gamma dose-rates, as in the quartz inclusion technique. Since potassium-40 has no alpha activity, if alpha counting is used to derive the beta and gamma dose-rates, chemical determinations have to be made of the potassium contents of sample and soil.

Various aspects of alpha counting are discussed in Chapter 4, including the 'overcounting' that may arise, particularly if there is escape of the radioactive gas radon from the sample.

2.2.4 ANOMALOUS FADING

Because the fine-grains comprise a mixture of unknown minerals, there is risk that thermoluminescence in the region of the glow-curve above 350° may not have the long-term stability expected of it. The first defence against this is the plateau test, but it is prudent to make a direct check as well. Two of the sample discs are used for this; one is irradiated and stored, and after several months its thermoluminescence is measured and compared with the thermoluminescence from the other disc immediately after irradiation. Although a few months might seem rather short in comparison with the millennia-long stability required, a proportion of samples do exhibit fading in this short time. The rate of fading initially is very much greater than subsequently so that any sample which fades in long term will fail the short term test. This is discussed further in Section 3.4 and Appendix F.

2.2.5 APPLICATION

In fine-grain measurements on the Romano–British pottery fragment of the previous section, a value of 9.3 Gy was obtained for the equivalent dose and a value of 0.3 Gy was obtained for the supralinearity intercept, giving a value of 9.6 Gy for the paleodose; the value obtained for k, the ratio of alpha-particle effectiveness to beta-particle effectiveness, was 0.18. There was no indication of fading in a two-month test. The alpha count-rate was 8.9 counts per kilosecond, indicating an alpha dose-rate of 14 Gy/kyear and hence an effective dose-rate of $0.18 \times 14 = 2.52$ Gy/kyear. In the case of alpha dose, the allowance for water content of 15% is made by dividing by 1.23, yielding 2.05 Gy/kyear. Using the previously derived values for the beta and gamma dose-rates, the age for the fragment is

$$\frac{9.6}{(2.05 + 1.63 + 1.40) \times 10^{-3}} = 1890 \text{ yr.}$$

2.3 Subtraction Dating

The age equation for fine-grain dating may be written as

$$P_{FG} = \text{age} \times (D'_\alpha + D_\beta + D_\gamma + D_c), \qquad (2.3)$$

where P_{FG} is the paleodose evaluated for the fine-grains. Similarly, for quartz inclusions,

$$P_I = \text{age} \times (0.90 D_\beta + D_\gamma + D_c). \qquad (2.4)$$

Hence after subtraction we have

$$\text{age} = \frac{P_{FG} - P_I}{D'_\alpha + 0.1 D_\beta}. \qquad (2.5)$$

This is independent of the external dose from gamma radiation and cosmic rays, and hence in principle the age can be derived from measurements on the sample only; that is, pottery for which there is no knowledge of burial circumstances (other than an estimate of burial wetness) can be dated. Unfortunately, in practice it is not common to find pottery which is suitable for both techniques; either there is some difficulty with the fine-grains, such as anomalous fading, or there is not enough quartz for an inclusion date. Additionally, there may be insufficient difference between the two paleodoses for the accuracy of the subtraction age to be useful.

If, in the example of the sample from Waddon Hill the error limits on the paleodoses are each taken as ±5%, then

$$P_{FG} - P_I = (9.6 \pm 0.4) - (5.7 \pm 0.3)$$
$$= 3.9 \pm 0.5 \text{ Gy},$$

that is, there is an error of 14% introduced into the paleodose by the subtraction and hence also into the age, which in this case turns out to be

$$\text{age} = \frac{3.9 \pm 0.5}{(2.05 + 0.16) \times 10^{-3}} = 1760 \pm 230 \text{ yr.}$$

Although the error limits do encompass the true age, this example illustrates the reduced accuracy inherent in a subtraction technique. However, there are circumstances when the annual gamma dose cannot be measured, and the technique is then rather valuable, as has been illustrated by Fleming and Stoneham (1973b) with respect to some Rennaissance terracotta. Obviously, the technique is most effective when there is a substantial difference between the two paleodoses—as would arise for a high k-value and alpha activity.

Other forms of subtraction dating are discussed in Section 6.6; a promising one is between large, potassium-rich feldspar grains and radioactivity-free quartz grains.

2.4 Accuracy and Error Limits

Accuracy is a particularly tedious aspect of thermoluminescence dating. Yet it has to be faced because if a date is to be useful some estimate of accuracy needs to be given. Inexperienced practitioners tend to be over-optimistic about accuracy, and ultimately this leads to disillusion on the part of the customer. A dozen or so quantities and components enter into the calculation of a thermoluminescence date, and the error limit derived for the date needs to reflect the combined effect of uncertainties in these quantities, appropriately weighted. Not all sources of error are quantifiable, and to assess whether such sources are causing significant interference it is necessary to look at the coherence of a group of dates for samples known to be contemporary (e.g., pottery all of the same style or all from the same stratigraphic level). An unfortunate characteristic of thermoluminescence dating is the variability of relevant conditions, from region to region or even from site to site within the same region. Thus to demonstrate experimentally on a particular site that thermoluminescence 'works' to the degree of accuracy predicted from the quantifiable uncertainties does not prove anything about its performance on a site where circumstances may be different. One example of a difficult-to-quantify interfering factor is leaching and deposition of radioactivity by groundwater during burial, which is discussed in Section 4.2.5. Such leaching and deposition is likely to depend on the compactness of the pottery fabric, so that it is advantageous if the samples have a variety of fabrics; agreement between individual dates is then a more powerful indication that all is well.

In Appendix B a summary is given of the system developed to deal with quantifiable uncertainties. In this system (Aitken and Alldred, 1972; Aitken, 1976) distinction is made between random and systematic errors. The former include measurement errors and other errors likely to be different from sample to sample. The latter include errors that would affect all samples, such as those due to uncertainties in basic radioactive-source calibrations, and errors that would affect all samples of a given site, such as those arising from uncertainty about moisture content during burial. Whereas the overall random error on a date can be reduced by averaging the results for a number of contemporary samples, this is not true for systematic error. Because of this the latter is the real barrier to better accuracy, and at present it is difficult to see real prospects of reducing the overall error limit (at the 68% level of confidence) to below ±5% of the age, except in particularly advantageous circumstances. More typically the error limit is in the 7–10% range, and there are some types of pottery fabric which have such poor thermoluminescence characteristics that a reliable date cannot be obtained.

2.4 Accuracy and Error Limits

As an example of the standard form of citation of a thermoluminescence date, we quote that obtained for context e of site 143 of the Oxford laboratory: 1070 BC (±100, ±220, OxTL 143e). The date given is the average for 7 samples from the same archaeological level of the site. The second error limit, ±220 years, is the overall predicted error limit calculated as indicated in Appendix B and taking into account all quantifiable sources of random and systematic error; it is this limit that is appropriate when making comparisons between thermoluminescence dates from different sites or between thermoluminescence dates and radiocarbon dates or other chronologies. The first error limit is the standard error on the mean value as derived from the scatter of the individual dates; that is, it is the standard deviation divided by $\sqrt{7}$. Thus it is an experimental estimate of the error ignoring the possibility of systematic errors. Hence it is useful for comparisons between similar contexts, for which the systematic errors are likely to be the same.

The first error limit has similarity to the standard deviation quoted for an uncalibrated radiocarbon date, whereas the second is more like the error given for a radiocarbon date after it has been calibrated and the calibration error has been included. Unlike thermoluminescence error limits, those for radiocarbon are not a fixed percentage of the age but are more or less fixed in terms of years. For the $\pm\frac{1}{2}\%$ measurement precision attainable with a good, modern radiocarbon installation, the corresponding error limit on the uncalibrated age is ±40 years, whereas after calibration this widens considerably to an extent depending on the form of the calibration curve in the period concerned; for the purpose of comparison, we take the not untypical value of ±150 years. The first error limit for a thermoluminescence date is usually of the order of ±3% of the age, as for the date cited; this is less than ±40 years until an age of 1300 years is reached. Taking ±7% as typical for the second error limit of a thermoluminescence date, we see that this is less than ±150 years until an age of 2000 years is reached. Thus it is in the last one or two thousand years that thermoluminescence dating can be expected to give a comparable or better accuracy than radiocarbon dating; in earlier periods its use should be restricted to sites on which there is no material for radiocarbon dating or on which there is ambiguity in the radiocarbon results. Of course, as discussed in Chapters 7 and 8, it is for sites that are beyond the circa 50,000-year limit of radiocarbon dating that thermoluminescence again becomes important.

In the citation of radiocarbon dates there is considerable confusion as to whether or not the date has been calibrated. Some authors and journals, unfortunately a minority, employ the convention that the lower case letters a.d., b.c., and b.p. signify uncalibrated dates, reserving A.D., B.C., and B.P. for calibrated dates, that is, the best estimate in terms of calendar years. Since thermoluminescence dates are all in the latter category, it is appropriate to make comparisons with calibrated radiocarbon dates and to use capital letters. Another source of confusion is that b.p. and B.P. are not to be taken literally as meaning 'Before Present' but have been defined as meaning 'before A.D. 1950'. It has been suggested that for thermoluminescence dates B.P. should mean 'before A.D. 1980', but this would seem to add to the confusion. It is strongly recommended by the present author that the A.D./B.C. system should be used for all dates of the last 10,000 years; beyond that the uncertainty associated with the use of B.P. becomes insignificant.

2.5 Authenticity Testing of Ceramic Art Objects

It is in the field of authenticity testing that thermoluminescence has had a revolutionary impact comparable with that of radiocarbon dating in archaeology. It is usually a question of deciding between an age of less than a hundred years and one of upwards of five hundred. Uncertainty about the annual gamma dose, which usually widens the error limit to around ±20%, is then unimportant and it does not matter that the 'burial circumstances' are unknown. The answer obtained is usually clear-cut, and the technique provides museum curators with a very powerful independent judgement of what on their shelves is genuine and what imitative. The result of a thermoluminescence test may alter the value of an object from an astronomical figure to a negligible one; consequently, although high accuracy may not be important, reliability is vital and tests need to be conducted with meticulous care. Reputable art dealers have doubtful pieces tested as a matter of routine, the usual fee for such a service being in the $100–200 range. The sample is obtained by drilling a small hole, 3 mm across by 3 mm deep, in an unobtrusive location using a tungsten carbide bit; this yields about 50 mg of powder, sufficient to make about four fine-grain discs for a restricted paleodose determination and leave a residue available for evaluation of annual dose. There is inevitably some degradation of large grains into fine-grains by the drilling, though not enough to upset the date at the level of accuracy expected with this technique; in most fabrics it seems that the large grains are detached rather than broken up. For porcelain (Stoneham, 1983) it is necessary to use thin ½-mm slices taken from a 3-mm core extracted by means of a diamond-impregnated corer and to evaluate the paleodose by means of the pre-dose technique described in Section 6.1; the pre-dose technique is also used for pottery of the past few hundred years and in cases where the type of clay is such that adequate suppression of spurious thermolumines-

2.5 Authenticity Testing

cence it is not possible. It is also possible to test the authenticity of bronze heads, etc., which have a core of clay; this core will have been heated in the casting process, and a sample for thermoluminescence dating is extracted through a convenient hole in the bronze.

Radioactivity measurements are made using alpha counting for the thorium and uranium and chemical analysis (flame photometry) for the potassium. Because of the restricted amount of material available, a scintillation screen of only 15 mm diameter is used for the alpha counting, and if necessary the fine-grain discs used for thermoluminescence can be utilised as sample, due allowance being made for the fact that the layer is less thick than the alpha particle range.[6] As indicated in Section 4.3.1, the alpha count-rate can also be used to derive the beta dose-rate due to thorium and uranium, that for potassium being derived from the chemical analysis. With regard to the overall accuracy obtainable in this application, the possible error due to dependence of the beta dose-rate on the thorium/uranium ratio is not significant.

Although a relaxed level of accuracy is acceptable in authenticity work, this does not mean that there can be a relaxed approach to measurement. Indeed, the opposite is true, and several years of thermoluminescence dating experience are necessary if wrong answers are to be avoided. Contamination of the sample by soil or encrustation adhering to the surface is one danger—an obvious one that is often underestimated by operators. Another is that the object being tested is a pastiche made from ancient and modern fragments; in such cases comprehensive multiple sampling is necessary.

Forgers of a scientific bent may employ clandestine gamma irradiation in an attempt to give the object seemingly authentic thermoluminescence. The ultimate defence against this is the use of a subtraction technique which gives the true age irrespective of any external irradiation or the zircon technique (see Section 6.3; for a fine example of authenticity application, to core material from the Bronze Horse of the New York Metropolitan Museum, see Zimmerman, Yuhas and Meyers, 1974). Both of these techniques require more sample than is commonly available, but there are other indicators of clandestine irradiation that an experienced practitioner learns to recognize.

Another possible way of fooling the thermoluminescence test is by reconstitution, that is, making the fake object from ground-up ancient brick by means of some chemical cementing agent. If no heat is used in the process, then the thermoluminescence age will be that corresponding to the ancient brick. Of course, there is much more to it than this; it is likely that the chemical agent will reveal itself by excessive spurious thermoluminescence, and in any case there are other techniques that can be applied, for example, thermogravimetric analysis and archaeomagnetic measurements—the randomization of the magnetic grain orientations inherent in grinding-up will have destroyed the sample's remanent magnetization.

2.5.1 AN EXAMPLE OF APPLICATION: HACILAR WARE

There are some styles of ceramic ware for which art historians have had doubts about the authenticity of the group as a whole. Application of thermoluminescence in such cases has led to a reassessment of the visual criteria on which these judgements are based; it has also had academic impact in the sense that previously the art historian may have been studying the forger's view of man's cultural and artistic development rather than actuality. An example of this is the so-called Hacilar ware, anthropomorphic vessels and figurines said to be 7000 years old and to have come from the renowned site of that name in southwest Turkey. Because their supposed origin was from a cultural phase following soon after the first appearance of pottery in that part of the world, the fineness of technique and beauty of form attracted particular interest. Out of a total of 66 pieces tested by thermoluminescence, 48 were found to be modern forgeries (Aitken, Moorey and Ucko, 1971). Of particular interest were the magnificent double-spouted vessels, each spout being in the form of a head with obsidian eyes (see Figure 2.6); out of seven such tested only one was genuine. Although these findings confirmed the doubts of some scholars, they were resisted by others. The argument was adduced that the thermoluminescence had been reset by recent heating, it being hypothesized that the pieces had been robbed out of marshy ground by peasants who subsequently baked them in preparation for burnishing, in the questionable belief that this would yield a higher price. Laboratory tests indicated that a day or two at 300°C would be necessary to erase the thermoluminescence, and there were two practical arguments against this having happened. First, risk of damage would dictate that the temperature used for drying out would be no higher than necessary to drive off the water. Second, it was unlikely that the available ovens—domestic bread ovens—would reach a temperature in excess of 200°C. In addition to this circumstantial reasoning, conclusive direct evidence against reheating was obtained by means of the pre-dose technique (see Section 6.1), which can be used to derive an upper limit for the age irrespective of whether or not the object has been reheated (as long as the heating did not exceed 700°C).

The arguments for rejecting the reheating hypothesis in other cases

2.5 Authenticity Testing

Figure 2.6 Double-headed 'Hacilar' vessel, recently fired according to thermoluminescence tests (photo courtesy of the Visitors of the Ashmolean Museum, Oxford).

follow similar lines to those given above. If the decoration is sophisticated, then the practical argument is stronger still—additionally one can ask, Why should the owner have wished to reheat?

2.5.2 THE GLOZEL CONTROVERSY

This archaeological *cause célèbre* of the 1920s became a thermoluminescence *cause célèbre* in the late 1970s; in this case the situation was the opposite of the last example, and baked clay objects presumed modern were found to be ancient. Beginning in 1924 a farmer living in an isolated farmhouse near Glozel in central France began to report the discovery of a wide range of objects, most of which had no parallels in excavated material anywhere else in the world: baked clay tablets with mysterious incised lettering in an unknown script, jars and other ceramics with and without inscriptions, phallic symbols in pottery, and bones and pebbles with animal inscriptions. The latter led to the claim that the site was Neolithic and that the tablets represented the earliest writing in the world. This view was hotly contested, mainly by non-French archaeologists who thought the whole thing was so unacceptable that it must be a hoax, the objects being forgeries. An acrimonious controversy continued more or less uninterrupted until the outbreak of war in 1939. During this time two international commissions of inquiry sat and pronounced exactly opposite decisions and some five law suits were fought with varying results.

Subsequently the questionable nature of the finds was generally accepted, and it was therefore a severe shock when thermoluminescence measurements made on some of them (McKerrell, Mejdahl, François and Portal, 1974) indicated that the ceramics were not recently fired but had probably been made about 2000 years ago. The majority of archaeologists still vehemently rejected the notion that the site was anything but a hoax, and the controversy was reactivated. This rejection was partly due to the history of the discoveries (though 'Glozelites' contest the accepted interpretation that this had been a shady business), and partly on grounds (Renfrew, 1975) that, even if the site was to be accepted as a unique manifestation of an otherwise unknown culture, the absence of run-of-the-mill pottery fragments typical of close-by sites of the period rules out its authenticity.

The archaeological rejection of the thermoluminescence finding led to the validity of thermoluminescence dating being called into question, at any rate in this particular instance (Daniel 1974, 1977; Aitken and Huxtable, 1975; McKerrell, Mejdahl, François and Portal, 1975). The proximity of a high-grade uranium mine seemed a strong coincidence, but thermoluminescence measurements on zircons extracted from several of the objects ruled out artificial irradiation, at any rate for those objects. The notion that the objects had been made by cementation of ground-up Roman tile was encouraged by the generally poor quality of the pottery fabric, and this possibility was investigated for six objects using archaeomagnetic intensity analysis (Barbetti, 1976). As mentioned earlier, the disorientation of the magnetic grains would give rise to a resultant magnetic moment negligible in comparison with that acquired by the sample upon heating in the laboratory. The first object tested, a bisexual figurine, did in fact show this sort of behaviour, indicating either that the

2.5 Authenticity Testing

original firing temperature had been less than 300°C or that reconstitution had taken place and that the temperature reached in this process had not exceeded 300°C. This finding was in accordance with the poorly fired appearance of the object. To explain why the thermoluminescence measurements on it indicated an age of about 2000 years rather than a geological age corresponding to unfired clay, in the case of the first alternative one must presume that the original firing had been sufficient to reset the geological thermoluminescence to zero.

The results for the other five objects indicated firing temperatures in excess of 500°C. For two of them, which were from tablets carrying the strange script, reliable determinations of the intensity of the ancient magnetic field in which they cooled were possible. The values obtained were 46 and 47 μT, suspiciously close to the present-day value at Glozel of 46 μT. At that time available evidence suggested that the field 2000 years ago was substantially higher, but subsequent research indicated that there had been a deviation from the general trend, reaching close to the present-day value (Shaw, 1979), thus removing what had at first seemed a strong discordance in the scientific evidence.

One problem in the Glozel investigation was the diverse nature of the objects and the tendency to forget that there could have been a number of different origins, some ancient, others modern. Also, in view of the conflicting accounts of the circumstances of excavation, there is the possibility that the ancient ones came from a site remote from Glozel (though the trace elements do correspond to local clay). Even so, the strange script remains a puzzle; according to Crawford (1977) the script does not have the characteristics of a natural language but is a collection of signs, of which there are at least 133. Another possibility is that the tablets themselves are ancient but the signs were added subsequently. Some tablets have small blobs of glass visible on and around the inscription surfaces, and consequently the thermoluminescence date that has been reported specifically for one such tablet is of particular interest: it lies within the last few hundred years. However, this does not conclusively prove a fairly recent origin for the inscription since there is the possibility that the tablet had been reheated through accidental association with glass-making activities that were going on at Glozel, perhaps as recently as the eighteenth century (François et al., 1977).

The Glozel affair has been discussed at some length for two reasons beside its intrinsic fascination: first, because the validity of thermoluminescence dating has been called into question on account of it; and second, because, as perusal of the cited references will illustrate, albeit with an extreme case, it shows the quagmire of embittered controversy in which a physical scientist is likely to have to wade once he involves himself with material of doubtful origin. Readers having it in mind to involve themselves in authenticity testing take warning!

Technical Notes

[1] The Franz magnetic separator consists of a flat chute positioned between the poles of an electromagnet. The chute is tilted both longitudinally and laterally. The grains are poured in at the top end and the chute is vibrated so that they move slowly downward. During this movement any grains that are slightly magnetic are pulled to the side of the chute which is highest because it is arranged that there is a magnetic field gradient such that the higher side of the chute is in a stronger field than the lower side. Hence at the bottom end of the chute one obtains weakly magnetic clay grains from the high side and non-magnetic crystalline grains from the lower side.

[2] A convenient practical procedure for the hydrofluoric acid treatment has been reported by Carriveau (1977), who recommends use of $AlCl_3$ at the end in order to remove precipitated fluorides; some workers use HCl for this purpose. Before using HF the laboratory safety officer should be consulted; a good fume cupboard is important as well as immediate availability of running water and burn jelly (containing, for example, calcium gluconate) against the possibility of skin contact. In this context concentrated HF usually means 40% strength, but some workers use 48%.

[3] Normalization be weight is not always satisfactory, because the light collection efficiency may differ due to slight variations in the way in which the sample is spread. Also, because it seems that only one grain in about 10 is 'bright' in its thermoluminescence, there is likely to be a significant statistical variation in the number of bright grains present in a 5 mg sample (the total number of 100-μm grains in a 5 mg sample is about 4000). Larger samples are undesirable because of the effects discussed in Section 5.2, where other methods of normalization are also mentioned: zero-glow monitoring (ZGM), equal pre-dose, and, the use of natural thermoluminescence for second-glow growth characteristics.

[4] It is important that the ratio be the same for all three elements; otherwise the correct one to be used will depend on the composition of the sample. For the arrangement of Figure 2.3, Bailiff (1982) found the individual ratios to be 0.287, 0.279 and 0.279 for K, Th, and U, respectively. This excellent 'element independence' cannot be assumed for other geometries; for further discussion see Section 4.3.3.

[5] Stainless steel discs may be preferred because of interference by spurious thermoluminescence that is presumed to come from aluminium oxide; this is only a danger if the disc is not well-covered with fine-grains. Some workers abraid the discs with sandpaper in order to improve the adhesion of the fine-grains. Analar grade acetone should be used in sample preparation and blank tests run on this

whenever interfering thermoluminescence is experienced; storage of acetone in plastic containers must be avoided.

[6] A difficulty in using grains carried on aluminium discs is that the aluminium itself tends to have a significant alpha activity. It is safer to deposit the residue on high-purity copper discs (diameter 15 mm). The thickness of the deposit is determined by weighing, and the alpha count-rate α_∞ that would be observed for a thick layer is derived from that actually observed (α) from a layer of thickness t mg/cm^2 by the relation

$$\alpha_\infty = \alpha/(2t/R - t^2/R^2),$$

where R is the average range, that is, 6.4 mg/cm^2 (25 μm) for equal thorium and uranium activities. The derivation of this formula is given in Appendix J.

CHAPTER 3

THERMOLUMINESCENCE

3.1 The Thermoluminescence Process

The details of the mechanism by which thermoluminescence is produced in any given mineral are not well understood, and in general it is only for crystals grown in the laboratory with strict control of impurities that these details can be elucidated. This is because thermoluminescence is very dependent on minute amounts of impurity and also on thermal history, both largely unknown in the context of dating. Investigation of the physical details of processes in particular minerals has not so far been able to assist significantly in dating; readers wishing to know more of this aspect should look elsewhere (e.g., Sankaran, Nambi and Sunta, 1983; McKeever, 1985). The situation is very different to that in radiocarbon dating; there the essential mechanism is the radioactive decay of the carbon-14 nucleus, which occurs at the same rate whether the sample is wood, shell, or human bone. In thermoluminescence dating different minerals have different mechanisms, and even for the same mineral each sample must be calibrated individually for sensitivity because it will be influenced by the sample's actual impurity content, etc.

However, the main features of thermoluminescence dating can be discussed usefully in terms of a simple model. It is convenient to base this model on the ionic crystal shown in Figure 3.1, though it should be realised that other types of insulators, such as covalent solids and glasses,

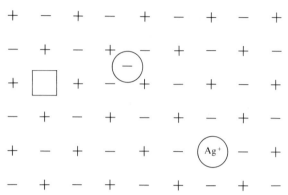

Figure 3.1 Simple types of defect in the lattice structure of an ionic crystal. From left to right: negative-ion vacancy, negative-ion interstitial, substitutional impurity centre.

also exhibit thermoluminescence; metals do not. An ionic crystal (e.g., calcium carbonate and sodium chloride) consists of a lattice of positive and negative ions; however, there can be defects in this regular order such as those due to impurity atoms, rapid cooling from the molten state, and damage caused by nuclear radiation. There are many types of defect that can occur, of which three simple ones are shown. A defect due to one of the negative ions being absent from its proper place, that is, a negative-ion vacancy, acts as an electron *trap* because the local deficit of negative charge attracts an ionized electron if it diffuses into the vicinity. Ionized electrons result from the action of nuclear radiation in detaching electrons from their parent nuclei. This ionizing effect of nuclear radiation, as opposed to the much less probable event of damage to the lattice, gives rise to the thermoluminescence.

Once in a trap an electron remains there until 'shaken out' by the vibrations of the lattice. As the temperature is raised these vibrations get stronger, and the probability of eviction increases so rapidly that within quite a narrow temperature range the situation changes from the electrons being firmly trapped to being free to diffuse around the crystal. A variety of fates awaits a diffusing electron. It can be re-trapped and re-evicted; it can be trapped at different types of defect better able to shield it from the lattice vibrations, that is, a deeper trap; or it can recombine with an ion from which an electron has previously been detached. This recombination can be of two types—radiative (i.e., with the emission of light) or non-radiative. Ions or atoms at which radiative recombination occurs are called *luminescence centres* and the light emitted is, of course, the thermoluminescence. Luminescence centres are a particular type of defect and are usually due to impurities—Ag^{2+} or Mn^{2+}, for instance. The colour of

3.1 The Thermoluminescence Process

the emitted light is characteristic of the impurity—blue/violet for silver, orange for manganese. According to this simple model evicted electrons have a chance of finding all types of luminescence centres that are present in the crystal. In reality a given type of trap is sometimes preferentially located close to a given type of centre, or they may both be part of the same complex defect. In this case the colour of the thermoluminescence emitted at the temperature corresponding to eviction from that trap may be different from the colour of the thermoluminescence at temperatures corresponding to eviction from other types of trap.

The above gives the essential basis of the thermoluminescence process. In summary, there is (i) ionization of electrons by nuclear radiation; (ii) capture of some of these electrons at traps, where they remain held as long as the temperature is not raised. Then centuries later in the measurement process, (iii) heating causes eviction from the traps, at a temperature characteristic of the type of trap. Almost instantaneously (iv) some of these evicted electrons reach luminescence centres and in the process of recombination there is emission of light, with a colour that is characteristic of the type of centre; the amount of light (i.e., the number of photons) is proportional to the number of trapped electrons, which in turn is proportional to the amount of nuclear radiation to which the crystal has been exposed.

Item (ii) requires qualification. There is a finite lifetime for an electron in a trap. Even at normal temperatures there is some probability of escape. This may be so slight that the lifetime is millions of years; such traps are 'deep' and the characteristic temperature at which rapid eviction occurs is correspondingly high—usually upwards of 400°C. In any crystal there are different types of trap, each with a different characteristic temperature; for the shallower ones lifetimes may be only an hour or less (with glow-peak temperatures below 100°C). Obviously, in dating it is only traps having lifetimes of tens of thousands of years and upwards, roughly corresponding to glow-curve temperatures of 250°C and upwards, that are of interest. In ancient pottery there is usually a variety of minerals, each with a variety of traps; as a result, the individual glow-peaks merge into each other and the thermoluminescence glow-curve is continuous—as illustrated in Figure 1.1; note that there is no thermoluminescence below 200°C, because the lifetimes of the associated traps are too short.

3.1.1 THE ENERGY-LEVEL DIAGRAM

A convenient way to represent the thermoluminescence process is shown in Figure 3.2. A trap is characterized by the energy E which an

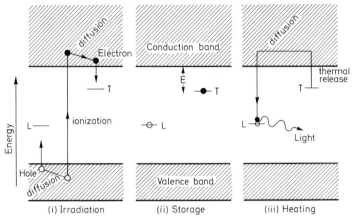

Figure 3.2 Energy-level representation of the thermoluminescence (i) Ionization due to exposure of crystal to nuclear radiation, with trapping of electrons and holes at defects, T and L, respectively. (ii) Storage during antiquity; the lifetime of the electrons in the traps needs to be much longer than the age span of the sample in order that leakage be negligible. This lifetime is determined by the depth E of the trap below the conduction band, and for dating purposes we are interested in those deep enough ($\sim 1\frac{1}{2}$ eV) for the lifetime to be the order of a million years or more. (iii) To observe thermoluminescence the sample is heated and there is a certain temperature at which the thermal vibrations of the crystal lattice causes eviction. Some of these evicted electrons reach luminescent centres, and if so, light is emitted in the process of recombining at those centres. Alternatively, the electron may recombine at a non-luminescent centre (a 'killer' centre) or be captured by a deeper trap.

electron must acquire from the lattice vibrations in order to escape from it and diffuse around the crystal; while diffusing it is described as being 'in the conduction band'. Before nuclear radiation detaches it from its parent lattice ion or atom, an electron is described as being 'in the valence band'. The traps and centres provided by defects are represented as intermediate between the valence band and the conduction band; electrons can be detached from these as well.

Implicit in the diagram is the concept of 'hole', already mentioned as a charge carrier. An atom from which an electron has been detached is said to be charged with a hole, and by gaining an electron from a neighbouring atom the hole can be passed on, so that it is convenient to regard it as a carrier of positive charge. Ionization can be thought of as the formation of electrons and holes, and luminescence as the recombination of electrons and holes at a luminescent centre; if the recombination occurs at a non-luminescent centre (sometimes called a 'killer' centre), there is no emission of light and the excess energy is dissipated as heat. Holes can be trapped in a similar manner to electrons, and the characteristic energy of a hole trap is represented in the diagram by its distance above the valence band; this is because it is in the valence band that holes can diffuse freely. In some crystals the storage of latent thermoluminescence is by means of trapped holes; however, this makes no practical difference, and it is convenient to presume in discussion that electrons are the trapped carriers.

It should be noted that the colour of the light emitted by a luminescence centre is not usually determined by the depth of the centre below the conduction band. An electron goes from the conduction band to an excited state of the centre, surplus energy being dissipated as heat; the emission of light occurs when, very quickly, the centre de-excites.

3.2 Other Types of Thermoluminescence; Prompt Luminescence; Spurious Thermoluminescence

If some types of crystal are subjected to pressure, there is emission of light on subsequent heating—called *piezo-thermoluminescence*. If grains are rubbed together or otherwise subjected to frictional effects, then on subsequent heating there is *tribo-thermoluminescence*. Similarly, the thermoluminescence resulting from prior exposure to light is termed *photo-thermoluminescence*. There is also emission of light while the agency is being applied, it then being termed, respectively, *piezoluminescence, triboluminescence,* and *photoluminescence*. 'Prompt' light is also emitted during exposure to nuclear radiation—*radioluminescence*—resulting, like the other types, from immediate recombination of electrons and holes at luminescence centres; *cathodoluminescence* specifically refers to exposure to beta particles or electrons. Light persisting after cessation of irradiation (or other agency) is termed *phosphorescence*. There are two mechanisms contributing to this: first, there is the release of electrons from very shallow traps because of their short lifetime at room temperature; and second, there is the tunnelling phenomenon discussed at the end of this chapter. An alternative terminology is *afterglow*.

The agencies able to produce latent thermoluminescence can also diminish it—an apparent contradiction discussed in the next paragraph. In the specific case of light as the agency, this diminution is termed 'bleaching', representing eviction of electrons from traps by photons; some of these electrons may recombine with holes at luminescence centres, so it is to be expected that photoluminescence will accompany bleaching. In general, the wavelength of the photoluminescence is longer than that of the exciting light, and this facilitates detection. It may also be noted that the photon energy necessary to achieve eviction from a trap is somewhat greater than the phonon energy required to achieve thermal eviction; this is because thermal eviction, though rapid, does give time for the lattice to relax into a new configuration, whereas photoeviction does not. Photoeviction is important both because of the need to process samples in subdued light once a surface layer has been removed and because it is the zero-setting mechanism in sediment dating.

There are two reasons why the same agency can both produce and diminish latent thermoluminescence. First, there is the obvious one that different types of traps, and therefore different parts of the glow-curve, may be affected differently. Second, for a given type of trap, filling and emptying may be going on simultaneously as competing processes—as with a leaky bucket being filled by a tap; if the traps are appreciably filled there will be a net eviction, whereas if they are nearly empty there will be a net input.

3.2.1 SPURIOUS THERMOLUMINESCENCE

In the context of dating it is the radiation-induced thermoluminescence that is the 'true' signal, and in general all *non-radiation-induced* (NRI) emission is spurious. Exactly what is encompassed by the latter term is not well defined and may depend on the context; the thermal radiation signal—red-hot glow or 'black body'—is something separate, and photothermoluminescence is usually excluded also. Certainly included is the *chemithermoluminescence* that accompanies some heat-stimulated chemical changes and phase transformations (e.g., aragonite changing to calcite); incidentally, for completeness in this section mention should be made of the *chemiluminescence* associated with some chemical reactions at room temperature, and of the *bioluminescence* as exemplified by the glow-worm, and indeed mentioned by Sir Robert Boyle in his 1663 report to the Royal Society.

Besides chemiluminescence, the other commonly experienced component of spurious thermoluminescence is tribo-thermoluminescence. It can be induced merely by 'stirring' already glowed grains on the heater plate, or even by blowing air across them. The mechanism has received sporadic attention, but without conclusive findings. For dating the important thing is to achieve suppression of all such spurious emission of light. As already emphasized in the first chapter, one essential condition is that the inert gas filling the glow-oven should contain less than a few parts per million of oxygen and water vapour. As might be expected with a surface phenomenon, small grains are more prone to exhibit spurious thermoluminescence than large ones.

3.3 Stability of the Thermoluminescence Record

3.3.1 PROBABILITY OF ESCAPE; FORMATION OF THE GLOW-PEAK

As mentioned earlier, the probability of escape from a trap rises very rapidly with temperature. This is illustrated in Figure 3.3, which also shows how the number of electrons remaining in traps of a given depth changes with temperature—as the probability of escape rapidly increases, the number of electrons remaining rapidly decreases. The intensity of the thermoluminescence being emitted (i.e., photons per second) is proportional to the probability of escape multiplied by the number remaining. Hence as the temperature is raised the thermoluminescence intensity rises to a maximum and then rapidly decreases to zero as traps of that depth become completely empty. The term 'characteristic temperature'

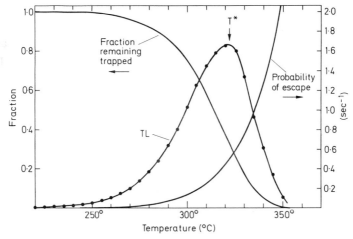

Figure 3.3 Formation of a thermoluminescence peak. The thermoluminescence intensity (scaled up ×10 in the diagram) is proportional to the rate at which the trapped electrons escape, and this is equal to the probability of escape multiplied by the number remaining trapped. The probability of escape rises very rapidly with temperature, but the thermoluminescence intensity does not increase for long, because the number of electrons remaining trapped is simultaneously reduced. Shortly after half of the electrons have escaped, the thermoluminescence intensity starts to fall, the temperature of its maximum being denoted by T^*. The diagram is based on equations given in the text and elsewhere,[4] using trap parameters $s = 10^{14}$ sec^{-1}, $E = 1.7$ eV, and a heating rate of 10°C/sec. At 15°C the calculated lifetime of electrons in such traps is 100 million years.

can now be replaced by the more usual term 'peak temperature,' T^*. Both peak temperature and peak width depend to a small degree on the rate at which the temperature is raised (see Figure 3.4). In practice, thermoluminescence peaks are often broader than the calculated ones shown in Figures 3.3 and 3.4; this is because they are composite, as illustrated in Figure 3.5.

Mathematically the probability of escape (per second) is described by the exponential expression

$$\lambda = s\, \exp(-E/kT), \tag{3.1}$$

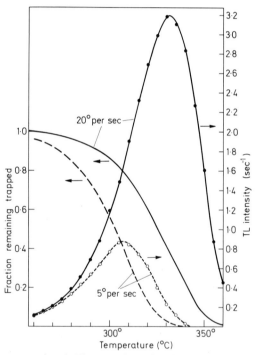

Figure 3.4 The same peak as in Figure 3.3 but at heating rates of 5°C/sec and 20°C/sec. At any given temperature the fraction remaining trapped is greater for the faster rate because there has been less time for emptying; hence T^* is higher by 35°C. Because the same number of electrons escape in a shorter length of time, the peak height for the faster rate is nearly 4 times that for the slower rate; it is not exactly 4 times because the glow-curve is slightly wider (43°C instead of 35°C). If time were used for the horizontal axis, then the scale for the faster rate would be contracted by a factor of 4 and the total area would be the same in both cases. The vertical scale on the right represents $10(n/n_o)s\, \exp(-E/kT)$.

3.3 Stability of the Thermoluminescence Record

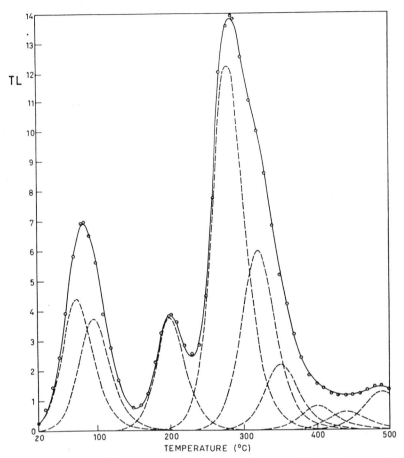

Figure 3.5 Analysis of the experimental glow-curve for natural fluorite into constituent peaks (Ganguly and Kaul, 1984). For the two peaks near 300°C the E values are 1.75 and 1.90 eV; consequently the lower one decays more rapidly if the sample is held at elevated temperature, and so the maximum of the composite peak then occurs at a slightly higher temperature—see Figure 3.6: Symbols: ⊙⊙⊙, from experimental glow-curve; ---, computer-generated peaks; ——, sum of computer-generated peaks.

where s is a *frequency factor* (which may be crudely thought of as the number of attempts to escape per second; its value depends on the type of trap, and is usually in the range 10^9 to 10^{16} sec^{-1}); E is the trap depth (in electron-volts)[1] and k is Boltzmann's constant, which relates the average energy provided by the lattice vibrations to the absolute temperature T; at 17°C the value of kT is 0.025 eV.

3.3.2 LIFETIME

As the temperature is raised in the course of the glow-curve, the increased lattice vibrations give sufficient energy to the trapped electrons for there to be a high probability of surmounting the barriers that retain them, and at the peak temperature eviction is rapid—in a fraction of a second. At the temperature of burial the probability of escape is many orders of magnitude smaller but it is not necessarily negligible: the average energy of the lattice vibrations is much lower but, although an electron is comparatively secure, in the end the random chance of an abnormally large vibration causes eviction. For dating the question is whether a significant percentage of the trapped electrons are likely to have suffered this random eviction during the sample's age span.

For a sample held at constant temperature, the rate at which electrons escape is given by the probability of escape multiplied by the number still trapped, that is, by the differential equation

$$-dn/dt = \lambda n, \qquad (3.2)$$

where n is the number remaining trapped at time t. It follows from this that the number of trapped electrons decays with time according to

$$n = n_0 \exp(-\lambda t), \qquad (3.3)$$

where n_0 is the number at $t = 0$. For such exponential decay it is convenient to talk of the *lifetime*,[2] the average residence time of electrons in a given type of trap; it can be shown that this is given by

$$\tau = \lambda^{-1} = s^{-1} \exp(E/kT), \qquad (3.4)$$

so that upon rewriting Equation 3.3 in terms of lifetime we have

$$n = n_0 \exp(-t/\tau). \qquad (3.5)$$

This equation deals with the case of there being n_0 trapped electrons at $t = 0$. In dating the situation is that there are no trapped electrons at time zero but they become trapped at a uniform rate thereafter; for this case it can be shown that the fractional loss of thermoluminescence due to escape during the sample's age span, t_s, is given by $\frac{1}{2}(t_s/\tau)$ to a good approximation[3] so long as t_s does not exceed one-third of τ. If we set 5% as the upper limit to the loss that can be tolerated, then the lifetime must be at least 10 times the age of the sample; if we allow 10% loss, then 5 times is enough. However, it is not prudent to place too much reliance on this because τ is strongly dependent on average burial temperature, and this is not usually known with sufficient precision. In practice prime reliance is placed on the plateau test used routinely for each sample dated; however, it is useful to have some idea of what is predicted by theory. In Table 3.1

3.3 Stability of the Thermoluminescence Record

TABLE 3.1
Rough Estimates of Lifetimes Corresponding to Various Glow-peak Temperatures[a]

Peak temperatures	100°C	200°C	300°C	400°C	500°C
Lifetime for burial at 10°C	2 hr	10 yr	600×10^3 yr	3×10^{10} yr	2×10^{15} yr
Lifetime for burial at 20°C	½ hr	2 yr	70×10^3 yr	3×10^9 yr	1×10^{14} yr

[a] Estimates are based on Equations 3.4 and 3.6, taking $s = 10^{13}$ sec^{-1} and $\beta = 20°$ sec^{-1}.

an estimate is given of the lifetimes for burial at 10°C and at 20°C of glow-peaks having various peak temperatures.

The actual values for any given type of trap, derived from measurement of its characteristic parameters E and s, are sometimes substantially different from these estimates; for instance, the thermoluminescence peak observed near 300°C in stalagmitic calcite has an estimated lifetime at 10°C of more than several million years; the same holds for the main peak in natural fluorite, also near 300°C. The purpose of this table is to illustrate first, the very sharp dependence of lifetime on glow-peak temperature, and second, the strong dependence on burial temperature. Estimated lifetimes for some other minerals relevant to dating are given in Appendix E, together with an outline of the methods available for measuring E and s.

3.3.3 KINETICS

Derivation of lifetime depends not only on trap parameters but also on the assumption of *first-order kinetics*. The essential meaning of the latter, sometimes termed *monomolecular decay*, is that the intensity of the thermoluminescence is determined solely by the rate of escape from the traps, as described by Equation 3.2. For this to be so both the degree to which available luminescence centres are 'used up' by electrons released earlier in the glow-curve and the probability of the electron being retrapped must be negligibly small. If not, then Equation 3.2 is not applicable and the situations are much more complex, of which the simplest approximation is *second-order kinetics,* or *bimolecular decay*. This daunting topic is briefly mentioned below, but no attempt at a full discussion will be made. Even for first-order kinetics there is no simple exact relationship between E and T^*, the various trap parameters, etc., being related together by the equation[4]

$$\frac{E}{kT^*} \exp\left(\frac{E}{kT^*}\right) = \left(\frac{sT^*}{\beta}\right), \qquad (3.6)$$

where β is the rate per second at which the temperature is raised. The value of E corresponding to a given T^* is only weakly dependent on s and β; for the values used in Table 3.1, $E \approx (T^*/400)$ eV, that is, for a peak temperature of 400°C the trap depth is 1.7 eV.

It is implicit in the foregoing equations for first-order kinetics that the peak temperature does not depend on the number of trapped electrons. This means that the peak position is independent of the radiation dose and of whether or not there has been partial decay. This is not the case for second-order kinetics: for a sample that has had only a small dose, the peak occurs at a higher temperature than for one that has had a large dose. Another difference is that it is no longer meaningful to talk of a lifetime; this is because the rate of decay with time is no longer proportional to the number of electrons remaining trapped; it is proportional to the square of that number in the case of second-order kinetics—hence the terminology. Compared with the first-order case, the rate of decay slows down as the traps get emptier, as would be expected with retrapping taking place.

Shift of peak position with degree of trap occupancy implies that samples containing minerals exhibiting second-order kinetics will fail to give a good plateau unless the artificial dose is roughly equal to the natural dose (McKeever, 1979); fortunately, this latter condition is normal practice, and so unnecessary rejection of samples is minimized. Conversely, for a

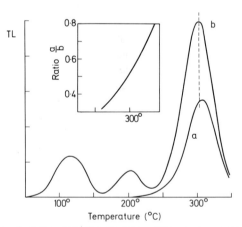

Figure 3.6 Failure of plateau test due to decay. The same sample of fluorite (natural calcium fluoride) was used for both glow-curves. In case (a) the sample was held for 2 min at 230°C subsequent to irradiation. Although the peak shift is only ~5°C, there is no hint of a plateau in the ratio a/b. This is a different type of fluorite than that of Figure 3.5, and evidently the higher-temperature component (of the main peak) is much less pronounced here. The dose was 1 mGy; the heating rate was 10°C/sec.

3.3 Stability of the Thermoluminescence Record

sample consisting of a single mineral with a single first-order peak there is the theoretical possibility that the plateau test can be passed despite decay having occurred. However, with natural minerals a true single peak is rare[5], and it is usual for peak shift to occur as decay proceeds, as illustrated in Figure 3.6 for fluorite; this peak shift is unlikely to be due to second-order kinetic behavior because there is no peak shift with increased dose. The presumed reason for peak shift in such cases is that the peak is composite and results either from several types of trap having different E values, as illustrated in Figure 3.5, or from a single type of trap having a finite spread in E value.

The kinetics of the thermoluminescence process have been discussed by a number of authors, beginning with Randall and Wilkins (1945) and Garlick and Gibson (1948); the reader is also referred to publications by Chen and Kirsch (1981) and by Levy (1979, 1982, 1983).

3.3.4 CORRECTNESS OF PREDICTED LIFETIMES; THERMAL QUENCHING

The predictions made for lifetimes are based on the trap parameters E and s. Inevitably the experiments by which these parameters are measured take place over a time many orders of magnitude shorter than the lifetime concerned; consequently, the validity of the predictions rests very heavily on the exactness of the equations used to describe the escape process, including, for instance, the assumption that s does not change with temperature. Checking of the predicted lifetimes by means of samples of known age or by looking for failure of very old samples to pass the plateau test does not turn out to be something that can be undertaken at all comprehensively. However, except for minerals affected by the phenomenon of anomalous fading discussed in the next section, there is no evidence of any of the predictions being incorrect.

One apparent failure was with respect to the 325°C peak in quartz. Using the value of E obtained with the initial rise method (see Appendix E), the predicted lifetime was 3000 years. Yet in a test (Fleming, 1970) using Romano–British samples of known age, the age deduced using this peak was within a few percent of the correct age, which was close to 1900 years; for the lifetime quoted the apparent age would be expected to be 25% below the true age. This discrepancy, albeit a favourable one, was investigated by Wintle (1975a), who found the cause to lie in not making allowance for *thermal quenching* when using the initial rise method. This term refers to a decrease, as the temperature increases, in the probability that a released electron will give rise to a photon of thermoluminescence; when the trap parameters were measured by means of a method not affected by thermal quenching, a value of 30 million years was obtained for the lifetime.

Specifically, thermal quenching is concerned with the decreased likelihood that an excited luminescence centre will de-excite by emission of a photon rather than dissipate its excess energy as heat in the form of enhanced lattice vibrations. The existence of thermal quenching in a mineral does not imply any breakdown in the kinetic equations describing escape. The temperature at which thermal quenching becomes significant depends on the type of luminescence centre. Early onset means that despite deep traps being present there is no high-temperature thermoluminescence because the available luminescence centres have effectively ceased to function.

3.4 Anomalous Fading

Another effect is shown in Figure 3.7, which gives glow-curves for a fine-grain sample of modern brick; one taken within a few minutes of irradiation and the other after overnight storage at $-18°C$. As will be seen,

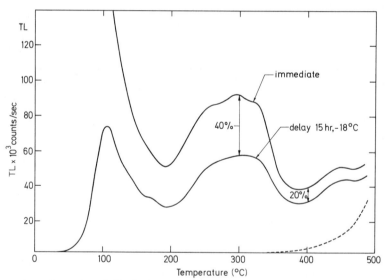

Figure 3.7 Anomalous fading in the thermoluminescence from a fine-grain sample of modern brick. For the lower glow-curve there has been a delay of 15 h between irradiation and measurement; despite storage at $-18°C$, there is substantial loss of thermoluminescence throughout the glow-curve.

3.4 Anomalous Fading

there is substantial fading in the glow-curve region where, on all expectations based on the kinetics of escape as just discussed, there should be stability. As we shall see, it can occur with a number of minerals likely to be present in clay, and so the effect calls into question the reliability of thermoluminescence dating. In the particular example shown the percentage loss does not vary much across the last 100° of the glow-curve; hence its occurrence will not necessarily be detected by the plateau test. This failure of the plateau test only happens in some cases; the test remains an essential minimum requirement.

In the context of thermoluminescence dating the effect was first noted[6] by Wintle (1973; see also Wintle, Aitken and Huxtable, 1971). In tests made on some twenty samples of various types of feldspar from lava flows, it was found that for four of them there was fading in the 350°–400° region of the glow-curve by 10–40% during overnight storage after artificial irradiation; for five of them the fading was less than 5% during a month, the rest being intermediate. Substantial overnight fading was observed in zircon and fluorapatite also, but was not detected in quartz and limestone stored for 2 years and 2 months, respectively.

A feature of this type of fading is that its dependence on storage temperature is rather weak; hence the alternative terminology *athermal* fading, sometimes *non-thermal*. Whereas in the case of thermal fading a 10°C elevation in the storage temperature is likely to cause an order of magnitude increase in fading rate (see Table 3.1), in the case of the anomalous fading exhibited by labradorite (a plagioclase feldspar) investigated in detail by Wintle (1977a) the storage temperature could be raised by more than 100°C before the same order of change occurred. The athermal character was even more marked at low temperature; although the fading was reduced by storage in liquid nitrogen (-200°C), it still remained appreciable, which would not have been the case for thermal fading; also, storage in liquid hydrogen (-250°C) gave no further reduction.

At present the commonly accepted explanation for anomalous fading is based on quantum-mechanical tunnelling, well-established as a mechanism responsible for similar effects in synthetic crystals (see Appendix F). The tunnelling process (see Figure 3.8) cannot be understood in terms of classical physics, only in terms of quantum mechanics: the wave function describing the trapped electron implies a small but finite probability that it will appear outside the energy barrier that retains it, and if this wave function overlaps that of a nearby centre able to accept an electron (i.e., a center that is positively charged), then a transition may occur, the electron thus escaping from the trap.

An explanation along these lines for the anomalous fading of thermoluminescence in natural minerals was first suggested by Garlick and Robin-

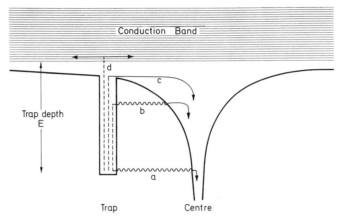

Figure 3.8 Escape routes from a trap: a, athermal tunnelling; b, thermally assisted tunnelling; c, d, thermal eviction (derived from Visocekas *et al.*, 1976).

son (1972) in the context of lunar samples, and it was one of the mechanisms proposed by Wintle (1977a) with respect to labradorite. However, other mechanisms may be operative as well, such as the two possibilities suggested by Wintle; one of these is concerned with slow diffusion of the defects themselves and the other with decay of effective luminescence centres. A centre is only effective when it is ready to accept an electron; if there is a slight but finite probability that electrons from the valence band can reach these centres, even though the temperature is not elevated, then the number of effective centres will slowly decay with time, with a consequent decrease in thermoluminescence sensitivity. There are certain difficulties in accepting this mechanism, which are discussed in Appendix F along with some other aspects of anomalous fading.

On the question of which minerals may be expected to be affected by anomalous fading, the tunnelling explanation suggests that one condition is close proximity of traps and centres. If these are randomly distributed, then fairly high concentrations are required, implying that minerals exhibiting bright thermoluminescence are more likely to be affected than others. Indeed, feldspars are bright in thermoluminescence, whereas quartz is rather dim. On the other hand, the very bright natural fluorite used in thermoluminescence dosimetry is immune, though equally bright artificial calcium fluoride doped with dysprosium is not. Also, feldspars not of volcanic origin, though bright, appear to be less prone; evidentally the state of crystallization has a strong influence too.[7] Whatever the reason, it is fortunate that the majority of mineral inclusions in clay are formed by the weathering of plutonic rocks rather than volcanic rocks.

3.4 Anomalous Fading

There can also be close proximity between traps and centres if both are part of the same complex defect; hence, conversely, it is not safe to assume that immunity is guaranteed by dimness.

It is also to be noted that within the same sample some thermoluminescence peaks may be affected and others not. Thus reliable dating of volcanic feldspars has been achieved by Guérin and Valladas (1980) using the 600°C region of the glow-curve despite strong fading in the 400°C region. With zircon and fluorapatite also there appear to be deep traps which are relatively immune (see Section 6.2) despite less deep traps in the same sample being strongly affected. These observations are consistent with another prediction from tunnelling theory: that the probability of transition decreases as the trap depth increases. On the other hand, no strong decrease of transition probability is indicated in the example of Figure 3.7, nor was it in some of the examples quoted by Wintle. In such cases, where it seems that all traps are affected more or less equally, there is an argument for thinking that the operative mechanism may be decay of luminescence centres.

3.4.1 IMPLICATIONS FOR DATING

When it was discovered that some samples of feldspar and of other minerals exhibited anomalous fading, there was a strong reaction in favour of using quartz rather than feldspar so far as inclusions were concerned, and considerable emphasis was placed on routinely testing each sample for fading when the fine-grain technique was employed. In retrospect it appears that the reaction against feldspars was overdone and that, at any rate, for Scandinavian pottery and burnt stones feldspar inclusions can be used to advantage (see Section 6.5); the same applies with respect to sediment dating (Chapter 8). However, the need for testing remains, as indeed it does with respect to the fine-grain technique. The incidence of the effect in pottery is highly variable from sample to sample and from site to site. There are some sites (see, for instance, Whittle and Arnaud, 1975; Whittle, 1975) where it is so prevalent that fine-grain dating is ruled out, but usually it is no more than 1 in 10 samples that are affected and then only to a small degree (the fading lowering the age by no more than 10%).

The upset to dating arises because the artificial thermoluminescence includes a short-lived component that has mostly disappeared from the natural thermoluminescence. One routine way commonly employed to minimize the extent to which the date is too young is to insert a delay between irradiation and thermoluminescence measurement; although the temperature dependence is weak, there is some acceleration in removal of the unwanted component if the storage is at elevated temperature, for example, overnight at 50°C. However this in no way guarantees that removal is complete and, except with respect to quartz and calcite, samples need to be given special storage tests to see if they are affected to a significant degree; if so, they should be rejected as not being reliable for dating unless it can be established that some thermal pretreatment such as just mentioned is effective.

Storage tests are inevitably short-term on an archaeological time scale, and the question of what constitutes an adequate duration is discussed in Appendix F. For these tests several portions of the sample, for example, several fine-grain discs, are needed. If first-glow artificial thermoluminescence is being used, an undrained portion is given an additive dose and the thermoluminescence measured after storage is compared with that from a similarly dosed portion which is measured immediately; to avoid interference by drift in equipment sensitivity, the measurements should be done at the same time. Unless discs of reliably equal sensitivity are used, normalisation between portions is required, for example, by weighing or by some form of thermoluminescence monitoring. Fading tests are sometimes made using second-glow artificial thermoluminescence, in which case the portions are irradiated after measurement of the natural thermoluminescence and the latter is used for normalisation. The second-glow technique has better precision, but there is evidence that for some samples it is not as effective as the first-glow test.

Fading is more rapid initially than subsequently, as might be expected; this is illustrated in Figure 3.9. Hence in testing for fading, if there is to be quantitative use of the data the 'immediate' measurement with which the delayed thermoluminescence is compared needs to be at a fixed time after

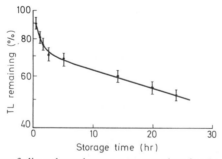

Figure 3.9 Anomalous fading: dependence on storage time for the plagioclase feldspar, labradorite, studied by Wintle (1977a). A number of portions were irradiated and then stored, at 10°C, for various lengths of time before measuring the thermoluminescence. The vertical scale is logarithmic, representing the percentage of the 300°C thermoluminescence remaining relative to the thermoluminescence from the portion which was measured immediately.

cessation of irradiation; obviously, the stringency of the test is increased by shortening this fixed time and by lengthening the storage time for the delayed measurement. In typical use these would be a few minutes and a few weeks, respectively.

The decay of thermoluminescence with time can be represented by several terms having an exponential dependence on time; on the other hand, for many samples a good fit to the data is obtained with a single term having logarithmic dependence on time. As discussed in Appendix F, this latter representation would be expected to be valid when tunnelling is the dominant mode of decay; in this case it is possible to set an upper limit to the amount of fading in antiquity using the result of the laboratory storage test. Also discussed in this appendix is the question of whether by sufficiently intensive thermal pretreatment (i.e., of longer duration and higher temperature than the overnight storage at 50°C mentioned earlier) it is possible to eliminate the fading component from the thermoluminescence of some minerals; optical bleaching is another pretreatment that may achieve elimination (Templer, 1985b).

Technical Notes

[1] The electron-volt (eV) is a unit of energy: 1 eV = 1.60×10^{-19} joule = 1.60×10^{-12} erg.

[2] It will be seen from Equation 3.3 that τ is the time at which the trap population has decreased by a factor e, that is, 2.7. It is related to the *half-life* $t_{1/2}$, the time taken for decay by a factor of 2, according to $\tau = (t_{1/2}/0.693)$.

[3] The precise expression for the fractional loss is

$$[1 - (\tau/t_s)(1 - \exp(-t_s/\tau))].$$

[4] The derivation of Equation 3.6 is as follows. From Equations 3.1 and 3.2 the rate at which trapped electrons are being evicted at a glow-curve of temperature T is given by

$$-dn/dt = ns \exp(-E/kT).$$

Since the thermoluminescence intensity is proportional to the rate of eviction, the maximum intensity occurs when the differential of this expression equals zero, that is, when

$$ns \exp(-E/kT) \left\{ (E/kT^2) \frac{dT}{dt} - s \exp(-E/kT) \right\} = 0.$$

After substitution of $\beta = dT/dt$ and T^*, this yields Equation 3.6.

[5] Even with the 110°C peak in quartz, there is a slight shift by 2 or 3°C (at a heating rate of 10°C/sec) for a peak that has suffered 50% decay.

[6] Comparable behaviour with respect to lunar material from the Apollo 11 and 12 missions was also reported [e.g., Dalrymple and Doell (1970), Hoyt *et al.*

(1970), and Garlick *et al.* (1971)]. The last named also noted strong anomalous fading in a terrestrial feldspar, labradorite.

In later lunar studies Hoyt *et al.* (1972) measured the thermoluminescence of single grains. Only one grain in about 15 carried natural thermoluminescence. Since a few of the grains that had no natural thermoluminescence showed only weak artificially induced thermoluminescence, it was hypothesized that whereas most grains were affected by anomalous fading, the occasional one was not. Grain-to-grain variation in anomalous fading has also been noted for zircons extracted from pottery; Sutton and Zimmerman (1976) found about half of the grains they examined were affected.

[7] The effect of recrystallization on the fading of labradorite and sanidine has been reported by Wintle (1974). Both faded at a slower rate and to a lesser extent after recrystallization, but the fading was still substantial. The minerals were subjected to a pressure of 1 kbar at 830°C for 3 days; they were cooled to 25°C within about 2 min, the first 100°C drop occurring in about 5 sec.

CHAPTER 4

NATURAL RADIOACTIVITY: THE ANNUAL DOSE

The measurement of radioactivity and its evaluation in terms of the annual dose that has been received by the sample is of equal importance to the measurement of thermoluminescence and the derivation of the paleodose. This takes us from solid state physics to natural radioactivity and to an evaluation of the energy absorbed by the sample from the nuclear radiations emitted; in a dating laboratory, expertise in these latter fields is probably a greater essential than in the former. It is also the area in which understanding by the excavator is most needed; this is because of the influence of radioactivity in the burial soil and the effect of variations in humidity of both sample and soil throughout the past.

To avoid misunderstanding it should be emphasized that in thermoluminescence dating we are concerned with radioactivity that is long-lived compared to the sample age span. The role that radioactivity plays is in providing a constant dose-rate for the production of thermoluminescence. This is in contrast to radioactive dating, such as by radiocarbon, where the half-life must be comparable to the sample age span, and the basis of dating is that the radioactive content of the sample decreases with age.

4.1 Natural Radioactivity

For the majority of samples, nearly all of the annual dose is provided, in roughly equal parts, by potassium, thorium, and uranium; the remainder, a few percent, comes from rubidium and cosmic rays. The radioactivity of potassium is due to its isotope potassium-40; when nuclei of this isotope undergo radioactive decay, beta particles and gamma radiation are emitted (see Table 4.1). With rubidium-87, there is emission of beta particles only. Natural thorium and uranium consist of *radioactive series* (or *decay chains*); the 'parent' of the thorium series is thorium-232, and when its nucleus undergoes radioactive decay, with emission of an alpha particle, the 'daughter' nucleus formed, radium-228, is also radioactive. This process continues through radioactive daughters until lead-208, which is stable, is reached. Natural uranium consists of two series, the parent of the principal one being uranium-238 and that of the minor one uranium-235; the latter accounts for only 0.72% of the atoms in natural uranium. The various members of the thorium and uranium series (see Table 4.2) emit a variety of alpha, beta, and gamma radiations. The amount of thermoluminescence induced depends both on the rate of emission and on the energy carried by the radiation. Typically the energy is on the order of several million electronvolts (1 MeV = 10^6 electron-volts = 1.602×10^{-13} J). More complete details of the decay series are given in Appendix G. In studying the decay series it should be remembered that all isotopes having the same name, for example, radium, are chemically identical but

TABLE 4.1
Radioactive Decay Schemes of Potassium and Rubidium[a]

Potassium-40 (natural abundance 0.0117%)		Rubidium-87 (natural abundance 27.8%)
potassium-40 (half-life: 1.25×10^9 years)		rubidium-87 (half-life: 48×10^9 years)
10.5%	89.5%	
γ(1.46 MeV)	β(1.36 MeV)	β(0.274 MeV)
argon-40 (stable)	calcium-40 (stable)	strontium-87 (stable)

[a] The value quoted for beta decay is the maximum energy of the beta particle spectrum. The average values are 0.583 MeV and 0.104 MeV for potassium and rubidium, respectively. The gamma emission is accompanied by electron capture.

4.1 Natural Radioactivity

TABLE 4.2
Radioactive Decay Schemes of Thorium and Uranium[a]

Thorium series		Uranium/radium series		Uranium actinium series (natural abundance 0.72%)	
Nuclide	Half-life	Nuclide	Half-life	Nuclide	Half-life
thorium-232	14.0×10^9 yr	uranium-238	4.47×10^9 yr	uranium-235	0.704×10^9 yr
↓ 1α		↓ 1α, 2β		↓ 1α, 1β	
radium-228	6.7 yr	uranium-234	245×10^3 yr	protactinium-231	32.8×10^3 yr
		↓ 1α			
↓ 1α, 2β		thorium-230 (ionium)	75×10^3 yr	↓ 2α, 1β	
		↓ 1α			
radium-224	3.6 d	radium-226	1600 yr	radium-223	11.4 d
↓ 1α		↓ 1α		↓ 1α	
radon-220 (thoron)	55 sec	radon-222	3.82 d	radon-219 (actinon)	4.0 sec
↓ 1α				↓ 1α	
polonium-216	0.16 sec	↓ 3α, 2β		polonium-215	1.8×10^{-3} sec
		lead-210	22 yr		
		↓ 2β			
↓ 2α, 2β		polonium-210	138 d	↓ 2α, 2β	
		↓ 1α			
lead-208	stable	lead-206	stable	lead-207	stable

[a] Full versions of these series are given in Appendix G.

differ in their radioactivity; thus whereas radium-228, the first daughter of the thorium series, has a half-life of 5.75 yr and is a beta emitter, radium-224 later in that series has a half-life of 3.6 days and is an alpha emitter whilst radium-226 in the uranium-238 series, also an alpha emitter, has a half-life of 1600 yr. The numbers 224, 226, and so forth, refer to the total number of neutrons and protons in the nucleus and are known as the *atomic weights* of the isotopes concerned, usually denoted by the symbol A; the scale is such that A g of an isotope contain 6.02×10^{23} atoms (Avogadro's number), that is, the mass of an atom of radium-224 is $224/(6.02 \times 10^{23}) = 37.2 \times 10^{-23}$ g.

Since an alpha particle consists of 2 neutrons and 2 protons and carries 2 units of positive charge, its emission causes a decrease of 4 in the atomic weight as well as a decrease in the charge of the nucleus, and hence a change in chemical behaviour. A beta particle carries one unit of negative charge but its mass is only 5.5×10^{-4} on the atomic weight scale; hence there is a change in isotope name but not in atomic weight. Gamma radiation consists of electromagnetic waves and in itself causes no change in either atomic weight or isotope name, the change that occurs being due to the alpha or beta emission to which it is an accompaniment; like light it has a dual nature and can alternatively be regarded as a stream of discrete photons.

4.1.1 RADIOACTIVE EQUILIBRIUM

The rate at which a group of N nuclei undergo decay is given by λN, where λ is the probability of decay per unit time related to the half-life by $\lambda = (0.693/t_{\frac{1}{2}})$. As is seen in Table 4.2 the half-lives of the members of a series differ widely, the longest half-life of each series being that of the parent. If a group of parent nuclei have been isolated by some process of chemical separation, then from being initially zero, the number of first daughter nuclei builds up exponentially, as determined by its own[1] half-life, until its rate of decay is exactly balanced by its rate of production, that is, $(\lambda_2 N_2) = (\lambda_1 N_1)$. Alternatively, if due to some process an excess amount of daughter is present the rate of decay will exceed the rate of production, and the amount will decrease until the equilibrium condition $(\lambda_2 N_2 = \lambda_1 N_1)$ is again reached. The daughter is said to be supported by the parent; if by some process the parent is removed, then the daughter becomes unsupported and its amount decreases exponentially with time as determined by its half-life.

The same considerations apply between successive members of the chain, so that in equilibrium

$$\lambda_1 N_1 = \lambda_2 N_2 = \lambda_3 N_3 = \cdots \tag{4.1}$$

Obviously the amount present of an individual member is given by $N_i = (\lambda_1/\lambda_i)N_1$; since the parent of each series has a very long half-life the amount of each daughter present remains effectively constant. For a short-lived daughter the amount, being proportional to its half-life, is very small; even for radium-226 with its half-life of 1600 yr, the amount in equilibrium with 1 g of its parent uranium-238 is only $1600/(4.5 \times 10^9) = 0.36$ μg. Thus rather than specify the amounts of the various members present it is better, and simpler, to specify the activity, λN, because of course for a series in equilibrium this is the same for all members. The internationally recommended unit is now the *Becquerel* and 1 Bq = 1 decay/sec; activity was formerly measured in *Curies* with 1 Ci = 37×10^9 Bq.

The activity per unit mass is known as the specific activity and measured in becquerels per kilogram (Bq/kg). The specific activity corresponding to a given concentration of parent can be calculated as is now illustrated in the case of thorium-232. The half-life of thorium-232 is $14 \times$

4.1 Natural Radioactivity

10^9 yr, so that the decay probability is $(0.693/14 \times 10^9 = 49.5 \times 10^{-12}$/yr. For a concentration of 1 ppm by weight, that is, 1 mg of Th-232 per kg of sample, there are $(6.02 \times 10^{20}/232) = 2.6 \times 10^{18}$ atoms of thorium-232 per kilogram. Hence the specific parent activity of a sample containing 1 ppm of thorium-232 is

$$\frac{49.5 \times 10^{-12} \times 2.6 \times 10^{18}}{365 \times 24 \times 3600} = 4.08 \text{ Bq/kg}. \qquad (4.2)$$

In natural uranium the parents of the two series are in the atomic proportions 99.28 to 0.72 and similar calculations show that 1 ppm of natural uranium corresponds to a combined parent activity of 13.0 Bq/kg of which 4.4% is contributed by uranium-235 (this percentage is higher than the corresponding one for weight because of the shorter half-life of uranium-235). For natural potassium a concentration of 1% by weight (of metal) corresponds to a specific activity of potassium-40 of 317 Bq/kg of which 89.5% is beta activity and 10.5% is gamma activity. For natural rubidium a concentration of 1 ppm by weight (of metal) corresponds to a specific activity of rubidium-87 of 0.88 Bq/kg, entirely beta.

4.1.2 RADON LOSS AND OTHER CAUSES OF DISEQUILIBRIUM

Midway through each of the three series there is a member which happens to be a gas—as determined by the atomic structure appropriate to the charge, of 86 units, on the nuclei concerned; these are radon-220 in the thorium series, called thoron in earlier terminology and radon-222 and -219 in the two uranium series, the latter being called actinon in the earlier terminology. Because it is a gas, some of the radon formed in a sample or in the soil is liable to emanate; this escape is particularly likely if the sample is porous rather than compact. If, say, 25% escapes, then not only is the radon activity 25% lower than the parent activity but the activities of all subsequent members are 25% lower also. Escape is most likely with radon-222, which has a half-life 3.8 days; although the other two emanate within the pottery, there is more likelihood that they will decay into a non-gaseous daughter before having time to diffuse out of the sample because they have shorter half-lives, 55 and 4 sec, respectively. Nevertheless, there can be substantial escape of thoron, particularly during measurement if the sample has been crushed.

There are other causes of disequilibrium too; for instance, radium may be leached out by ground water, and then all subsequent members of the series are depleted too. Another example is in stalagmitic calcite; uranium is available to be incorporated into calcium carbonate at formation but not thorium. Hence in young stalagmites and stalactites there is uranium without daughters; then as time elapses there is a gradual build-up, governed by the 75,000 yr half-life of thorium-230 throughout the rest of the series. This of course is the basis of the uranium series method of dating.

4.2 Annual Dose

In evaluating the rate at which the latent thermoluminescence builds up, we need to know the rate at which ionization is being created in the crystals of the sample, and this in turn means the rate at which energy is being deposited by the various radiations. One way of approaching this would be to make detailed calculations based on the penetrating powers of the different radiations and the energy absorption coefficients of the constituents of the sample. Fortunately we are saved from this complex procedure by use of the 'infinite matrix assumption'. Conservation of energy requires that within a volume having dimensions greater than the ranges of the radiations the rate of energy absorption is equal to the rate of energy emission; if we go further and assume that the matrix is uniform both in radioactivity and absorption coefficient it follows that the absorption of energy per unit mass is equal to the emission of energy per unit mass.

As an example let us derive the annual alpha dose in a sample containing the thorium series in equilibrium. Suppose the specific parent activity is c_h Bq/kg; this then is the number of parent disintegrations per second in a kilogram of sample and since the series is in equilibrium it is also the disintegration rate of each member of the series. To obtain the rate of energy emission we need to multiply by the energy carried by the alpha particles, and from Appendix G (which is based on experimental measurements collated in nuclear data tables) we see that for the 6 alpha particles emitted per 'chain disintegration' the total energy is 36 MeV; hence the dose-rate is $(36 \times c_h)$ MeV/sec/kg. Remembering that 1 MeV/kg equals 1.6×10^{-13} Gy, the annual alpha dose, expressed in μGy/a, is given by

$$D_{\alpha, h} = 36 \times 1.6 \times 10^{-7} \times c_h \times 365 \times 24 \times 3600$$
$$= 182\, c_h. \tag{4.3}$$

In the case of the main uranium series, the energy carried by its 8 alpha particles is 43 MeV, and the annual alpha dose is

$$D_{\alpha, u} = 217 c_u, \tag{4.3a}$$

where the c_u is the specific parent activity of the uranium series.[2] Proceeding along similar lines, the other components of the annual dose can be

4.2 Annual Dose

TABLE 4.3
Components of Annual Dose for Unit Specific Activity[a]

	Alpha	Beta	Gamma
Potassium	—	2.68	0.79
Rubidium	—	0.53	—
Thorium: full chain	182.2	7.06	12.69
pre-thoron	76.1	2.54	5.13
Uranium: full chain	217	11.42	8.98
pre-radon[b]	98	4.77	0.44

[a] Annual dose components are given in micrograys per year (μGy/a) for samples having a specific parent activity of 1 Bq/kg. In the case of potassium, this refers to the combined beta plus gamma activity, and in the case of uranium the combined activity of both chains in their naturally occurring ratio (i.e., 0.956 Bq/kg from uranium-238 plus 0.044 from uranium-235).

[b] Radon-219 and its daughters have been included in calculating the pre-radon components, that is, it is assumed that gas loss only affects the uranium-238 series; this is because the short half-life of radon-219 makes its escape unlikely.

obtained and these are given in Table 4.3; further details are contained in Appendices G and J.

If the results of radioactive analysis are expressed in elemental concentrations, then it is convenient to have the annual dose data as in Table 4.4. To give some idea of the relative importance of the different components

TABLE 4.4
Components of Annual Dose from Potassium, Rubidium, Thorium, and Uranium for Given Concentrations[a]

	Alpha	Beta	Gamma
Potassium	—	830	244
Rubidium	—	23	—
Thorium: full chain	738	28.6	51.4
(pre-thoron)	(308)	(10.3)	(20.8)
Uranium: full chain	2779	146.1	114.9
(pre-radon)[b]	(1260)	(61.0)	(5.6)

[a] Values are quoted in micrograys per year and are for concentrations (by weight) as follows: 1% of natural K, 50 ppm of natural Rb, 1 ppm of natural Th, and 1 ppm of natural U.

[b] As with Table 4.3 the pre-radon components include radon-219 and its daughters.

in practice, in Table 4.5 they have been expressed as percentages of the total annual dose for 'typical' composition of pottery and soil, though it must always be emphasized that wide deviations from this composition can occur, and also in the k-value by which the actual alpha dose has been multiplied in order to allow for the reduced thermoluminescence effectiveness of alpha particles. Subject to these reservations, one may note that in fine-grain dating the three most important components are the potassium beta dose and the thorium and uranium alpha doses; in quartz inclusion dating the potassium beta dose is the largest component thereby giving some reduction in the vulnerability to interference from gas escape and other disequilibrium effects. The component most seriously affected by gas escape is the uranium gamma dose: complete loss of radon would

TABLE 4.5
Components of Annual Dose for 'Typical' Pottery and Soil Expressed as Percentages[a]

	Effective Alpha	Beta	Gamma	Total
(a) Fine-grain dating				
Potassium	—	16	5	21
Rubidium	—	$\frac{1}{2}$	—	$\frac{1}{2}$
Thorium: full chain	21	6	10	37
(thoron and after)	(12)	(4)	(6)	(22)
Uranium: full chain	24	8	7	39
(radon and after)	(13)	(5)	(7)	(25)
Cosmic[c]	—	—	3	3
	45	30	25	100
(b) Quartz inclusion dating[b]				
Potassium	—	30	8	38
Thorium: full chain	—	10	19	29
(thoron and after)	—	(6)	(11)	(16)
Uranium: full chain	—	16	12	28
(radon and after)	—	(9)	(11)	(20)
Cosmic[c]	—	—	5	5
	—	56	44	100

[a] The percentages given correspond to pottery and soil having 1% K, 50 ppm Rb, 10 ppm Th, 3 ppm U, and a value of 0.15 for k, the alpha effectiveness. The total annual dose will be 5180 μGy/a for fine-grains and 2820 μGy/a for quartz inclusions (exclusive of beta attenuation). The given concentrations correspond to specific activities (in Bq/kg) of 310 for K, 44 for Rb, 41 for Th and 39 for U. The percentages given have been rounded off. If precision is required reference should be made to Table 4.3 or 4.4.

[b] No entry is made for rubidium under quartz inclusion dating because of the low energy of the beta particles emitted.

[c] The cosmic ray contribution has been taken as 150 μGy/a corresponding to a burial depth of one or two metres at low altitude (see Appendix I).

4.2 Annual Dose

result in the loss of 95% of this component. The cosmic ray contribution is just significant whereas the rubidium contribution is negligible.

4.2.1 INHOMOGENEITY EFFECTS

Absorption Coefficient. For pottery, and most other types of samples, the assumption that the constitutent minerals, clay, and soil are similar in their absorption coefficients turns out to be a good approximation; this is because quartz, feldspar, and clay have effective atomic weights and atomic numbers that are close, being contained within a span of about 5%. However, this is not the case for the water that will have soaked into the pottery in all but arid conditions, and as already mentioned in Chapter 2, water absorbs more than its fair share of energy from nuclear radiations; compared to the constituents of pottery the absorption coefficient per unit mass of water is higher by 50% for alpha particles, by 25% for beta particles, and by 14% for gamma radiation. Allowance for water content is discussed in Section 4.2.3.

Radioactivity. The quartz inclusions in pottery are compartively free of radioactivity (but see Appendix C) and the resulting inhomogeneity in the alpha dose leads to the necessity of extracting grains of a given size range before measuring the thermoluminescence, that is, one uses the techniques of quartz inclusion dating and fine-grain dating. For beta particles the dose within pottery is approximately uniform, because the grains are well mixed and usually much smaller than the ranges of the beta particles (up to a few millimeters); a small correction does have to be made for attenuation within a 100-μm grain and of course for larger grains the attenuation is more serious. (An abundance of large inclusions would mean also that apart from those inclusions the beta dose is higher than average).

Radioactive inclusions such as zircon, apatite, and potassium feldspar are another type of inhomogeneity. Fortunately these are not usually abundant enough to upset the validity of the standard methods; if such interference is suspected the extent to which uranium-rich zircon and apatite grains are present can be estimated by means of the induced fission track and alpha autoradiography techniques mentioned later in this chapter, and potassium feldspar can be estimated by mineral separation. On the plus side, radioactive inclusions give the possibility of using one of the special techniques discussed in Chapter 6.

When the radioactivity of the pottery is different from that of the burial soil there is a gradient in beta dose extending about 2 mm into the pottery as the dose changes from that corresponding to half pottery and half soil to that corresponding to pottery; this transition layer must be sawn off and discarded. This puts a lower limit on the thickness of the sample that can be processed since it is rather difficult to saw off a 2 mm layer from both surfaces of a fragment that is less than 8 mm thick. Estimates of the error resulting from failure to discard the surface layer is given in Appendix H.

4.2.2 EXTERNAL DOSE

Unless the sample is abnormally large the gamma dose is mainly determined by the radioactivity of the soil; this is because of the much greater range of the gamma radiation, on the order of 0.3 metre. A useful concept here is the self-dose percentage p; this is the average gamma dose in the sample due to its own radioactivity, expressed as a percentage of the gamma dose in a sample of the same radioactivity having infinite extent (i.e., the infinite matrix dose). The average gamma dose in the sample due to soil radioactivity is $(100-p)\%$ of the soil infinite matrix dose. (This follows from the principle of superposition: if both sample and soil have the same radioactivity the sum of the two must equal 100%). Hence for sample and soil having infinite matrix doses D_γ^i and D_γ^e, respectively, the average gamma dose in the sample is given by:

$$D_\gamma = \frac{p}{100} D_\gamma^i + \left(1 - \frac{p}{100}\right) D_\gamma^e$$

$$= D_\gamma^e + \frac{p}{100} (D_\gamma^i - D_\gamma^e). \qquad (4.4)$$

Figure 4.1 shows, for 2-MeV gamma radiation, the self-dose at the centre of a sphere as a function of radius. The average self-dose in the sample will be substantially less than the self-dose at the centre; also we are concerned with a wide spectrum of gamma energies, mostly below 2 MeV (see Figure M.3). As discussed in Appendix H rough rules of thumb are: for a spherical sample of density 2 g/cm³, $p = 2d\%$ where d cm is the diameter of the sample; for a pottery fragment of the same density for which the thickness t cm is small compared to lateral extent, $p = 100\{1 - \exp(-0.06-0.07\ t)\}$. Apart from these rules being approximate, there is the difficulty that sample shapes do not usually conform to either model. Fortunately the associated error in the date due to neglecting the effect is barely significant in most circumstances; thus even for a sample having $p = 30\%$ (i.e., a sphere of 15 cm or a flat sherd of thickness 4 cm) buried in soil such that $D_\gamma^i = 1.5\ D_\gamma^e$ we see from Equation 4.4 that

$$D_\gamma = 1.15\ D_\gamma^e. \qquad (4.4a)$$

4.2 Annual Dose

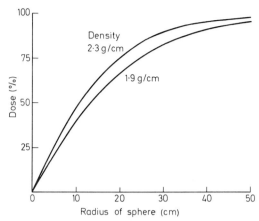

Figure 4.1 Self-dose at centre of sphere for gamma photons of energy 2 MeV, expressed as a percentage of the dose for an infinitely large sphere (from Murray, 1981). For photons of lower energy a given percentage is reached for a smaller radius; thus for the composite spectrum from thorium and uranium (for which the gamma energies spread from 2.2 MeV down) the percentage reaches 75% at radii of only 15 and 18 cm for the above densities (17 and 20 cm in the case of potassium); these data have been derived from Fleming (1979), and like the figure, are based on the approximate treatment given by Evans (1955:739) which includes both primary and secondary flux.

For D_γ constituting about 15% of the total effective annual dose, the error in age due to taking $D_\gamma = D_\gamma^e$ would only be about 2%.

On the other hand if we are concerned with a sample that is of negligible radioactivity, as is sometimes the case with flint or calcite, the attenuation of the soil gamma dose may be important; this is because the latter will be the main component of the annual dose. In a case where the soil gamma dose forms, say, 80% of the annual dose, the age evaluated for a sample having $p = 30\%$ will be low by 24% unless correction is made.

Of more immediate concern in practical application is the question of uniformity of radioactivity in the surroundings of the sample. It is implicit in Figure 4.1 that a lump of rock will contribute to the external gamma dose if it lies within about 0.3 metre. This limit is on the cautious side unless the proportion of rock present is high, but ideally samples for dating should only be accepted from situations in which the soil is uniform to a distance of 0.3 metre; if this requirement is not met additional uncertainty is introduced, to a degree depending on the extent to which the radioactivity of the rock, or other soil, is different. Figure 4.2 illustrates the situation for a pit; note that if the sample was close to the upper surface of the pit and the overburden was slow in accumulating we have the adverse situation in which half of the gamma dose-rate was missing for an appreciable part of the burial time. Figure 4.3 shows the fall-off of

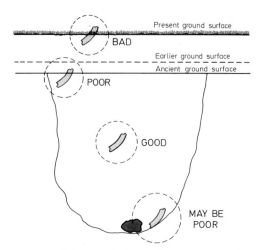

Figure 4.2 Gamma rays reach the sample from a distance of up to about 0.3 metre. Hence for a reliable assessment of the gamma contribution to the annual dose to be made the sample should be in a uniform surround of soil, such as obtains in the middle of a pit or ditch. The situation for the sample at the bottom will be poor if the subsoil, or the lump of rock, has an appreciably different level of radioactivity then that of the filling of the pit.

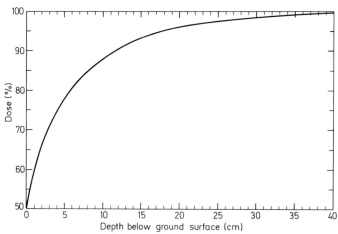

Figure 4.3 Variation of gamma dose in soil as ground surface is approached, expressed as a percentage of the dose at infinite depth (based on computation by Løvborg, see Table H.1). The radioactive composition was taken as close to 1% K, 10 ppm Th and 3 ppm U. The density taken for the soil, inclusive of 25% by weight of water, is 2 g/cm^3; data for density ρ can be obtained to a reasonable approximation by multiplying the values on the depth axis by $2/\rho$), that is, for $\rho = 2.5$ the percentage shown above for 25 cm becomes that for 20 cm. Actual soils have a *dry* density of 1.4–2 g/cm^3, sedimentary rocks 1.8–2.4, and igneous rocks 2.4–2.8.

4.2 Annual Dose

gamma dose as the ground surface is approached. The rule of thumb is that the sample should have been buried to a depth of 0.2 metre or more for at least two-thirds of the burial time.

The soil gamma dose-rate, D_γ^e, can be evaluated either by radioactive analysis in the laboratory or, preferably, by on-site measurements. The use of thermoluminescence capsules for the latter has already been mentioned in Chapter 2 and further discussion is given in Section 4.3.3; on-site measurements can also be made with a portable gamma spectrometer as discussed in Section 4.3.4. If either of these on-site techniques is used then it is possible to deal with some types of non-uniform surroundings by utilizing an intact profile elsewhere on the site that replicates the one from which the thermoluminescence samples have been obtained; successive soil layers of differing radioactivity is one such situation. As illustrated in Figure 4.4 there is variation of dose near to boundaries and so it is important to know the position of the sample. Data enabling calculations to be made for other layer thicknesses are given in Appendix H. However it should be stressed that these apply strictly only to soil having the density and water content stated. Undoubtedly it is better to avoid the need for computation by obtaining samples from uniform surroundings if at all possible.

Although the importance of uniform surroundings needs to be stressed, the archaeologist who does not have them should not necessarily despair.

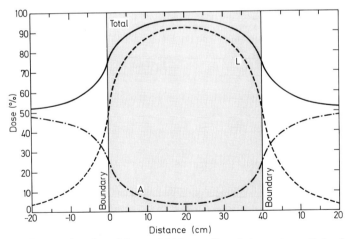

Figure 4.4 Gamma dose for soil layer of thickness 40 cm between two soils having half as much radioactivity, expressed as a percentage of the dose in the layer if it was of infinite thickness. Curve L shows the contribution from the layer itself and curve A the contribution from the adjacent soils. Densities and compositions are as for Figure 4.3.

First, although there may be different types of soil and rock within range of the samples they may not differ in their radioactivity; this can be checked by preliminary laboratory analysis. Second, whereas for the 'typical' sherd and soil of Table 4.5 the gamma component is a substantial percentage of the whole dose-rate (25% for fine-grain dating), if the radioactivity of the sample is very much higher than that of the soil the importance of the gamma component is greatly diminished; again this can be found out by preliminary laboratory analysis. Of course the converse is also possible—weak sample radioactivity—and this is often the case for burnt flint and stalagmitic calcite on paleolithic sites. Third, one of the radioactive inclusion techniques discussed in Chapter 6 may be applicable and these greatly reduce dependence on the gamma component.

Cosmic Radiation. In many circumstances the cosmic ray contribution is sufficiently small that a standard value of 150 μGy/a can be assumed. The value at ground level is about twice this and there is rapid initial fall off with depth in the first 30 cm; thereafter the fall off is much slower and further halving occurs at a depth of 5 metres or so.

It is only in special circumstances such as weak sample radioactivity, high altitude, or very thick overburden (as is often the case in a paleolithic cave) that adjustment of the standard value needs to be considered. Further information is given in Appendix I. Of course, the on-site methods—thermoluminescence capsules and gamma spectrometry—automatically include the cosmic contribution and no explicit use is then made of the standard value. However, for sample contexts over which there is overburden amounting to several metres or more, the question arises as to whether that overburden has been present for a substantial fraction of the burial time; if not then the average cosmic contribution may be higher than as measured today.

4.2.3 EFFECT OF MOISTURE

Water in the pores of pottery or in the soil absorbs part of the radiation that would otherwise reach the thermoluminescent grains. Alternatively, on the infinite matrix basis, one can think of the water as decreasing the radioactivity per unit mass compared to the dry situation. Although the effect corresponding to the degree of wetness measured 'as found' can be calculated, uncertainty as to how the wetness may have varied throughout the burial period is a serious barrier to better accuracy. Of course, for arid environments that have always been arid there is no problem; similarly for many sites in NW Europe there is a fair degree of certainty that pottery and soil have always been fully saturated—though the possibility

4.2 Annual Dose

of interference by change in water table is sometimes present. However, the ultimate way to lessen or even remove the dependence on moisture content is by means of one of the radioactive inclusion techniques discussed in Chapter 6. The saturation level (*porosity*) sets an upper limit to the effect on the age; for pottery the saturation level may be as low as 5% for compact fabric but as high as 25% for cruder, more porous types; for soil, values in the range 20%-40% are likely. Estimation of how near to saturation soil and pottery will be in any given situation is a matter for a soil specialist, and also a matter of knowing various parameters such as pore size distribution. However, the uncertainty about the situation in the past, whether due to climatic change or due to natural or human interference with the site's drainage and so forth, vitiates too much effort in this direction. From measurement of 'as found' water contents, it seems that nearness to saturation is common except in obviously dry circumstances; excluding such situations it is usual to assume that the water content has averaged $(0.8 \pm 0.2)W$, where W is the saturation content expressed as (weight of water/dry weight). In what follows F will be used to denote the fraction of saturation to which the assumed average water content corresponds, and δF the uncertainty in that value, that is, in the above $F = 0.8$ and $\delta F = 0.2$.

As mentioned before, water absorbs more than its fair share of radiation on a weight for weight basis. From nuclear tables it has been estimated (Zimmerman, 1971) that averaged over the relevant spectra the absorption coefficient for water is 50% higher than for pottery in the case of alpha radiation, and 25% and 14% higher for beta and gamma, respectively. Consequently, if the dose-rates have been evaluated from measurments on dry material, the actual values to be used in the age equation are given by:

$$D_\alpha = \frac{D_{\alpha,\text{dry}}}{1 + 1.50\ WF}, \tag{4.5}$$

$$D_\beta = \frac{D_{\beta,\text{dry}}}{1 + 1.25\ WF}, \tag{4.6}$$

$$D_\gamma = \frac{D_{\gamma,\text{dry}}}{1 + 1.14\ W_1 F}, \tag{4.7}$$

where W refers to sample and W_1 to soil. Table 4.6 shows the percentage reduction in annual dose for the 'typical' compositions of Table 4.5 for several assumed saturation levels. The quoted uncertainties correspond to the ± 0.2 assumed for the uncertainty in the value of 0.8 taken for the fractional uptake F; since they are expressed as percentages they also represent the percentage of uncertainty resulting in the calculated age—thus for the inclusion technique, the above mentioned uncertainty in F

TABLE 4.6
Moisture Content[a]—Percentage Reduction in Annual Dose

	Alpha	Beta	Gamma	Total for 'typical' compositions[b]	
				Fine grain	Inclusion
$W = 0.1, W_1 = 0.2$	12±3	10±2	18±5	13±3	13±4
$W = 0.1, W_1 = 0.3$	12±3	10±2	27±7	15±4	17±5
$W = 0.2, W_1 = 0.4$	24±6	20±5	36±9	25±6	26±7

[a] W and W_1 are the saturation water contents expressed as (weight of water/dry weight) for sample and soil respectively. F has been taken as (0.8±0.2). This is the fractional uptake, that is the actual weight of water is FW times the dry weight of the sample.

[b] The 'typical' compositions are given in Table 4.5.

produces a 7% uncertainty in the calculated age for pottery and soil having, respectively, 20% and 40% porosity. It should be noted that failure to make any allowance for the moisture effect, or underestimation of it, causes the calculated age to be erroneously too recent: thus if $W = 0.1$ and $W_1 = 0.3$ are used when the true average values during the burial time were 0.2 and 0.4, respectively, the calculated age will be about 10% too recent. Similarly an underestimate of F will give rise to an erroneously low age: if F is taken as 0.4 when the true value was 0.8 then in the case of $W = 0.2$ and $W_1 = 0.3$ the date will again be about 10% too recent.

Because of the short range of alpha particles, about 20 μm, the question arises as to whether any water does in fact intervene in their path; if the tracks lie entirely in impervious material the water will have no effect. However, the pores in pottery are very fine, a substantial proportion of them being less than 10 μm in diameter; nevertheless the possibility that the moisture effect on the alpha dose is less than that given by Equation 4.5 cannot be completely ruled out.

4.2.4 RADON EMANATION

The escape of the gas radon from pottery is a two-stage process. First there is movement out of solid material into water-filled or air-filled pores; second there is diffusion along the pores and hence out of the pottery. The first stage is accomplished when the radon nucleus is formed by emission of an alpha particle from the radium nucleus; conservation of momentum requires that the nucleus recoil in the opposite direction with a kinetic energy of about 0.1 MeV, giving it a range in pottery of about 0.02 μm. If it ends up in a pore then the second stage takes place—as long as it has

4.2 Annual Dose

not undergone radioactive decay while still within the pore; if so then the next member of the chain, being non-gaseous, will deposit on the wall of the pore. For radon-222 of the uranium-238 series the half-life is 3.8 days, and even for a water-filled pore the diffusion length before decay is the order of a few centimetres; for an air-filled pore it is much longer. However, for thoron (radon-220 of the thorium series) the half-life is only 54 sec thereby reducing the diffusion length by a factor of 6000; for actinon (radon-219 of the uranium-235 series) with a half-life of 4 sec the diffusion length is substantially shorter still. Hence we might expect that only the escape of radon-222 is of interest; however, it seems that in alpha counting at any rate, the escape of thoron makes itself felt. The diffusion lengths and other data given above are from Tanner (1964).

Effect on Annual Dose. From Table 4.5 we see that for 'typical' pottery and soil, 100% escape of radon leads to a 25% decrease in the annual dose for the fine-grain technique and a 20% decrease for the quartz inclusion method. As we shall see in due course there are at present a number of factors in thermoluminescence dating which contribute an error of the order of 5%, and if we set this as the threshold at which the effect on the annual dose becomes significant we might conclude that for pottery and soil near to the typical we need only concern ourselves with radon escape when it exceeds around 20%, that is, when 20% of the radon atoms born in the pottery and soil succeed in escaping before they decay into the non-gaseous daughter. However, as is discussed later (Section 4.3.2), it seems that in measuring radioactivity by alpha counting lower levels of escape may upset things, and escape of thoron would then have importance.

The Recoil Process. Since the recoil range is only 0.02 μm it would seem unlikely that anything approaching 20% of the recoiling nuclei would even reach a pore, quite apart from coming to rest in it and diffusing out. However for pottery from some regions the escape does reach this level and beyond—even 50%. One possible explanation is that the pottery is permeated with very fine pores to such an extent that there is not really any solid material in it, that is, although the uranium is uniformly distributed nearly all of it lies within 0.02 μm of a pore. An alternative explanation is that the decaying radium nuclei are preferentially located near the pore wall; this could be through deposition of radium salts from groundwater, an unwelcome possibility to be discussed shortly.

Effect of Moisture. A suprising effect is that pottery emanates more strongly when wet than when dry. This is because a recoiling nucleus reaching an air-filled pore is likely to overshoot and embed itself within the wall opposite; if there is water in the pore there is a much higher probability that it will be stopped within the pore (see Figure 4.5). Of course, diffusion is slower out of a water-filled pore but as long as the diffusion length is sufficient the enhancement of the first stage will be dominant. For thoron and actinon the reduction of diffusion length in a water-filled pore will be dominant and the overall escape thereby reduced.

Radon in Soil; Back Diffusion. In the above discussion it has been assumed that the radon loss can be equated to the amount diffusing to the surface of the pottery. This is true for a piece of pottery in air but if it is buried in soil there is the possibility that the level of radon at the surface, due to emanation from the soil, will be such that the outflow is balanced by back-diffusion into the pore or even that the latter predominates. The level of radon in the soil is controlled by two processes; first, diffusion to the ground surface and escape into the air; second, transportation as dissolved gas in groundwater movement. Both of these are highly depen-

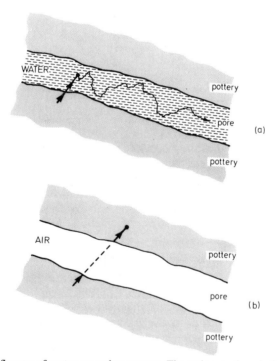

Figure 4.5 Influence of water on radon escape. The radon nucleus is formed when its parent emits an alpha particle. In the recoil from this emission the nucleus travels about 0.02 μm in pottery but several thousand times more in air. Hence it is likely to traverse an air-filled pore and embed itself in the opposite wall, as in (b). But if the pore is filled with water, as in (a), it is likely to end its recoil in the water, and then eventually escape by diffusing to the outlet of the pore.

dent on local conditions, though with respect to the former, there are some generalizations that can be made. In wet soil the diffusion length is only a few centimetres and it is only from a surface layer of about that thickness that appreciable loss of radon will occur; in dry sandy soil the diffusion length, and hence the thickness of the layer from which loss occurs, may be of the order of several metres. Since for other reasons we only accept pottery for thermoluminescence dating if it is buried to a depth of 30 cm or more, we see that in wet soil back diffusion is likely to be important, depending on the relative levels of radioactivity of soil and pottery and on the emanating power of the soil grains relative to the pottery.

In dry sandy soil there are two effects resulting from the diffusion length of several metres that should be noted. First, in a stratum of weak radioactivity overlaying one of strong radioactivity, the radon may be much higher than corresponds to the stratum itself. Second, when an excavation trench is dug there is loss of radon from the exposed section to a distance of several metres; hence a capsule at 30 cm from the exposed surface will record less than the true annual gamma dose unless the trench has been refilled with the original soil.

What to Do about Radon?   There are three aspects to be considered:

1. influence on the annual dose,
2. upset to *measurement* of radioactivity, particularly alpha counting,
3. radon escape as an indicator of other types of radioactive disequilibrium.

The last is discussed in the next section and (2) later in the chapter. With regards to (1) the first approach tried was to make laboratory measurements of the amount of radon escaping, both on the pottery fragment and on the soil, at the level of moisture content presumed during burial. Apart from the experimental difficulties, including the variability of the emanating power with environmental conditions, this approach has the drawback that it does not tackle the question of back diffusion; highly emanating pottery buried in highly emanating soil does not necessarily suffer a loss of annual dose, quite apart from the question of the extent to which radon in the soil has been removed by groundwater. The better approach is to evaluate the radon *retention* by measurement of one of the daughters supported by it, either lead-210 by gamma spectrometry, or polonium-210 by alpha spectrometry. The half-life of lead-210 is 22 yr and so although the degree of radon escape may be drastically altered when the sample is removed from the ground, the level of lead-210 does not follow but remains for several years at the level corresponding to the degree of radon retention while buried; polonium-210 is supported by lead-210 and so it too remains at that level. Of course, this still does not tell us anything about variation during antiquity but it is very much more reliable than the first approach. Unfortunately, the facilities required for either type of spectrometry are somewhat more sophisticated than alpha counting; they are discussed in Sections 4.3.2 and 4.3.5.

The same approach can be used for radon retention in the soil, though for humid conditions the capsule technique can be expected to be reasonably reliable—as long as excavation has not interfered with groundwater movement. Although, for fine-grain pottery dating the decrease in overall annual dose due to 100% loss of radon from the soil is only 7% for the 'typical' circumstances of Table 4.5, for burnt flint and calcite the gamma component is usually dominant with consequent accentuation of dependence on radon escape.

Discussion so far has concentrated on radon as opposed to thoron or actinon; this is on the basis that the diffusion lengths of both the latter are too short to allow escape. However, escape of thoron does occur substantially from ground-up pottery and hence is relevant to (2) above; it may also be relevant to (3).

Although one of the spectrometry techniques needs to be used if any attempt is to be made at correction of the annual dose, the more simple approach of laboratory measurement of the amount escaping can be used for rejection of samples that may be unreliable on this account (for instance using the gas cell mentioned in Section 4.3.2; there are also other indicators of strong emanation). We have noted that for the 'typical' circumstances of Table 4.5 there is unlikely to be a significant effect on the calculated age until the escape reaches 20%; however circumstances are often non-typical and the limit at which the effects of escape become significant should be calculated in each case—for instance, for a sample in which the radioactivity is predominantly due to uranium, the limit should be placed much lower, at about 10%. There are also points (2) and (3) to be considered, and as a working rule it is advisable to discard, or put aside for special investigation, any sample for which the radon or thoron escape exceeds 10%.

As to which regions of the world are likely to yield pottery exhibiting strong radon loss, the outstanding examples seem to be from sites on igneous geology.

4.2.5 THE CONSTANCY OF THE ANNUAL DOSE; DISEQUILIBRIUM AND LEACHING

It is implicit in the age equation that the radioactivity of the sample (and soil) has remained constant since time zero, thereby providing a constant

4.2 Annual Dose

annual dose. When it comes to the dating of stalagmitic calcite by thermoluminescence (Section 7.5) allowance needs to be made for the fact that when calcite crystals are formed only uranium-238 and uranium-234 are incorporated so that initially the rest of the chain is missing, from thorium-230 onwards; subsequently the full chain 'grows in' at a well-defined rate as determined by the thorium-230 half-life of 75,000 yr, the annual dose increasing correspondingly. The question is whether we may expect any similar effects in pottery. On the one hand the clay from which the pottery was made may have been in disequilibrium; on the other hand initial equilibrium may have been disturbed either during firing or during burial. One type of disequilibrium known to occur in clay from geochemical studies is an excess of uranium-234 relative to uranium-238; however, the 250,000 yr half-life of the former is so much longer than the time period of not more than 10,000 yr with which we are concerned in pottery, that there is effective constancy. As far as pottery is concerned the same is more or less true for thorium-230.

On the other hand, disequilibrium of radium-226, half-life 1600 yr, is a very different matter. If the radium was in disequilibrium after firing then the degree of disequilibrium measured today will be very much less than the initial degree; thus if the radium is 10% away from equilibrium today, for a sample which is 1600 years old it would have been 20% away initially, and for a sample which is 3200 yr old, 40%. However it is not usually known whether the disequilibrium was present initially or whether it has been the result of leaching during burial. If the latter, then assuming the groundwater conditions have not changed during the burial period, the degree of disequilibrium will have been more or less constant. Thus it is not possible to make correction for radium disequilibrium, but only to calculate the widening of the error limit on the basis of the alternative hypotheses.

Substantial degrees of radium disequilibrium have in fact been observed in some pottery (e.g., Meakins, Dickson and Kelly, 1978, 1979; Carriveau and Harbottle, 1983; Murray and Aitken, 1982; Mangini, Pernicka and Wagner, 1983). Because of dilution by other components of the annual dose the effect would not be too serious if the disequilibrium concerned only radium and its daughters. Unfortunately uranium also can be leached from pottery; this has been demonstrated by ion exchange measurements in the laboratory (Hedges and McLellan, 1976) and evidenced in ancient pottery from the Hong Kong region using gamma spectrometer measurements (Murray, 1981, 1982). Fortunately samples exhibiting disequilibrium usually show appreciable emanation of radon, and detection of appreciable escape in the course of routine measurements must be regarded as a warning signal. On geochemical grounds, mobility (and consequent disequilibrium) is to be expected in regions of high salinity such as occur in Australia, Spain, and the Middle East.

4.3 Measurement

In Chapter 2, alpha counting was given as the technique for evaluation of the annual alpha dose and direct dosimetry by means of an ultrasensitive thermoluminescence phosphor as the technique for the beta, gamma, and cosmic doses. We shall now examine these techniques in more detail and then discuss alternative methods.

Because different methods are affected differently by interfering effects, such as sample inhomogeneity and radioactive disequilibrium,—particularly radon escape—use of methods in parallel is advantageous as a reliability test; in addition there are ancillary techniques which can be used to check to what extent these interferences are present.

4.3.1 ALPHA COUNTING; DERIVATION OF ANNUAL DOSE

This technique is remarkable both for the high ratio of sample count to background count—around 100:1 for typical pottery—and for its simplicity and inexpensiveness, at any rate in basic form. It is more sensitive than other practical methods, including neutron activation—despite the latter's high-powered technology. Alpha counting was adopted for thermoluminescence dating in the early days, having been developed for detecting low levels of alpha radioactivity in biological tissue (Turner, Radley and Mayneord, 1958), and it retains a central place in most thermoluminescence laboratories despite the complications that have developed. Initially it was used not only for alpha dose-rate but also for beta and gamma dose-rate in conjunction with chemical analysis of potassium (usually by flame photometry); however, as already mentioned in Chapter 2, in the conversion of count-rate to dose-rate there is dependence on the sample's thorium to uranium ratio in the case of these two latter; this ratio can in fact be determined using the 'pairs' technique discussed later, although rather long counting times are required. For authenticity testing, when only a few tens of milligrams may be all the sample that is available, the combination of alpha counting and chemical analysis for potassium remains the preferred one; the error limits are wider in any case and uncertainties due to the effects just mentioned do not increase them significantly.

4.3 Measurement

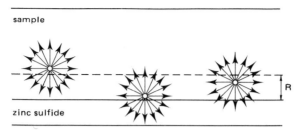

Figure 4.6 Thick source alpha counting: idealized sketch. For emitting nuclei which are close to the zinc sulphide screen nearly half of the alpha particles produce scintillations; but for nuclei further away the proportion is less and for nuclei beyond the range R none of the alpha particles reach the zinc sulphide. On average one quarter of the particles emitted by nuclei within the layer of thickness R reach the screen; this is derived in Appendix J. For the thorium and uranium series the average value of R is 25 μm for a density of 2.6 g/cm^3, but for the most energetic alpha emitter of the series the value is twice this; thus for the standard screen of diameter 42 mm the within-range layer contains about 0.2 g. However because of the difficulty of spreading evenly it is not practical to use less than about one gram of sample for this size of screen.

Theory. Following on from Figure 2.5, we show in Figure 4.6 an idealized version of the layer of sample as it lies on the zinc sulphide screen; both are thicker than the range R of the alpha particles being emitted by the sample; for the moment it is assumed that all are emitted with the same initial energy. Of the alpha particles emitted by nuclei within a distance R of the screen, only a quarter actually reach the screen; this is because the particles are emitted isotropically and many of the tracks are rather oblique—the derivation is given in Appendix J. It follows that if c_α is the number of alpha particles emitted per unit mass in unit time, the count-rate will be

$$\dot{\alpha} = \tfrac{1}{4} AR\rho c_\alpha, \qquad (4.8)$$

where ρ is the density of the sample and A the area of the screen (the diameter of the screen is very much larger than R). We see that the count-rate is proportional to R as well as to the rate of emission. The higher the energy of emission E for the nucleus, the greater is R (though the dependence is not linear, see Figure 5.7) so the count-rate from a given activity of a high energy emitter will be greater than that from the same activity of low energy emitter. Since the alpha dose-rate is proportional to the product (Ec_α) this is advantageous; the dependence on E of the conversion factor from count-rate to dose-rate is weakened, and as a consequence the factor is almost the same for the thorium and uranium series.

If the sample contains a radioactive series of n alpha emitters in equilibrium and c is the activity per unit mass of the parent then

$$\dot{\alpha} = \tfrac{1}{4} A(R_1 + R_2 + R_3 + \cdots)\rho c \qquad (4.9)$$
$$= \tfrac{1}{4} A\bar{R}\rho nc, \qquad (4.9a)$$

where $R_1, R_2, \ldots$ etc. are the ranges of the alpha particle emissions from the individual members of the series and $\bar{R}$ is the average range. The individual alpha ranges for the members of the thorium and uranium series are given in Appendix J. It is convenient to state these in terms of the product $R\rho$: for the 6 alpha emitters of the thorium series the average value of this product is 70 μg/mm² and for the 8 alpha emitters of the main uranium series it is 58 μg/mm²; for $\rho = 2.5$ g/cm³ these values correspond to 28 and 23 μm, respectively. On substitution into Equation 4.9 we obtain

$$\dot{\alpha} = \tfrac{1}{4} A \{(70 \times 6 \times c_h) + (58 \times 8 \times c_u)\} \times 10^{-6}, \qquad (4.10)$$

where c_h and c_u are the specific activities (Bq/kg) for the thorium and uranium series[2] respectively, A is measured in mm², and $\dot{\alpha}$ is measured in counts per kilosecond (ksec). It is convenient from here on to substitute the value of A appropriate to the standard 42-mm diameter screen and to introduce the respective electronic threshold fractions of 0.85 and 0.82 for the thorium and the uranium series to be discussed shortly. The equation then becomes

$$\dot{\alpha} = 0.123\, c_h + 0.132\, c_u. \qquad (4.11)$$

Conversion to Annual Dose. From Equations 4.3 and 4.11 we see that for a sample containing only the thorium series, the alpha dose-rate (in μGy/a) is given by $D_\alpha = 1480\, \dot{\alpha}$, and for uranium only, $D_\alpha = 1640\, \dot{\alpha}$. The value corresponding to equal activity of the two series is

$$D_\alpha = 1560\dot{\alpha}, \qquad (4.12)$$

where, as in Equation 4.11, $\dot{\alpha}$ is the count-rate (per ksec) for a 42-mm diameter screen and electronics are adjusted for the standard threshold fractions. If Equation 4.12 is used without regard to the actual ratio of activities in a sample, the error even for a sample that is all thorium or all uranium is only 5%. When it comes to *effective* dose-rate, the upper limit to the error is not more than 1–2%; this is because the alpha effectiveness (k-value) appropriate to a thorium-only sample is about 8% higher than that appropriate to a uranium-only sample.

As regards the beta component of the annual dose we see from Table 4.3 that the dose-rate for unit specific activity of the thorium series is 0.62 times that for the uranium chain; from this table and Equation 4.11 we obtain $D_\beta = 57\dot{\alpha}$ for a thorium-only sample and $D_\beta = 87\, \dot{\alpha}$ for a uranium-only sample. The best we can do from simple alpha counting is to assume,

4.3 Measurement

as is often the case, that the sample contains equal activities of the two series and use the average of the two conversion factors, that is, for the combined thorium and uranium contribution to the beta dose-rate we take

$$D_\beta = 72\dot{\alpha}. \qquad (4.13)$$

Although the majority of pottery samples do have the two series present at about the same activity level, this is not always the case; one possibility is to obtain an estimate of the activity ratio by means of the 'pairs' technique described shortly, despite the rather long counting times required for this. If this is not done, then we have to assume equal activities and we use Equation 4.13; if in fact there is no uranium in the sample this equation will give a dose-rate that is 20% higher than the true value, whereas at the other extreme (no thorium) it will be 20% lower. The associated error introduced into the date will be substantially less than 20% because of dilution by other contributions to the overall dose-rate; however it is preferable to use direct thermoluminescence dosimetry for the beta dose-rate particularly as this obviates the need for determination of the potassium content by chemical analysis.

In the case of gamma radiation it is the thorium series that is relatively stronger for a given specific activity. Again using Table 4.3 and Equation 4.11 we obtain $D_\gamma = 103\dot{\alpha}$ for a thorium-only and $D_\gamma = 68\dot{\alpha}$ for a uranium-only sample; hence for the case of equal activities of the two series the combined thorium and uranium contribution to the gamma dose-rate is

$$D_\gamma = 85\dot{\alpha}. \qquad (4.14)$$

Thus the dependence on thorium/uranium ratio is about the same (20%) as in the case of beta radiation but in the opposite sense. Tempting though it might be to rely on compensation, this would not be justified because it cannot be assumed that the soil has the same thorium/uranium ratio as the sample. As for the beta component, thermoluminescence dosimetry is the preferred technique; there are other reasons too, one being that thermoluminescence dosimetry samples a very much larger volume of soil.

The Effective Alpha Dose-Rate; The a-value System. As discussed in Section 2.2 alpha radiation is less effective in inducing thermoluminescence than beta and gamma radiation, and the effective alpha dose-rate D'_α is equal to the value of D_α given in Equation 4.12 multiplied by the k value as measured for the sample concerned. In this system (Zimmerman, 1971) there is complication because the measured value of k is strongly dependent on the energy of the alpha particles used for measurement; this drawback is avoided in the a-value system, described in Appendix K, which makes use of the experimental observation that for alpha particles the thermoluminescence per unit length of track is approximately independent of particle energy. In this system the effective alpha dose-rate (μGy/a) is obtained directly from the alpha count-rate as

$$D'_\alpha = 1280 a \dot{\alpha}. \qquad (4.15)$$

Alternatively, if the alpha dose-rate has been obtained by some method other than alpha counting, the effective dose-rate can be obtained by using $k = 0.86a$ for the thorium contribution and $k = 0.78a$ for the uranium contribution, or $k = 0.82a$ if the thorium/uranium ratio is unknown and equal activities are assumed. Also mentioned in Appendix K is the b-value system; if b is measured in gray micron², then $b = 13\ a$.

4.3.2 ALPHA COUNTING IN PRACTICE

A container in common use for alpha counting in shown in Figure 4.7; the sample is in direct contact with the zinc sulphide screen so that a new screen is needed for each sample. Screens can be made by sprinkling zinc sulphide onto Sellotape, using a ring of stainless steel or perspex to carry the latter; alternatively, ready made plastic screens impregnated with zinc sulphide can be purchased.[3] The usual diameter of the screen is 42 mm, but for the limited amount of sample available in authenicity testing a lesser diameter may be necessary, for example, 15 mm. It is vital to use materials that are free from alpha radioactivity in the sample holder and to clean with a suitable detergent, or methanol, before each use; in measuring the no-sample background the screen should be covered with a disc of

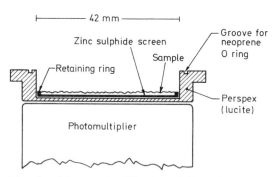

Figure 4.7 Container for alpha counting. The zinc sulphide scintillation screen is retained by a stainless steel ring which is in turn held down by a split, slightly sprung, stainless ring (not shown). Normally the sample, which is crushed and placed in direct contact with the zinc sulphide screen, is counted unsealed, but as a crude test for radon escape the container can be sealed by means of a lid and O ring.

perspex or copper so as to avoid alpha particles from radon in the air; aluminium must be avoided because even when it is nominally 'super-pure' it usually contains residual radioactivity. The no-sample count-rate should be less than 0.01 counts/ksec; the electronic background, that is with the scintillation screen removed, should be substantially less.

Unlike a beta or gamma counter no lead shielding is required. This is because of the short range of alpha particles and because of the comparative insensitivity of zinc sulphide to beta and gamma radiation. Of course, the sample plus photomultiplier must be enclosed in a dark box because of the latter's sensitivity to light.

Electronics. Each scintillation of light produced by an alpha particle in the zinc sulphide screen causes photoelectrons to be emitted from the photocathode of the photomultiplier (e.g., EMI type 6097); through the multiplying action of the dynode chain, by a factor of around a million overall, sizeable electrical pulses are obtained at the anode—one for each scintillation. These pulses are then counted by means of suitable electronic circuitry (Figure 4.8). After amplification to a pulse height of a few volts, the pulses are fed to a discriminator unit which accepts only pulses of greater voltage than the threshold that has been set; in this way pulses corresponding to beta particles and gamma rays are rejected, as are noise pulses from the photomultiplier dynode chain. The output pulses from the discriminator are fed to a counter and the number of recorded counts

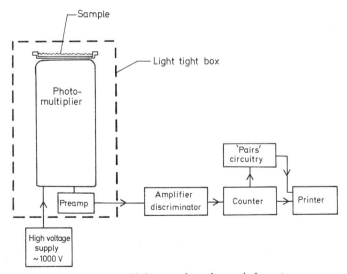

Figure 4.8 Alpha counting: electronic layout.

received is printed out at regular intervals (e.g., every kilosecond) as a check that the count-rate is constant and that there have been no bursts of electrical interference pulses.

The Threshold Fraction. There is a wide spread in the spectrum of pulse sizes from the photomultiplier anode. In part that is due to the various energies with which alpha particles are emitted by different members of the radioactive series, 4–9 MeV, but the dominant effect is due to the 'thick source geometry'—an alpha particle which dissipates most of its energy in a long path within the sample does not have much left for creating the scintillation in the zinc sulphide screen.

A typical spectrum of the pulse height distribution due to a sample is given in Figure 4.9. In routine counting, the pulses due to beta particles, gamma rays, and photomultiplier noise are rejected by setting the threshold as indicated. The fraction of alpha particles that are rejected is estimated from the shape of the spectrum from a source having a high concentration of thorium (or uranium); this fraction is then used to correct observed counts to total counts in routine low-level counting. It is usual to set the threshold so that for a sample containing only the thorium series 0.85 of the alpha pulses are above the threshold. It turns out that, with the same setting, for a sample containing only the uranium series 0.82 of the

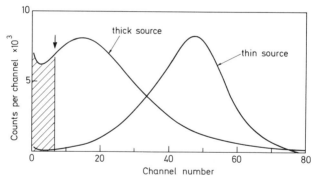

Figure 4.9 Pulse height distribution for alpha particles incident on zinc sulphide scintillator. This is measured using a multichannel analyser; each channel on this corresponds to a given height of pulse and this is proportional to the energy deposited by the particle in the phosphor. The wide distribution in the case of the thin source, which was americium-241, arises because of poor resolution in scintillator and photomultiplier; in the case of the thick source, which was pitchblende, the incident particles have energies spreading from maximum energy of alpha emission in the uranium series, 7.7 MeV, down to 0 for particles emitted at a distance from the scintillation equal to their range. The rise below channel 3 represents pulses for beta particles and photomultiplier noise. By setting the threshold as indicated by the arrow, these (as well as about 16% of the alpha pulses) are ignored in routine counting.

4.3 Measurement

alpha pulses are passed; hence for a sample containing equal activities of the two series the threshold fraction is 0.835. An alternative way of setting the threshold is to use a calibration sample; some suitable ones are mentioned in Appendix J.

The pulse shaping circuitry at the photomultiplier anode is such that the pulse duration is about 1 μsec; hence there is negligible likelihood of two pulses being superimposed ('piled up') until count rates of the order of 10,000 counts/sec are reached; sample count rates are rarely greater than 50 per *kilo*sec.

Screen-to-Screen Reproducibility. As indicated in Figure 4.4 the sample is in intimate contact with the screen. Although it is possible to interpose a thin sheet of aluminized plastic, in practice there is appreciable reduction in count rate for any thickness that is sufficiently robust for routine use because of the short range of alpha particles. Hence a new screen is needed for each sample. Experience with screens made by sprinkling zinc sulphide onto Sellotape indicates that for a given operator the screen-to-screen reproducibility is better than ±3% in terms of the count rate from a standard sample; the same is true of a given batch of ready-made plastic screens. When variation does occur it appears to be due to an excessive amount of zinc sulphide on the screen causing attenuated pulse height because of optical absorption; usually loadings are in the range 5–10 mg/cm^2.

For checking reproducibility the source used should, as far as possible, replicate the sample, that is, it should be one of the natural radioactive series, it should be alpha-thick, and it should be of the same diameter as the screen. A suitable one can be made using pitchblende dispersed in resin; for a 42-mm diameter screen a count-rate of 300 counts/sec can be obtained so that a statistically precise evaluation can be made in a minute. The pitchblende source must not be in contact with the screen because otherwise the background is likely to become substantial.

Reflectivity; Colour of Sample. Of the scintillation light reaching the photocathode some is direct and some is reflected. The proportion of the latter will depend on the irregularity of the sample/screen interface and on the colour of the sample. Taking extremes, the average pulse height for a bright white sample would be twice that for a dull black one, the whole spectrum being stretched or contracted, respectively. Thus the fraction of alpha pulses rejected at the electronic threshold for the white sample would be half that for the black sample, so if the threshold fraction had been set at 0.835 for an intermediate sample, implying a fraction rejected of 0.165, the fraction rejected for the white sample would be about 0.11

and for the black 0.22, corresponding to actual threshold fractions of 0.89 and 0.78, respectively. This implies that the count-rate would be about 6% higher and 6% lower, respectively. Since in practice samples are rarely extreme in colour the effect is not usually significant. It can be reduced by lowering the threshold (i.e., increasing the threshold fraction) but this should be done with prudence because of the risk of accepting beta pulses and the increased vulnerability to electrical interference.

The 'Pairs' Technique. Typically the alpha count rate from a sample is around 10 counts/ksec, that is, the average time interval between pulses is 100 sec. However, in the thorium chain about 3% of the counts occur in pairs—that is, within about 0.2 sec of each other. This is because the alpha emitter polonium-216 has a half-life of only 0.145 sec and it follows immediately after another alpha emitter. If the number of pairs occurring with a spacing of less than 0.21 seconds is recorded by an electronic coincidence circuit, the 'pairs rate' observed is a measure of the thorium activity; it is shown in Appendix J that the pairs rate is given by

$$\dot{p}_h = 0.0048\, c_h, \qquad (4.16)$$

where the conditions and symbols are the same as for Equation 4.11. For the 'typical' composition of Table 4.5, $c_h = 41$ Bq/kg, and hence $\dot{p}_h$ is only about 0.2 pairs/ksec, that is, it takes a week to accumulate 100 pairs, implying a statistical uncertainty of $\pm 10\%$ even then. Thus the counting of pairs is hardly a practical proposition for determining thorium content routinely though it is useful for special samples. However, even in a day's counting, the occurrence of grossly abnormal thorium contents can be detected, thereby signalling the need for special investigation.

It should be noted that random coincidences make a significant contribution to the pairs actually observed and this must be subtracted. The fraction of true singles which get counted as pairs is (0.21 sec × count rate), that is, an actual rate of 0.02 counts/ksec for a singles count rate of 10 counts/ksec, but rising to 2 counts/ksec for a singles count-rate of 100. Since the 'true' pairs rate for the latter is likely to be only about the same, we see that the technique ceases to be useful for abnormally active samples. In addition to the random pairs there is also a contribution to the observed pairs rate by 'fast pairs' due to the 0.002-second half-life of polonium-215 in the uranium-235 series. For 'typical' composition these amount to about 10% of the thorium pairs. The full formula for deriving the contribution made to the total alpha count-rate $\dot{\alpha}$ by the thorium series is

$$\dot{\alpha}_h = 28\left(\dot{d} - \frac{0.21\,\dot{\alpha}^2}{10^3}\right) - \frac{\dot{\alpha}}{12}, \qquad (4.17)$$

4.3 Measurement

where $\dot{d}$ is the 'doubles' rate (per kilosecond), that is, the total pairs rate inclusive of random pairs. The $\dot{\alpha}^2$ term subtracts these random pairs and the final term makes allowance for the fast pairs from the uranium series. This formula applies for a coincidence window of 0.21 sec. In *Daybreak* equipment the window is open from 0.02 to 0.4 sec after each count; the correct formula is then

$$\dot{\alpha}_h = 21 \left(\dot{d} - \frac{0.38 \, \dot{\alpha}^2}{10^3} \right). \tag{4.17a}$$

Thoron and Radon. The first manifestation of the troublesome effects of these gases met by an experimenter is difficulty in achieving a low background count-rate for the scintillation screens. The need to shield the screen from alpha particles emitted by thoron and radon in the air has already been mentioned but there is also a more subtle effect; particularly inside buildings of brick and granite there is enough thoron and radon in the air for significant absorption on the screen to occur if it is left exposed; this gives rise to a background that decays with time and quite often the 10-h half-life of lead-212, which occurs in the thorium chain later than thoron, can be observed. The remedy is of course to store the screen sealed in its container for a few days before use.

The next manifestation is that for samples from which there is escape of gas the count-rate builds up with time if the container is sealed (using the O ring and lid shown in Figure 4.5). The reestablishment of the thorium chain is governed by the 10-hr half-life just mentioned and of the uranium chain by the 3.8 day half-life of radon itself (though the lead-210 and polonium-210 will remain effectively constant at the level established during burial because of the 22-yr half-life of the former). In early days this build up was considered as a means of evaluating the degree of escape, and three regimes of counting were distinguished: the unsealed count-rate $\dot{\alpha}_0$, the average count-rate during the first 24 hrs after sealing $\dot{\alpha}_1$, and the count-rate 10 days after sealing $\dot{\alpha}_2$. The excess of $\dot{\alpha}_1$ above $\dot{\alpha}_0$ is indicative of thoron build-up and the further increase to $\dot{\alpha}_2$ indicative of radon build-up. Although this is now seen to be a simplistic approach due to the problem of 'overcounting' discussed below, an increase of the sealed count-rate above the unsealed does remain a useful semiquantitative indicator that a significant degree of gas escape is occurring; a 10% increase is usually taken as the warning level though a lesser increase does not guarantee that the absence of interfering effects such as overcounting is assured (Pernicka and Wagner, 1982). The difficulty of retaining radon in the container should not be underestimated; radon easily escapes through a thin plastic container or through a seal made with Sellotape.

Overcounting. Early practice was to count the sample in the rubbly form in which it was left after 'vicing' in the course of fine grain sample preparation for thermoluminescence measurements. This was done to minimize error due to the possible presence of high-radioactivity inclusions such as zircon; for grains of 100 μm or more a substantial proportion of the alpha particles would fail to emerge and hence would not reach the fine grains on which the thermoluminescence measurements were being made; by counting the sample in the 'viced state' the zinc sulphide is in the same situation as the fine grains. However it has now been established (Pernicka and Wagner, 1982; Murray, 1982) that it is not uncommon for the observed count-rate, even for unsealed samples, to be substantially higher than corresponds to the thorium and uranium contents as evaluated by other methods; this is termed 'overcounting'. The samples concerned were pottery, soils, and sands. On the other hand, tests using a known amount of artificial alpha emitter mixed in an inert matrix have indicated no discrepancy (e.g., Bowman, 1977; Huntley and Wintle 1981), although these were usually finely powdered samples.

Because predicted count-rates are based on average thorium and uranium contents, undercounting rather than overcounting would be expected if radioactive inclusions are present. Another possible cause is radioactive disequilibrium; there would be apparent over-counting if the daughters are in excess, though of course, as with radioactive inclusions, the alpha count-rate gives a more relevant evaluation of the dose-rate than chemical analysis of parent concentration. There are however several other possible causes. First there is the possibility of the converse situation to the presence of zircon grains—that the alpha radioactivity is preferentially located in the surface layer of large grains so that the effective radioactive concentration, as seen by the zinc sulphide screen, is higher than the average; again the alpha count-rate is the more correct evaluation. Second, there is the possibility that the fine grains are higher in radioactivity than the coarse ones and these are preferentially located directly on the screen in the gaps between larger grains. Third, there is the possibility that thoron and radon accumulates in such gaps and in grain fissures adjacent to the screen enabling alpha particles to reach the screen from a greater quantity of gas than if it remained within the grains. This is the cause emphasized by Pernicka and Wagner (1982) who quote examples of thermoluminescence ages for radon emitting pottery which are gross underestimates of the true age if uncorrected alpha count-rates are used; however, correlation between overcounting and radon escape is not yet well established, there being no dependence shown in the data of Murray (1982).

Jensen and Prescott (1983) have found that overcounting is dramatically

4.3 Measurement

reduced if the samples are finely powdered, but they recommend fluxing with lithium tetraborate into a glass disc as the ultimate solution—any effect arising from air spaces or inhomogeneity then being completely ruled out. Murray (1982) obtained good results by casting in polyester resin. All of these remedies get away from the situation in which the zinc sulphide screen sees the same alpha count-rate as the fine grains used for dating, and consequently, they are vulnerable to upset by radioactive inclusions and inhomogeneous distribution of activity within the grain; however, the balance of evidence suggests that serious upset in this way is uncommon and use of one of the remedies mentioned is therefore recommended. The presence of inclusions can be checked by means of fission track mapping or autoradiography.

The Gas Cell. A better estimate of degree of radon escape than is obtainable by sealing an alpha counting container can be obtained by means of the gas cell shown in Figure 4.10. The sample, first in solid form as when buried, and then, in a second experiment, in the crushed condition, is placed in the tray so that the only alpha particles reaching the zinc sulphide screen are those from radon that has escaped and from radon daughters that have been deposited on the walls or the bottom of the tray.

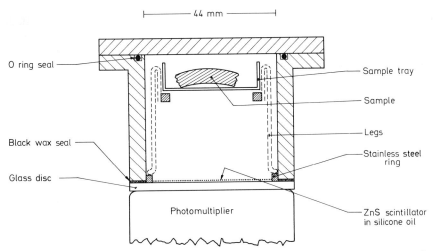

Figure 4.10 Gas cell for detection of radon escape. The only alpha particles that can reach the scintillation screen are those from radon that has escaped from the sample and from deposited daughters. Absorption in the air reduces the count-rate but the rate of radon emission can be evaluated by calibration with a source of known emanating power. Obviously all materials used should be of low radioactivity, for example, borosilicate glass for the disc carrying the zinc sulphide; brass is satisfactory for the body of the container.

The cell is placed on top of a photomultiplier in the same way as the perspex sample container in direct alpha counting. By employing a suitable counting schedule (Aitken, 1978) the escape of thoron and radon can be separately evaluated utilizing the different half-lives involved. The amount by which the escape observed in the second experiment exceeds that in the first allows evaluation of the 'lost counts' when α_0 is being measured.

Because gas is being collected from a quite large sample (several grams instead of the 0.2 g that contribute to the direct alpha count) the gas cell is a highly sensitive device; escape of more than 1–2% of the radon from typical pottery is detectable. However, although sensitive, the device is subject to irregular shifts in counting efficiency, making it only semi-quantitative.

Alpha Spectrometry. Because the sample is thick compared to the alpha range, and in any case because the detection system does not have good resolution, it is not possible to use the zinc sulphide method as a spectrometer and distinguish different energies of alpha emission. For high resolution silicon surface barrier detectors are available with the capability of separating emission energies differing by more than around 0.03 MeV; with such a detector, and using a multichannel analyser to sort the pulses received according to amplitude, the count-rates from individual members of the thorium and uranium series can be obtained. But, of course, the source must be thin compared to the alpha ranges and even if it was feasible to obtain a thin enough layer of sample the amount of radioactivity present would be much too low to give a measurable count-rate. Hence, to make use of the high resolution, it is necessary to employ chemical separation[4] and obtain a thin deposit of the radioisotopes. This can be done, as is in fact done routinely for uranium series dating, but it is laborious and a well-developed expertise is necessary for reliable results.

A more feasible use of alpha spectrometry routinely is to restrict it to the measurement of polonium-210; although acid dissolution is required the need for chemical separation is avoided. This radioisotope is the daughter of lead-210 which is subsequent to radon in the series; hence, its activity is equal to the activity of the radon retained in the sample. Of course, when the sample is dug out of the ground the degree of gas escape changes due to the different conditions (moisture content, absence of back diffusion), but because of the 22-yr half-life of lead-210 the activity of that radioisotope, and also of the polonium-210 which it supports, remains equal to that corresponding to the degree of retention in the ground for several years.

4.3.3 THERMOLUMINESCENCE DOSIMETRY

The use of thermoluminescence itself for measurement of dose-rate seems highly appropriate—and attractive in its low cost—because no extra electronics are required. Through the development of thermoluminescence dosimetry (TLD) for use in medical radiotherapy and radiation protection a number of phosphors are available which can measure doses as low as 10 μGy accurately and without difficulty; from Table 4.5 it is seen that this is about a day's effective dose in typical pottery. However, it is by no means as simple as it might seem at first sight, and for accurate results careful attention must be given to exactly what is being measured. Ideally a direct measurement would be made by direct mixing of the phosphor into the sample, but this is not valid because of the difficulty in replicating the alpha particle effectiveness appropriate to the sample; in addition there would be the problem of replicating grain size. Instead the alpha, beta, and gamma plus cosmic components are measured separately; the last being the most straightforward and the former the most difficult.

Two phosphors are in common use in this context: natural calcium fluoride, *fluorite,* and synthetic calcium sulphate activated with dysprosium, $CaSO_4:Dy$. Some practical details about the former are given in Appendix M. For further information about other thermoluminescence phosphors, the reader is referred to Mejdahl (1978) and to Portal (1981); in respect of non-dating applications there are a variety of sources (e.g., Cameron *et al.,* 1968; McKinlay, 1981; Oberhofer and Scharman, 1981).

The first thing to appreciate when utilizing thermoluminescence dosimetry is that the dose absorbed from a given radiation field by the thermoluminescence phosphor does not necessarily equal the dose absorbed from the same field by the material being dated; and that in the case of gamma radiation the ratio between the two depends on the energy spectrum of the gamma rays.

If the phosphor is close to the dating material in effective atomic number (Z) the ratio is near to unity and for this reason lithium fluoride and lithium borate are favoured for radiobiological application as being nearest to 'tissue equivalent'. Neither of these are a good match to quartz and feldspar; fluorite and calcium sulphate are not a good match either but they are preferred because of higher sensitivity, by a factor of about 20 relative to lithium fluoride. In most circumstances the mismatch does not introduce significant error, being accommodated by means of calibration. The sensitivity of magnesium silicate is higher still, and this has the additional advantage of being close to quartz and feldspar in effective Z.

A characteristic of nearly all synthetic phosphors is that they are prone to a small amount of anomalous fading, and it is usually necessary to make correction for this. The extent of the fading seems to vary with production technique; for $CaSO_4$: Dy fading by 8% in 6 months is typical for beta and gamma radiation, the effect being stronger for alpha. For fluorite the fading is negligible, only 1–2% per yr, but on the other hand the best quality fluorite so far found as regards thermoluminescence characteristics has the drawback of a small but non-negligible self-dose due to radioactive impurities, amounting to 100 μGy/a.

Both of these phosphors are sensitive to light, particularly the ultraviolet component of sunlight, daylight, or fluorescent light. Storage needs to be in opaque containers, and handling for measurement should be in subdued red light. Exposure to light usually causes fading but there can also be enhancement, as discussed in Appendix M.

For measurement, the standard equipment for dating can be used; however the wavelength regions of the thermoluminescence from $CaSO_4$: Dy are around 480 (blue-green) and 570 nm (green-yellow) so that the blue-violet filter, normally fitted inside the photomultiplier housing, needs to be removed. This is not necessary in the case of fluorite because its thermoluminescence emission extends over the wavelength range 300–450 nm (ultraviolet to blue).

The measurement procedure usually consists of comparison of the first-glow thermoluminescence (i.e., the thermoluminescence that accumulated during storage) with the second-glow thermoluminescence induced by beta irradiation of the same portion (see Figure 4.11); this gives higher precision than the additive dose technique used for dating. It is satisfactory only as long as there is no sensitivity change between first and second glow such as might be caused by the pre-dose effect; this effect only becomes significant for doses of 1 Gy and upwards so it is not usually relevant in the context of annual dose evaluation. Another advantage of using a one-portion procedure is that by adjusting the duration of the beta irradiation approximately the same level of thermoluminescence can be obtained in the second glow as in the first. This avoids possible error due to non-linearity of equipment response, such as due to 'pile-up' of photomultiplier pulses or inadequacies in the ratemeter.

Calibration of radioisotope sources and other aspects of artificial irradiation are discussed in Chapter 5. It is worth emphasizing here that the dose delivered is strongly influenced, not only by source-sample distance, but also by the backing material on which the phosphor is carried; the thickness of the portion being irradiated is also important, and for precise measurements a monolayer of grains is recommended.

4.3 Measurement

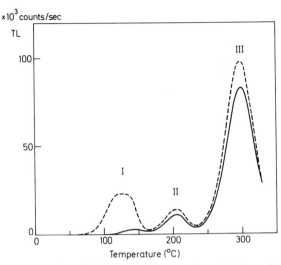

Figure 4.11 Glow-curve from fluorite. The full curve shows the thermoluminescence due an irradiation administered some weeks before measurement so that peak I has decayed away. Evaluation of dose is by comparison with the thermoluminescence due to a calibrating beta dose, shown dashed, that is administered to the same portion. By expanding the scale for peak II, that peak can be used for the comparison as well as peak III. If the dose is of the order of 1 Gy or more, there is risk of sensitivity change between the first glow and the second glow due to the pre-dose effect; to minimize this, the first glow-curve should be terminated as soon as a temperature sufficient to erase peak III has been reached.

Gamma Dose-Rate from the Burial Soil. A convenient capsule for the phosphor has already been shown in Figure 2.4. One essential feature is that the 0.7 mm thick copper wall is sufficient to stop beta particles from reaching the phosphor, another is that the capsule material is free of radioactivity—copper, stainless steel, and plastics are usually satisfactory, but as has been emphasized in other contexts aluminium is to be avoided. After filling the capsule with about 0.1 g of phosphor it is made watertight by hard soldering. On site, shortly before burial, it is heated to about 400°C for a few minutes so as to set the thermoluminescence to zero; this can be conveniently done in the flame of a camping gas cooker; indication that high enough temperature has been reached is obtained either by noting colour change in a patch of a special temperature sensitive paint[5] or by observing a dull red glow in the dark. The capsule is inserted into the soil in as similar a situation as possible to that from which the sample was removed; it must of course be at least 30 cm from any exposed surface (see Figure 4.3) and this is achieved by poking it along a hole made with an auger; a string or stainless steel wire is attached for retrieval. If a plastic capsule is used, a convenient means of insertion is to pack the container within a steel tube (Mejdahl, 1970, 1978), the latter being driven into the ground and left there until retrieval.

The annual gamma dose is usually in the range 500–1500 μGy/a. Although this is sufficient for a measurable dose to be accumulated in the phosphor in the course of a week or so, it is preferable to leave the capsule buried for a year in order to even out seasonal variation in water content; in the types of site studied by Mejdahl the effect of climatic variation was not strong but this is not always the case (Martini, Piccinini and Spinolo, 1983). Another reason for avoiding a short burial time is to avoid losing accuracy due to the 'travel dose' acquired during transit to the laboratory for measurement; it should be appreciated that the above-ground gamma dose-rate is comparable to the burial dose-rate. Hence it is usually desirable to employ a 'travel monitor'; this capsule is set to zero on the day that the buried capsules are retrieved and then kept close to those capsules until measurement. This procedure also deals with the dose received from X-ray security checks, though this is unimportant as long as the check is 'film-safe'. There may be a small contribution from cosmic rays during flight, this dose being of the order of ten times the ground level cosmic dose; however it is still only about 10 μGy/day, so unless the flight is long it will usually be negligible.

The finite thickness of the capsule wall, necessary to stop beta rays, causes some attenuation of the gamma flux too. As discussed in Appendix L, for the 0.7-mm wall copper capsule of Figures 2.4 using fluorite as phosphor, the dose for unshielded quartz is estimated to be 1.10 times the dose actually received by the phosphor. Of course the external gamma dose received inside a sample is attenuated to some extent as well; this has been discussed in Section 4.2.2 and in Appendix H.

The numerical value of the correction factor relating infinite matrix dose to measured dose depends on the phosphor used as well as on capsule wall. This is because different phosphors have different relative responses to gamma rays in the energy region below 0.1 MeV where attenuation is appreciable. Fluorite overresponds substantially in this region and were this not the case, the observed depression of dose would be greater. The response of a given combination depends also on the energy spectrum of the gamma rays being received, but fortunately the spectra from potassium, thorium, and uranium are sufficiently similar that there is no significant dependence on which is dominant.

For $CaSO_4$: Dy in a copper capsule of wall thickness 1 mm, the correction factor has been calculated by Valladas (1982) to be 1.08. For $CaSO_4$: Dy in a steel tube of wall thickness 1.5 mm Mejdahl (1978) found experimentally that the dose received by aluminium oxide in a quartz tube of wall thickness 2.5 mm is 0.95 times the phosphor dose; aluminium oxide is close enough in absorption characteristics to be considered as equivalent to quartz.

The amount of phosphor put in a capsule should be 0.1 g or more. Although less would be adequate in providing measurement portions, there are dosimetric effects which then become significant.[6]

Beta Thermoluminescence Dosimetry. In measuring the gamma dose it is important that the capsule wall is thick enough to prevent beta particles from reaching the phosphor; this is because, except for the discarded outer 2 mm of the sample, the beta dose from the soil does not contribute to the thermoluminescence. In the case of beta dose measurement it is necessary to prevent alpha particles from reaching the phosphor, because of the need to make separate measurement of the components mentioned earlier.

The most direct method of beta thermoluminescence dosimetry is that developed for utilization with the quartz inclusion technique (Fleming, 1969, 1979). Fluorite grains having the same size range (around 100 μm) as the quartz used for dating are mixed into a gram of powdered pottery and stored for several weeks shielded by lead from environmental gamma radiation. The grains are specially treated before use so as to make them insensitive to alpha radiation; by controlled exposure to water vapour at elevated temperature (600°C) it is possible to desensitize the outer layer of the grains to an appropriate depth. Hence in terms of absorbed dose fluorite grains are a good replication of the quartz grains used for dating, that is, 100 μm grains of which the alpha-irradiated outer layer has been removed by etching with hydrofluoric acid; etching is not possible with fluorite but this prior desensitization serves the same purpose—elimination of the alpha contribution to the thermoluminescence. After storage, the accumulated thermoluminescence of the fluorite grains is measured, recovery being by sieving and washing; hence the dose-rate is evaluated. This does include a small contribution of gamma dose from the sample itself, and correction[7] must be made for this.

Although elegant, the grain substitution method is tedious in execution, particularly because of the difficulty of getting well-controlled diffusion depths in the desensitization of the fluorite grains, and it is not used routinely. An alternative approach is the external dosimeter method first developed by Mejdahl (1969). A later version of this approach is shown in Figure 4.12 and a scaled-down unit of this type needing only 1 g of sample has already been illustrated in Figure 2.3; an intercomparison of results from different methods of beta dose evaluation has been given in a report by Haskell (1983). In all of these, alpha particles are stopped by a plastic

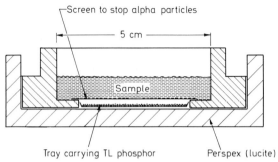

Figure 4.12 Measurement of beta dose-rate by thermoluminescence dosimetry (Bailiff and Aitken, 1980). After the phosphor has been exposed to the beta flux from the sample for a few weeks its thermoluminescence is measured. About 0.5 g of phosphor (e.g., fluorite) are used and it is fixed on the tray (of copper, 0.25 mm thick) with silicone resin. A 'lamp oven' is used for measurement of the thermoluminescence: the tray is placed on a stainless steel plate which is heated from beneath by means of a 100 W pre-focussed projector lamp.

absorber, usually 0.18 mm thick, and the external phosphor grains receive only the beta dose from the sample together with some gamma contribution from the sample itself and from the surroundings (despite lead shielding), as well as cosmic ray background.

Although it might be expected that the beta dose experienced by the phosphor grains is half that within the sample itself, it is found, by using standard samples of known composition, to be substantially less; this is on account both of attenuation in the plastic and of the finite distance between sample and phosphor. For the unit of Figure 4.12 the ratio between phosphor dose and internal sample dose is 0.18 and for the unit of Figure 2.3, in which the dosimeter tray is only 0.75 mm from the plastic absorber, it is 0.28. Thus for a sample having the 'typical' composition of Table 1.1, the phosphor dose-rate in the former unit is only about 0.8 μGy/day. However a thermoluminescence phosphor has adequate sensitivity for a precise measurement to be made after a few weeks of storage; a more serious limitation is that the background due to external gamma radiation and cosmic rays contributes a comparable dose-rate despite the 2.5-cm thick lead shielding, and for samples substantially weaker in radioactivity than 'typical', the precision becomes poor. In any case careful measurement of the background dose-rate using radioactivity-free quartz is necessary. Substantially thicker shielding, of the order of 8 cm, would be advantageous although expensive. Another consideration here is 'cross talk' between samples, for example, if a weak sample is stored close to a strong one, the gamma flux from the latter will contribute significantly to the dose measured for the former; hence individual shield-

4.3 Measurement

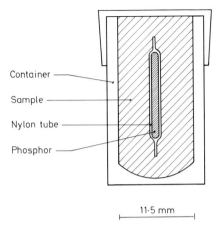

Figure 4.13 Measurement of beta dose-rate by thermoluminescence dosimetry (Valladas and Valladas, 1983). The phosphor is contained in a nylon tube (internal diameter 1.1 mm) which is sealed by applying pressure with hot pliers. This tube is immersed in the powdered sample; typically a few weeks is long enough for an accurately measurable level of thermoluminescence to be reached. The sample needs to be pressed into the container firmly enough for it to be effectively infinite as far as beta particles are concerned. Alpha particles are prevented from reaching the phosphor by the ¼-mm thick wall of the nylon tube. The mass of phosphor, $CaSO_4:Dy$, is about 18 mg.

ing is advisable. There is also a contribution to the measured dose due to gamma rays from the phosphor's own sample. It is not necessary to evaluate this and make subtraction because it is automatically allowed for in calibration[8].

In the unit developed by Valladas and Valladas (1983) the phosphor grains are contained in a thin-walled narrow tube of nylon which is inserted into the powdered sample, as shown in Figure 4.13. This has the advantage that the ratio between the phosphor dose and the dose within the sample is higher; for the dimensions shown, the ratio is 0.51, being less than unity because there is attenuation of beta particles in the wall of the nylon tube. Also the small size of the container makes this unit easier to shield and hence on both accounts the background is relatively less important. On the other hand, it does not have the convenience of the phosphor grains being fixed on a tray ready for the oven plate; however it does allow several portions of grains to be measured for each storage and hence give greater precision.

In all types of unit it is desirable for the sample to be 'beta thick', that is, several millimetres or more.[9] Besides giving maximum dose to the phosphor this also removes dependence on weight of sample as far as beta dose is concerned; however for the gamma contribution there is rough proportionality to weight.

Alpha Thermoluminescence Dosimetry. The problem in adapting any of the beta thermoluminescence dosimetry methods is that because alpha particles have such a short range the sample needs to be in direct contact with the phosphor, as in alpha counting; similarly, if mixing-in of grains is considered these must be fine (10 μm or less) and the problem of recovery is formidable. However, encouraging results have been obtained by Wang and Zhou (1983) using a thin layer[10] of fine grain phosphor deposited on 10-mm discs of aluminium foil immersed directly in the ground-up sample; after a suitable storage period the foil is withdrawn, adhering sample grains blown off, and the thermoluminescence is measured by placing the foil directly on the heater plate. The beta dose is obtained with a similar dosimeter shielded on both sides by 80-μm thick polyethylene; the alpha dosimeter is shielded only on the aluminium foil side and so the alpha dose is obtained by doubling the difference between the individual dose recorded by the alpha and beta dosimeters. The alpha dose so obtained is the effective dose in the phosphor, so to evaluate the full alpha dose the effective dose needs to be divided by the alpha effectiveness of the phosphor.

Although the method is attractive in its simplicity it is not as convenient as direct alpha counting (several weeks storage being required) nor does it necessarily avoid the various interfering effects such as overcounting, though it will avoid risk of the small error that there is in alpha counting arising from extremes of sample reflectivity.

4.3.4 GAMMA SPECTROMETRY USING A SCINTILLATION CRYSTAL

An alternative to burying capsules for evaluation of annual gamma dose is to take on-site a gamma spectrometer or a gamma scintillometer. These have the advantage that the measurement takes less than half an hour but the disadvantage is that the diameter of the auger hole, which must of course be 30 cm long (the same as for the capsule) has to be substantially wider, typically 70 mm rather than 10 mm. In the case of the gamma spectrometer, the individual contributions from potassium, uranium, and thorium are determined, such information being helpful when assessment is being made of the likely errors due to variation in radon escape and leaching by ground-water.

The scintillation crystal currently used in these devices is of sodium iodide doped with thallium as an activator; in the present application the

4.3 Measurement

crystal is typically a cylinder of size 44 mm × 51 mm. Gamma rays produce scintillations in the crystal and these are viewed by a photomultiplier which, as in the alpha counter, produce electrical pulses (one pulse per gamma ray) as a consequence. However, because the gamma ray may undergo various types of interaction in the crystal, the size of the pulse is not always the same. This is illustrated in Figure 4.14 for the 1.46-MeV gamma ray from potassium-40. Each pulse in the *photopeak* corresponds to the conversion of a 1.46-MeV gamma photon into an electron having the same energy, the electron being ejected from an iodine atom in most cases; the electron dissipates all of its energy in the crystal (except those which escape through being produced too near the surface) resulting in scintillations produced at luminescence centres formed by the thallium atoms. The *Compton continuum* is due to photons that interact with the atoms of the crystal by a different process such that only part of their energy is deposited in the crystal, the rest continuing as a photon of lower energy; the amount deposited varies from event to event with a fixed probability. A typical spectrum from soil is shown in Figure 4.15; in the thorium and uranium series there are a variety of energies of gamma emission and consequently the spectrum is a melange of photopeaks and Compton continua.

A qualitative difference between alpha particles and gamma radiation should be noted. While passing through matter a group of alpha particles initially of the same energy lose energy according to the distance trav-

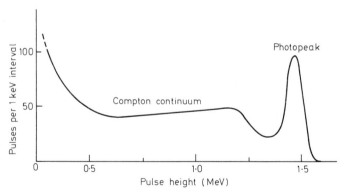

Figure 4.14 Pulse height spectrum for potassium-40 gamma rays in sodium iodide crystal. Gamma photons produce scintillations in the crystal; each scintillation is converted into an electrical pulse by the photomultiplier. For some the pulse height corresponds to the full energy of the photon (1.46 MeV), giving the photopeak, but for the remainder only some of the energy contributes to the scintillation, giving the Compton continuum. The larger the crystal the smaller is the height of the continuum relative to the photopeak.

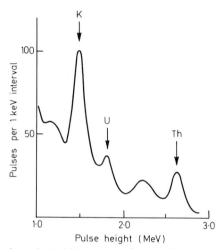

Figure 4.15 Composite pulse height spectrum from sodium iodide crystal for potassium, thorium, and uranium gamma rays from soil. Whereas for the thorium series a pulse height region ('window') spanning the 2.6 MeV peak will contain only pulses due to the thorium series gamma rays, the regions spanning the potassium peak at 1.46 MeV and the uranium series peak at 1.76 MeV will contain pulses from the Compton continua associated with higher energy peaks, as well as minor interfering peaks; hence the need for 'spectrum stripping'.

elled, finally reaching zero energy and coming to a halt. On the other hand, a beam of gamma ray photons is *attenuated* and *degraded* as it passes through matter; some travel without interaction and hence retain their full initial energy, others disappear completely through photoelectric interaction (as is responsible for the photopeak in the scintillator crystal) and some lose only part of their energy, through Compton interaction, continuing on as gamma photons of lower energy; there is also another process, the production of pairs of electrons and positrons, which is not of much importance at the gamma energies concerned here. Thus it is possible, for gamma photons, unlike alpha particles, to give rise to a scintillation pulse corresponding to the full energy of emission even though the site of the emission is some distance away. Of course because of the degradation process there are photons of lower energy too and these contribute to the continuum of the spectrum observed from a thick source, or from a discrete source when there is appreciable amount of absorber interposed.

The Portable 4-Channel Gamma Spectrometer. It will be seen from Figure 4.15 that there are some high energy photopeaks that are distinguishable. Fortunately there is one suitable for determination of each of the three

4.3 Measurement

main radioactive series with which we are concerned: there is the peak at 1.46 MeV for potassium-40, the peak at 1.76 MeV from bismuth-214 of the uranium series, and the peak at 2.61 MeV from thallium-208 of the thorium series. Evaluation of the two decay series in this way does of course assume that there is radioactive equilibrium.[11]

After amplification the pulses are fed to the pulse height analyser section of the electronics. In this there are 4 channels, corresponding to potassium, uranium, thorium, and cosmic rays. Each channel has an acceptance 'window' that takes a range of pulse heights centred on the peak concerned. The actual width of the window can be selected to suit circumstances, but typically the potassium window would accept pulse heights corresponding to the energy range 1.38–1.53 MeV, the uranium window 1.69–1.84 MeV, and the thorium window 2.46–2.76 MeV; the cosmic ray channel is set to accept all pulses above 3 MeV. The pulses received by each channel are stored in its memory and at the conclusion of the count time the number stored in each is read out sequentially onto a digital meter. Reference to Figure 4.15 will show that whereas the thorium channel will receive only thorium gamma rays, the uranium channel will receive some gamma rays of the thorium Compton continuum, and the potassium window will receive rays of both the thorium and the uranium continua. Correction is made for these interferences by the 'spectrum stripping' procedure described in Appendix L, which also indicates how the instrument is calibrated so that count-rates can be interpreted in terms of the individual dose-rate components (and incidentally in terms of weight-for-weight concentrations of potassium, thorium, and uranium).

Several commercial versions of this instrument are available, it being developed for mineral prospection. For archaeological use an important consideration is the outer diameter of the housing that carries the scintillation crystal, photomultiplier and preamplifier; in the instrument that the author has found convenient to use over several years, the diameter is 64 mm, the length 27 cm, and the weight 1 kg; the electronic package weighs 5 kg and it is powered by eight 1.5-volt dry batteries. Because of temperature dependence an important feature is spectrum stabilization (see Appendix L).

The Gamma Scintillometer. In the gamma scintillometer the whole spectrum of pulses above a certain threshold is used to evaluate the dose-rate. This threshold must be chosen with care because otherwise the calibration of the instrument will depend on the relative proportions of potassium, thorium, and uranium; this is because the ratio of the energy absorbed in the sodium iodide crystal to that absorbed in pottery is different for the different spectra; also in talking about count-rate we must remember that the energy represented by a count is proportional to the scintillation size giving rise to it. An instrument suitable for thermoluminescence use has been developed by Murray (1981; Murray *et al.*, 1978a, b) who finds that for a sodium iodide crystal of the size 44 mm diameter by 51 mm long, the use of a threshold equivalent to 0.43 MeV ensures that the dose-rate is proportional to the count rate (above threshold) with negligible dependence on whether the radioelement concerned is potassium, thorium, or uranium; it is 30 μGy/a per 1 count/sec. So for typical soil the count rate is about 30 counts per sec and a reliable measurement can be completed in a few minutes; this is substantially less than the time required to make a measurement with the gamma spectrometer. (Note however that a separate, longer, measurement is needed to get the cosmic ray contribution.) The detector head is the same and electronics package similar though less complex; for accurate measurements spectrum stabilization is desirable though it is not so vital as with the gamma spectrometer.

Laboratory Use. Both of these instruments can be used for soil samples brought to the laboratory but this is a second best alternative. The 30 cm radius sphere from which gamma rays reach the detector when it is used on-site contains about 300 kg of soil; obviously it is only practical to bring about one-hundredth of this weight to the laboratory quite apart from the question of whether or not the archaeologist would like to see a big hole in his site. Consequently in laboratory use the count-rate from the sample is much lower and substantial lead shielding, at least 5 cm thick, needs to be employed to avoid being swamped by the background from the surroundings. However, when on-site measurements are not possible this is a better way of evaluating the soil gamma dose-rate than alpha counting both because it measures gamma activity directly and because it avoids the problems of overcounting. It has also been used for pottery samples (Meakins, Dickson and Kelly, 1979).

4.3.5 HIGH RESOLUTION GAMMA SPECTROMETRY

In crystals used as semiconductor detectors the electrons ejected from atoms by gamma rays are directly collected as an electrical pulse without the need for a photomultiplier; as a result the resolution (for a 0.6 MeV photopeak) is at least an order of magnitude better, the width of a photopeak being only 1 or 2 keV (i.e., 0.001–0.002 MeV) compared to about 50 keV for a sodium iodide crystal. The advantages are twofold; first, the better resolution allows individual measurement of isotopes that have

4.3 Measurement

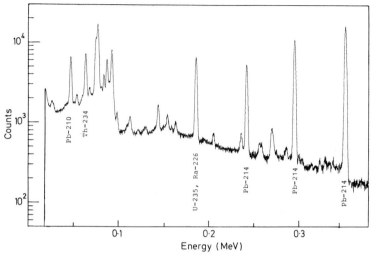

Figure 4.16 Part of the spectrum from a uranium-rich sand as detected by a high resolution germanium crystal (from Murray, 1981).

gamma energies separated from each other by only a few keV; second that the sharpness of the photopeaks makes it possible to detect those from low energy gamma rays despite the presence of the Compton continuum from higher energy gamma emitters in the same sample (see Figure 4.16). These advantages mean that measurement can be made of individual members of the thorium and uranium series and information can be obtained about the degree of radioactive disequilibrium. It is not as incisive a tool as alpha spectrometry, nor is it as sensitive, but it has the very strong advantage that is does not require the painstaking chemical separation procedures of the latter. Of course, there is the difficulty of poor sample to background ratio already noted with the scintillation-based spectrometer, but the importance of the information gained makes it attractive nevertheless, not only for soil but also for pottery.

Two types of detector crystal are in current use: germanium doped with lithium, Ge(Li), the so-called 'jelly' detector, and intrinsic germanium. Both types need to be cooled by liquid nitrogen ($-196°C$) when in operation, and the former has the further restriction that it must stay at that temperature whether in operation or not, otherwise expensive reconditioning is necessary. Because of the need for cooling they are not at all convenient for on-site work, although portable instruments are in fact available. The cost of a crystal suitable for archaeological dating is in the range $10,000–20,000 and the associated electronics cost about the same again; to make full use of this type of detector comprehensive peak fitting programmes are necessary, preferably using a mainframe computer.

To get maximum count-rate from a limited quantity of sample, a well-type detector is used and in the system developed for archaeological use by Murray and Aitken (1982) the sample size is only 7 g, thereby making it a practical proposition for pottery. The crystal is shielded against laboratory background by 7.5 cm of low activity lead; the duration of the counting period is usually 24 hours. An important feature is that the 46-keV gamma ray from lead-210 can be measured, and as mentioned earlier, this is a measure of the radon retention while the sample was buried. Much useful information about radioactive disequilibrium has been obtained with this instrument (e.g., Murray, 1982); however, because of the limited sample size the number of counts acquired in 24-hr period is not sufficient for good statistical accuracy unless the level of radioactivity is higher than average, and in practice after peak fitting and subtraction of background typical error limits for most of the radioisotopes are around ±10%. Where application is limited to soil measurements a detector suitable for a much larger sample size is to be preferred.

4.3.6. FISSION TRACKS AND AUTORADIOGRAPHY

The alpha particles from a radioactive inclusion, such as zircon, will largely dissipate their energy in the inclusion itself rather than in the fine grains used for thermoluminescence measurement; also, in zircon dating (Section 6.3) where thermoluminescence from the zircon is used, there is difficulty due to zoning of the radioactive contents and the thermoluminescence properties; this holds in the case of stalagmitic calcite also. Hence there is the need to study the spatial distribution of the radioactivity in samples, with a resolution sufficient to detect inhomogeneities on a scale of the order of magnitude of alpha tracks, that is, 10–20 μm.

In *fission track counting* (see, for example, Fleischer, Price and Walker, 1975) a polished section of the sample is covered with a polycarbonate foil (e.g., 'Makrofol') and exposed to the neutron flux of a nuclear reactor. Neutrons induce fission in the thorium and uranium nuclei in the sample and some of the resulting nuclear fragments plough into the foil leaving tracks of damaged structure. After exposure the foil is etched with sodium hydroxide and since the damage tracks are vulnerable to this they become visible under an optical microscope. The tracks are short and hence a photograph of the foil reveals the spatial distribution of the thorium and uranium; radioactive inclusions are revealed by distinctive clusters of tracks. Fast neutrons are used for thorium, and thermal neutrons for uranium. The technique can also be used for determination of bulk concentration by using a powdered sample. Extensive use of this technique in thermoluminescence dating has been made by the group at Heidelberg (see, for example, Mangini, Pernicka and Wagner, 1983). With *autoradiography* (Sanzelle, Fain and Miallier, 1983) tracks of natural alpha particles are used and a nuclear reactor is not required. However, because these tracks are longer the resolution is not as good as with fission tracks; exposure times of several weeks are required.

4.3.7 NEUTRON ACTIVATION AND CHEMICAL ANALYSIS

In neutron activation (often referred to as INAA—Instrumental Neutron Activation Analysis) the sample is exposed to the neutron flux of a nuclear reactor; artificial radioactivity is thereby induced in the constituent elements, and by analyzing the resulting spectrum of gamma rays with a gamma spectrometer (of the high resolution type mentioned in Section 4.3.5) the concentrations of most elements can be determined, including potassium, rubidium, thorium, and uranium. It is a well-established technique and given easy access to the necessarily comprehensive facilities it has strong advantages. Of course only the parents of the radioactive series are measured; as explained in Section 4.1 the concentrations of the daughters are far below the limits of detection by this means. Thus ancillary checks need to be made concerning the question of disequilibrium—the annual dose derived on the assumption of equilibrium will be much more seriously in error if that assumption is unjustified than for any of the methods in which the evaluation is based on measurement of natural radioactivity. Measurement of polonium-210, as discussed at the end of Section 4.3.2, is one such check which is effective and which does not involve much extra effort (Mangini, Pernicka and Wagner, 1983).

The simplest and least expensive method of determining the potassium content is by the well-established technique of *flame photometry* (Suhr and Ingamells, 1966). This has good sensitivity and only a few tens of milligrams are required, making it very useful in authenticity testing; however, this small sample size does make the analysis vulnerable to upset due to inhomogeneity. Potassium can also be measured by the techniques of *X-ray fluorescence* (XRF) and *atomic absorption* (AA).

4.4 Summing Up

Evaluation of annual dose is no less a hydra-headed monster than thermoluminescence itself. This chapter has not done more than indicate where the problems—and some solutions—lie; hopefully it provides a jumping off ground for the deeper waters of the specialist literature, in particular the proceedings of the *Specialist Seminars on Thermoluminescence and ESR Dating*. As regards choice of measurement method two principles can be suggested. First, each practitioner should be biased towards the methods in which he or his colleagues have best expertise. Second, each component of the annual dose should be measured in as simulative a way as possible, implying: *alpha for alpha, beta for beta, and gamma for gamma*. As regards the more fundamental problems of radon escape, disequilibrium and leaching, two things need to be emphasized. First, particularly in igneous regions and where there is high salinity of the groundwater, it is rash to assume these difficulties are not present; at the least a check for radon escape should be made and where it is significant a thorough investigation of disequilibrium must follow—or the site must be abandoned. Second, there are many sites on which these problems do not intrude; the difficulties encountered on other sites should not be allowed to detract from the success of thermoluminescence on these and it is here that available effort is most effectively utilized.

Technical Notes

[1] It is not always realized that the half-life of the daughter determines the time taken to reach equilibrium in the case of build-up. It follows from the differential equation $dN/dt = \lambda_1 N_1 - \lambda_2 N$, where N is the number of atoms of daughter present at time t, N_2 being reserved for the number present at equilibrium as in the text. If $N = 0$ at $t = 0$ then the solution to this equation is $N = N_2 [1 - \exp(-\lambda_2 t)]$ since it is assumed that $\lambda_2 \gg \lambda_1$, so that effectively the number of parent atoms remain constant.

In terms of activity, putting $A = \lambda_2 N$, $A_1 = \lambda_1 N_1$, and remembering that at equilibrium $A_2 = A_1$, we have $A = A_1 [1 - \exp(-\lambda_2 t)]$. In the case of a daughter that becomes unsupported the differential equation is: $dN/dt = -\lambda_2 N$, and since in this case $N = N_2$ at $t = 0$ it follows that $A = A_1 \exp(-\lambda_2 t)$.

[2] Although Equation 4.3a was specifically derived for the uranium-238 series, the energy release in the uranium-235 series is sufficiently close that allowance for the minor contribution from this series (which provides 4.4% of the activity) does not alter the numerical factor significantly. This means that when substituting into Equation 4.3a the combined activity of the two uranium series should be used for c_u. This applies to Equation 4.10 and all other equations and Tables also unless the contrary is explicitly stated.

[3] One suitable type of zinc sulphide is G345 from Levy West, Enfield, Essex, U.K. This is nickel-killed and silver-activated with a grain size in the range 20–40 microns. Suitable ready-made scintillation screens for alpha counting are obtainable from W. B. Johnson, Montville, New Jersey, U.S. They are available as 30 cm × 30 cm sheets or as circles of various sizes.

[4] An outline of the steps necessary has been given by Mangini *et al.* (1983). The samples, approximately ½ g, have to be decomposed using $HF/HClO_4$ or by fusion with Na_2O_2 or $Li_2B_4O_7$. Then after adding appropriate spikes, the thorium and uranium are separated by two ion exchange columns and electroplated on platinum or stainless steel discs for counting with the silicon surface barrier detector. The radium-226 is determined by allowing the daughter radon to accumulate in solution and transferring it, by bubbling nitrogen through the solution, to a pulse ionization chamber for counting. The polonium-210 does not need chemical separation because polonium has the property of depositing spontaneously from solution onto a silver disc; this is then counted with the surface barrier detector.

An alternative approach is to use a large area ionization chamber, the sample being introduced inside the chamber as a very thin layer of fine particles (~1 μm diameter) deposited from suspension in acetone or by means of an aerosol spray. Direct measurement without chemical enrichment is possible because of the large area employed, but the same comments apply: it is laborious and requires expertise. A further problem in this case is to achieve a low enough background and very careful selection of materials for the chamber is necessary. Martini, Spinolo and Dominici (1983) have reported the use of a chamber with a deposited area of 200 cm^2, having an instrumental resoluton of 0.024 MeV; in the chamber reported by Murray and Heaton (1983) the deposited area was 1 m^2 and the instrumental resolution was 0.05 MeV. The resolution obtainable from a sample of finite thickness is somewhat less, being of the order of 0.1 MeV in the latter case; this requires the use of very thin layers, the order of 2 $\mu g/mm^2$.

[5] A suitable paint is Tempilaq (343°C) obtainable from Tempil Division of Big Tree Division, 2901 Hamilton Boulevard, South Plainfield, New Jersey 07080 (U.K. agents: Bayer U.K. Ltd., Bayer House Richmond, Surrey TW9 1SJ).

[6] If the amount is too small the dose-rate recorded is dependent on the quantity used (Murray, 1981); this is because electrons ejected from the wall are then responsible for a substantial part of the total thermoluminescence and so the greater gamma absorption coefficient of copper makes itself felt. If phosphor and wall are more similar, such as is the case for aluminium oxide or obsidian in a glass capsule, then the effect will be negligible.

[7] For one gram of typical pottery the correction is about 4% of the infinite matrix gamma dose-rate; it would rise to 7% in the case of a sample of which the radioactivity was entirely due to thorium. The gamma contribution would be reduced by using a smaller sample but this would make more acute the practical problem of avoiding fluorite grains getting within the 2-mm outer layer of the sample in which the beta dose-rate falls below the infinite matrix beta dose-rate.

[8] The pitchblende and monazite standards diluted in silica provided by the New Brunswick Laboratory, referred to in the section on Calibration Standards in Appendix J are suitable together with a non-hygroscopic potassium salt such as KCl; somewhat higher activity sands than used for alpha counting are recommended, such as 1% for Th and 0.1% for U.

As mentioned in note 4 of Chapter 2 it is important that the calibration should be independent of whether the dose is from K, Th, or U. If there is a substantial gamma contribution from the sample, this element-independence may be upset because the ratio of infinite matrix gamma dose to beta dose is 0.3 for K, 0.8 for U and 1.8 for Th. This and other aspects of calibration have been discussed by Bailiff and Aitken (1980). In respect of the unit of Figure 4.10 it appears that the higher relative gamma contribution from Th is compensated by greater attenuation of beta particles in the plastic window on account of their lower average energy, and vice versa for K; the gamma contributions, expressed as a percentage of the dose to the phosphor, are estimated to be 5%, 20%, and 10% for K, Th, and U respectively, but the ratio of phosphor dose to beta dose in the sample is within 2% of the average value irrespective of composition.

Another consideration in calibration is the beta stopping power of the standards relative to the samples being measured. Essentially the effect of a higher stopping power is to reduce the depth from which the phosphor receives beta particles, similarly to thick source alpha counting; hence the flux received is inversely proportional to stopping power. If silica-based standards are used for Th and U, no correction is needed for a pottery sample, but in respect of KCl the relative stopping power is 0.865; hence the phosphor dose from KCl is greater than would be obtained from pottery having the same K content.

[9] Obviously the bulk density of the sample is important here; this can be as low as 1.5 g/cm^3. The most energetic beta particles are from protactinium-234m in the uranium series; they have a maximum energy of 2.26 MeV and carry 40% of the dose from the series. From calculations for water by Berger and Seltzer (1982) 95% of the dose from a 2.26 MeV beta emission will be contained in a sphere of radius equivalent to about 0.6 g/cm^2, that is, 4 mm for a density of 1.5 g/cm^3. Experimentally, for the unit of Figure 4.10 it was found that reducing the thickness of sample from 6 mm to 3 mm decreased the phosphor dose by 4% on average.

[10] The thickness of the phosphor layer was about 10 μm, and since this is about one-third of the average alpha particle range, it might be expected that there will be significant attenuation (even for 10 μm grains the dose is down by about 10%, see Figure C.2, and for a layer irradiated from one side only the depression will be at least twice this). However Wang and Zhou (1983) found reasonable agreement with direct alpha counting, the tendency being for TLD to overestimate rather than underestimate.

[11] Note however that both of these gamma emitters are the highest energy ones of the chain and so for evaluation of gamma dose-rate they are the most relevant ones to use. In the case of the uranium series note that bismuth-214 comes after radon and that 95% of the gamma dose-rate is post radon. Hence even though there is radon loss from the soil, evaluation by means of bismuth-214 is a good measure of gamma dose-rate.

CHAPTER 5

ARTIFICIAL IRRADIATION

In evaluating the paleodose, the natural thermoluminescence is compared with that induced in the same sample by exposure to an artificial radioisotope source. We now come to the question of how the dose absorbed by the sample during that exposure is evaluated; it is a far more complex matter than calibrating the activity of the source, because the dose delivered to the sample is influenced by its composition, its thickness, and its surroundings as well as, of course, by its distance from the source. As with thermoluminescence dosimetry we shall proceed in the order of increasing complexity: gamma, beta, alpha.

As mentioned in earlier chapters, assessment of the thermoluminescence contribution from alpha particles is complicated by their short range and by their poor thermoluminescence effectiveness relative to beta and gamma radiation on the basis of thermoluminescence per unit dose. Except possibly at low energies, the two latter have the same effectiveness.[1]

The energy absorbed by a medium from a given flux of radiation (i.e., photons or particles per unit area, as the case may be) depends on the atomic composition of the medium. However with exceptions such as zircon, the minerals encountered in thermoluminescence dating are contained within quite a narrow range, and the calibration of a source in terms of dose delivered is only weakly influenced by composition; there is no significant difference between the calibrations appropriate to common pottery minerals such as quartz and feldspar and so forth. Thermoluminescence phosphors such as fluorite or calcium sulphate are often used for intercalibration of sources because of the high precision then attainable; if the comparison is between a beta source and a gamma source, then a small correction factor is necessary in order to obtain the dose delivered to a pottery mineral; from now on we shall refer only to quartz and it can be assumed that the dose for other minerals (except heavy ones) is the same. Anticipating the discussion later in this chapter some relevant factors are given in Table 5.1.

Whereas the influence of composition on dose received is weak, the effect of the sample and its container on the radiation flux itself, through attenuation and scattering, can be strong and this is an important aspect of quantitative artificial irradiation.

There is extensive specialist literature relevant to this chapter, daunt-

TABLE 5.1
Beta Stopping Powers and Gamma Absorption Coefficients (Ratio to Quartz)[a]

	Beta[b,c]	Gamma[c,d]
Al_2O_3	0.98	0.99
$CaCO_3$	0.99	1.00
CaF_2	0.95	0.98
$CaSO_4$	0.98	1.00
Water	—	1.12

[a] As an example of using a gamma beam to calibrate a beta source, suppose that the dose to water from a caesium gamma beam is specified as 1 Gy/min. Then, neglecting attenuation in the container wall, the dose to CaF_2 is (0.98/1.12) = 0.88 Gy/min. If 2 min exposures with the beta source are required to induce the same level of thermoluminescence as 1 min exposure in the gamma beam, then the dose to CaF_2 from the beta source is 0.88 × 0.5 = 0.44 Gy/min; the dose to quartz will be (0.44/0.95) = 0.46 Gy/min. Remember that the dose from the beta source is dependent on distance, substrate, and grain size.

[b] The beta column shows the mass stopping power relative to quartz (SiO_2) appropriately averaged over the yttrium-90 beta spectrum (see Figure 5.3) that is emitted by the standard strontium-90 source. Some values for quartz are: 3.39 MeV/gm/cm^2 at 0.1 MeV; 1.71 at 0.5 MeV; 1.57 at 1 MeV; 1.59 at 2 MeV. These values include the small component due to bremsstrahlung production: 2% at 2 MeV, and less at lower energies. Data are based on Pages et al. (1972) and Berger and Seltzer (1982).

[c] The ratio values for beta stopping power do not change by more than a few percent throughout the yttrium-90 spectrum; for gamma absorption the variation is appreciable, becoming strong below 0.1 MeV; see Figure M.2.

[d] The gamma column shows the mass absorption coefficient relative to quartz for the 0.66-MeV gamma emission from caesium-137; values for the gamma emissions at 1.17 and 1.33 MeV from cobalt-60 are within 0.01 of these values. The values for quartz are 0.0293 per gm/cm^2 for caesium-137, and 0.0266 for cobalt-60.

ing in its complexity. A concise account such as those by Boag (1971) or by Greening (1981) is recommended.

5.1 Gamma Irradiation

A gamma source capable of delivering an absorbed dose to a sample that is of the same order as the paleodose is a much more expensive item than an appropriate beta source, and much more comprehensive shielding and source handling facilities are required. Thus unless a thermoluminescence laboratory is situated within an institute where a gamma source has been set up for other purposes, evaluation of paleodose is done using a beta source. However, there are considerable problems in calibrating a beta source for different presentations of sample, and because such calibration is done by comparison against a gamma source some understanding of how to carry out a gamma irradiation is important. Suitable gamma sources are likely to be found in hospitals employing radiotherapy, in radiobiological units, and in atomic energy establishments; the calibration of these is usually derived by means of an ionization chamber which is in turn calibrated against a national standard. Obviously, high precision is required in all these successive steps and one can see the attraction of the thermoluminescence dating systems that are independent of calibration, as discussed in Sections 6.3.2 and 6.4, and in the section on random errors in Appendix B.

Two radioisotopes are in common use as gamma sources: cobalt-60 and caesium-137. The former has a half-life of 5.3 years and emits gamma rays of 1.17 MeV and 1.33 MeV; the latter has a half-life of 33 years and emits gamma rays of 0.66 MeV. To obtain a dose-rate convenient for thermoluminescent dating of, say, 1 Gy/min, with a source-sample distance of, say, 10 cm requires a source activity of about 10^{12} Bq (about 40 curies) in the case of cobalt-60, and about four times as much in the case of caesium-137. The thickness of lead required to make it safe for people to be within a metre of such a source is considerable, and remote control is an essential feature. However the details of gamma source construction and so forth are unlikely to be the concern of the thermoluminescence practicioner; on the other hand, an understanding of the right way to present a thermoluminescence sample to the gamma beam is important because otherwise the actual dose received can easily be different from what is assumed.

5.1.1 BUILD-UP

Gamma rays do not induce thermoluminescence directly but only through the intermediary of the low energy electrons into which the energy of the gamma beam is gradually degraded on passage through matter. This degradation occurs through photoelectric, Compton, and pair production interactions of the gamma photons with atoms, followed by successive collisions with atoms of the secondary electrons produced; at each collision the incident electron loses some of its energy, this being transferred to further secondaries ejected from the atom. Thus on passing through matter the 0.66 MeV of energy carried by a gamma photon from caesium-137 is gradually transformed into a large number of low energy electrons, almost a hundred thousand, which are eventually travelling slow enough to be captured into atomic orbits. These secondaries are the agency by which the thermoluminescence is induced—upon capture into an orbit that forms an electron trap. As the secondaries slow down they are more easily deflected at each collision and so they soon become isotropic in direction, like the molecules of a gas.

The general picture of what happens when a beam of gamma rays enters matter is shown in Figure 5.1. There is attenuation of the primary beam—the number of photons still having full energy decreases by a few percent per mm (this is for gamma rays in the range 0.1–3 MeV incident on quartz

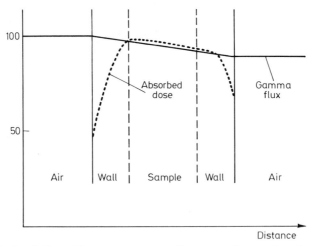

Figure 5.1 Irradiation with a gamma source. The gamma beam enters the sample container from the left, producing secondary electrons. These ultimately produce the absorbed dose, mostly after scattering and degradation in energy; the dose builds up in a distance comparable with the range of the most energetic secondary. For 0.66-MeV gamma rays from caesium-137 this distance is about 1 mm in materials such as glass or aluminium. The dose diminishes in the exit wall because the backscattered component of the secondary electrons gradually falls to zero. The primary gamma flux is attenuated in the wall and sample by about 2% per mm.

5.1 Gamma Irradiation

and other pottery minerals)—and there is build up of the secondary electrons resulting from the interactions that have occurred. At a depth into the solid roughly equal to the range of the highest energy electrons produced (about 1 mm for 0.66-MeV gamma rays incident on the minerals just mentioned) the electron flux reaches a steady level ('charged particle equilibrium'). At the exit face there is again depletion of the secondaries because the backscattering from the air is insignificant compared to the diffusion out of the solid. Obviously the thermoluminescence sample should only be present in the region of charged particle equilibrium; hence the walls of the container should not be too thin. If so, then additional build-up material should be placed in front and behind. A correction, usually only a few percent, needs to be made for attenuation by the material in front and within the sample itself.

The container walls and any build-up material should have the same gamma absorption and electron stopping characteristics as the sample. Perspex (lucite) is commonly used with biological materials, but for minerals relevant to dating glass is preferable[2], with aluminium as next best.

5.1.2 DOSE CALCULATION

The dose received by a sample depends on the *energy absorption coefficient* μ_{en} which specifies the fraction of the beam's energy absorbed from it per unit length of path.[3] Because this coefficient is proportional to density, ρ, it is more convenient to use the *mass energy absorption coefficient* (μ_{en}/ρ), which specifies the fraction absorbed per unit weight, independent of density. For gamma energies in the range 0.2–3 MeV and substances for which the average atomic number is below 20, as is the case for minerals relevant to thermoluminescence dating (except for zircon), the mass absorption coefficient does not vary much from element to element, except for hydrogen for which the coefficient is about twice the average for other elements.

It is common for gamma beams to be calibrated in terms of the dose absorbed by a sample of water, this being representative of biological material; it is important that the thermoluminescence practitioner should ascertain exactly what is being quoted. Whereas, for 0.66-MeV gamma rays, the mass energy-absorption coefficient is 0.0327 per gm/cm^2 for water, it is only 0.0293 for quartz; hence for a given gamma beam and sample configuration the dose to quartz will be 0.90 times the dose to water. Alternatively, the beam may be calibrated in terms of *exposure*, using the now obsolete *roentgen* units; for an exposure of 100 R the dose to quartz is 0.866 Gy. For cobalt-60 gamma rays the numerical values just quoted are 0.0296, 0.0266, 0.90, and 0.867, respectively.

Gamma irradiation is often required for fragments before they are powdered, for example, for testing whether crushing has any adverse effects on the thermoluminescence. A suitable container for this purpose is a glass petri dish with the spare space filled with glass balls. Other effects, such as the contribution of secondary photons scattered sideways, may then become significant; advice should be sought from a dosimetry specialist. In the case of fine grains irradiation should be in bulk rather than already deposited on disc.[4]

5.2 Beta Irradiation

The commonly employed radioisotope for beta irradiation in thermoluminescence dating is strontium-90, half-life 28 years; however, it is the daughter nucleus supported by it, yttrium-90 (half-life 64 hours), which provides the beta dose[5], the maximum energy of beta emission being 2.26 MeV. Strontium-90 is used in cancer therapy and the plate applicator (see Figure 5.2) made commercially[6] for this purpose is convenient for use. In

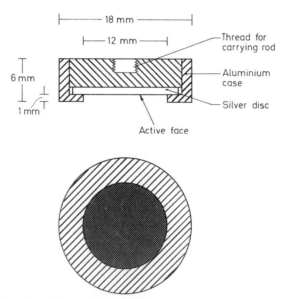

Figure 5.2 Strontium-90 beta source. Upper part shows section and lower part the front face (which must never be looked at directly, only via a mirror or with a transparent absorber interposed). The radioisotope compound is incorporated into the front surface of the silver disc in a very thin layer (about 0.02 mm for a 1.5 GBq source); there is then a 0.1 mm screen of silver on which is a protective coating of gold or palladium.

5.2 Beta Irradiation

this, a thin layer of the radioisotope compound is incorporated in a 1-mm thick disc of silver, its front face being screened by 0.1 mm of silver and protected against corrosion by a thin coating of gold or palladium. The standard source in use for thermoluminescence has an active area of 12 mm diameter and an activity of 1.5 GBq (40 millicuries); typically it is positioned at a distance of 16 mm above the sample. There is considerable health hazard from such a source; precautions and safe handling are discussed in Section 5.7.2.

The dose delivered to a sample from a beta source is influenced quite strongly by effects discussed shortly such as build-up, attenuation, and backscattering; as a result, the thickness of the sample, the substrate on which it is carried, and even the optical transparency can affect the effective calibration. Derivation of the calibration from measurements with an ionization chamber is not easy and the preferred procedure is to make intercomparison with a calibrated gamma beam, that is, the thermoluminescence induced in a sample by a known gamma dosage is measured and compared with the thermoluminescence induced in an identical sample by a given duration of irradiation from the beta source; the calibration applies only to the precise conditions of the beta irradiation and subsidiary intercomparisons need to be made for other sample thickness, substrates and so forth. In round terms the dose delivered per minute by a source of the strength quoted above is on the order of 1 Gy for the 16-mm source-sample distance; the backscatter effect is particularly strong, for example, the dose to a sample carried on silver is twice that to one carried on perspex (lucite).

In making intercomparison it is good practice to adjust the irradiation times so as to match the thermoluminescence levels obtained in each case; this avoids error due to non-linearity of thermoluminescence response, such as occurs with quartz, and of detection equipment response (e.g., due to 'pile-up' of photomultiplier pulses). Matching of thermoluminescence levels implies use of the same portion for both irradiations, which in any case gives higher precision than the multi-portion additional dose technique, and it is then necessary to check for change of TL response between first and second glow, making correction if appropriate. When a phosphor is used for the intercomparison with the gamma source the beta source calibration for quartz is obtained by multiplying by the *mass stopping power* ratio between quartz and the phosphor concerned, that is, multiply by 1.05 in the case of calcium fluoride and by 1.02 in the case of calcium sulphate (see Table 5.1).

Although calibration is a tedious business, for routine use beta irradiation is usually much more convenient than gamma. The beta irradiation facility can be located in the same room as the thermoluminescence oven, within easy reach; it is also common practice to position the beta source (temporarily) above the oven heater plate so that the irradiation is administered without disturbing the sample. Alternatively an automatic facility can be used in which a dozen or so samples are loaded onto a turntable which is programmed so as to position each sample in turn beneath the beta source for a predetermined length of time. In the system developed at the Risø National Laboratory (Bøtter-Jensen, Bundgaard, and Mejdahl, 1983) the whole measurement sequence is automatic, each sample being moved sequentially from beta source irradiation to thermoluminescence oven.

5.2.1 ATTENUATION

Whereas alpha particles and gamma rays are emitted from the nucleus with a discrete energy, beta particles have a continuous spectrum of energies[7]—see Figure 5.3. From the point of view of absorbed dose we are concerned with E_{av} (0.93 MeV in the case of yttrium-90), but when we are considering attenuation we have to take into account the fact that the energies spread from zero up to E_{max}.

On passing through matter a beta particle may suffer a variety of interactions: some result in scattering in direction, some in loss of energy with the production of energetic secondary electrons, and some in the produc-

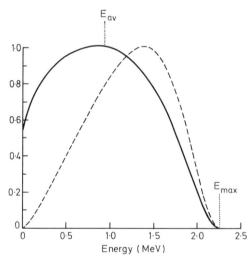

Figure 5.3 Spectrum of beta particles emitted by yttrium-90 (from Cross, 1983). The full curve shows the number of particles emitted per unit interval of energy and the dashed curve shows the energy carried; both are normalized to unity at maximum.

5.2 Beta Irradiation

tion of X-rays ('bremsstrahlung').[8] The slower the electron the greater the probability of the first two types of interactions. However, for a beam of beta particles having the continuous spectrum shown in Figure 5.3 the low energy end does not become depleted because electrons that have been slowed down and brought to rest are replaced by secondaries generated in the interactions of higher energy electrons; as a result, the shape of the spectrum remains roughly the same as absorption proceeds. For the spectrum of yttrium-90 with E_{max} = 2.27 MeV the number is halved by approximately 0.4 mm of aluminium[9]; each additional 0.4 mm causes a further halving so that 3 mm of aluminium reduces the intensity to less than 1% of its initial value.

For minerals relevant to thermoluminescence dating, the degree of attentuation is close to that of aluminium. For a fine-grain sample the attenuation is negligible but for sample thickness upwards of 0.1 mm it begins to be significant (see Figure 5.4).

5.2.2 BUILD-UP

As with gamma irradiation there is a build-up of dose as the beta beam enters matter, though to a lesser degree and by not quite the same mechanism; also the build-up layer is only about one-tenth as thick. From a 11-mm diameter source positioned 16 mm above the sample the paths of beta particles at entry into the sample are largely perpendicular to its plane (to the extent of all lying within 30° of the perpendicular). As scattering proceeds their directions rapidly become isotropic, rather like the molecules of a gas. If we think in terms of thermoluminescence per unit path length we see that for a layer of given thickness the thermoluminescence in the surface layer will be less than that within the interior just because the path length within it is less. More succinctly, the effect is due to the paths becoming oblique. Experimental evaluation of the effect indicates that for a source-sample distance of 16 mm the build-up is almost complete within the first 0.1 mm of sample, being about 20% lower at the surface (see Figure 5.4). The average dose received by 100-μm grains spread in a monolayer is consequently somewhat less, by about 6% than the dose at full build-up; experimentally it has been confirmed by Wintle and Murray (1977), as predicted from the data of Figure 5.4, that the dose to a monolayer of 100-μm gains is 1.16 ($\pm$ 0.04) times the dose to a thin layer of fine grains (1–10 μm), both with aluminium as the backing material. As elsewhere in this chapter the data given are for the standard strontium-90 source at the standard distance. If the source is close to the sample the build-up effect is likely to be much less pronounced because the incident electrons have a much greater degree of obliquity.

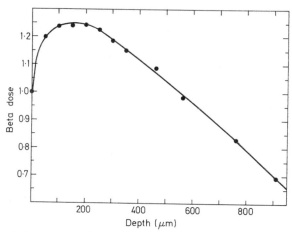

Figure 5.4 Beta dose within slice as a function of depth, for material of density 2.7 g/cm^3 having about the same beta absorption characteristics as aluminium (i.e., quartz and most other dating materials) irradiated by a strontium-90 source in standard geometry. The data points were obtained by irradiation of fine-grain CaF_2:Dy deposited on a 0.45-mm thick aluminium disc with aluminium absorber of the indicated thickness interposed. The source was positioned 16 mm above the Nichrome plate on which the disc was placed; for smaller source-sample separation the build-up will be less because a substantial degree of obliquity in the paths of the beta particles then exists at incidence.

In order to obtain the true variation within a slice of specified thickness a small correction, by 9% per mm, needs to be made to the curve shown to allow for the greater proximity to the source of the upper parts of the slice (from Wintle and Aitken, 1977a,b).

5.2.3 BACKSCATTERING

So far we have not considered the substrate on which the sample is carried. If this has the same scattering characteristics as the grains of the sample then it has no effect. For pottery minerals this is approximately the case for aluminium and so it is advantageous in this respect to use aluminium discs or planchettes to carry the sample grains. On the other hand, aluminium has the disadvantage that it can give rise to a parasitic thermoluminescence signal, and if the sample does not block off all light from the disc it may be preferable to use stainless steel; also, there are occasions when the grains are irradiated directly on the heater plate of Nichrome (or tantalum). Use of these materials gives rise to an enhanced dose because a greater proportion of the electrons crossing the interface are scattered back into the grains than if the substrate was aluminium. The effect is dependent on atomic number, and Figure 5.5 shows the experimentally observed magnitude of the effect for several materials in

5.2 Beta Irradiation

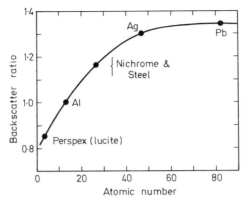

Figure 5.5 Backscatter enhancement of beta dose from strontium-90 source at a distance of 16 mm. The ratio shown is the dose received by a monolayer 100-μm grains of fluorite on the substrate indicated, relative to that with the grains on aluminium. The backscattered contribution may be stronger still for fine grains because the backscattered spectrum has lower average energy than the incident one (from Murray and Wintle, 1979, with additions).

the case of 100-μm grains; with a substrate of Nichrome the dose is enhanced by about 20% whereas with a plastic material it is diminished by 20%.

For a very thin layer of substrate the backscattering does not reach its full value; thus grains on a thin aluminium sheet resting on plastic material acquire a lesser dose than if the aluminium is thick. According to the data of Murray (1981) the backscattering contribution should reach 95% of full value for an aluminium thickness of 0.35 mm, that is, a thickness comparable with that which would attenuate the yttrium-90 spectrum by a factor of 2; it is relevant to note here that the standard disc used for fine-grains has a thickness of 0.5 mm.

5.2.4 SPATIAL VARIATION OF DOSE

In medical use it is the surface dose-rate that is relevant and it is the value for this that is quoted by the manufacturers. For thermoluminescence work there has to be a gap between sample and source, and for a separation of a few millimetres there is substantial diminution in the dose received at the outer part of a 10-mm diameter sample compared with that received in the centre (as elsewhere a 11-mm diameter active area of source is assumed). According to the data of D. W. Zimmerman (1970), for a separation of 2.5 mm the outer 10% of the area receives 30% less dose than the central 10%. By increasing the separation to 16 mm the diminution becomes 10% and this is considered small enough to standardize on this separation.[10] Uniformity of dose across the sample is only obtained at the cost of lower dose-rate; thus the dose-rate for a separation of 16 mm is about one-tenth of that for 2.5 mm. Another reason for avoiding a close position is acute dependence on distance; this is likely to entail error, particularly if the irradiation is carried out with the sample on the heater plate, which is liable to warp slightly. Even for the 16-mm position plate-warp can significantly change the dose received; the dependence on separation is approximately 10% per millimetre for this position. This dependence, and the variation across the sample, could be avoided by using a source substantially wider than the sample; however this would entail the total activity of the source being greater with the disadvantages of increased cost and greater difficulty in handling without health hazard.

5.2.5 SAMPLE SPREADING; THE MONOLAYER TECHNIQUE

Obviously it is important to have the sample spread evenly over a specified area for even at a source-sample separation of 16 mm there is still some difference of dose between centre and periphery. For coarse grains there is an additional reason—the influence of variations in sample thickness.

Consider a sample of 100-μm grains irradiated on a stainless steel disc which in places is bare and which elsewhere carries a layer which is two or three grains thick. The grains underneath receive a greater dose than those on top because they are in the region of the build-up maximum (see Figure 5.4) and to some extent because they receive a bigger backscattered dose. If the same weight of sample covered the disc in a monolayer, then no grains would be in the region of the buildup maximum. Another effect with a multilayer sample is that, due to scattering and absorption, the thermoluminescence emitted from grains underneath makes a diminished contribution to the measured signal. To some extent these various effects would be expected to cancel out, but in practice it has been found that the reproducibility is much improved if a monolayer of grains is used.

The recipe for making a monolayer sample is as follows. The disc (10-mm diameter) is coated with silicone oil using an aerosol spray[11]; the grains are then sparingly poured onto the disc as evenly as possible; finally the disc is inverted and tapped so as to remove any grains not adhering to the silicone-wetted disc. A refinement is to use a brass annulus of diameter $6\frac{1}{2}$ mm so as to avoid having any grains on the outer part of the disc where the dose is diminished. The weight remaining on the disc when using 100-μm grains (in practice they are obtained by sieving and lie

5.2 Beta Irradiation

in the range 90–125 microns) is between 2 and 3 mg, roughly 2000 grains. Obviously this monolayer technique has the disadvantage that the amount of sample is rather small and in some applications the level of thermoluminescence may be unacceptably low.

5.2.6 LARGER GRAINS AND THICKER SAMPLES; SLICES; OPTICAL ATTENUATION

The build up curve of Figure 5.4 may be used to predict, by integration, the average dose in thick samples. This is shown in Figure 5.6, expressed as a ratio to the dose that would be received by a thin layer of fine grains; after going through a broad maximum with values in the range 1.20–1.23 for slice thicknesses in the range 0.2–0.5 mm the ratio falls again, reaching 1.09 for a slice of thickness 0.9 mm. For a monolayer of 100-μm grains the average dose predicted is 1.16 times the fine-grain dose and this has been confirmed directly (Wintle and Murray, 1977).

Also, as mentioned above, there may be an effect due to optical attenuation; for grains that have poor transparency the deeper parts are ineffective as far as thermoluminescence received at the photomultiplier is

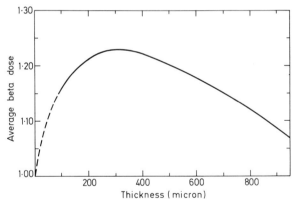

Figure 5.6 Average beta dose within slice irradiated on aluminium substrate as a function of slice thickness, normalized to zero thickness (i.e., to fine-grains on an aluminium disc). The curve applies to material of density 2.7 g/cm³ having about the same beta absorption characteristics as aluminium for irradiation by a standard strontium-90 source from a distance of 16 mm. The thermoluminescence received from a beta-irradiated slice depends also on the optical transparency; consequently except for highly transparent materials calibration against gamma irradiation is necessary.

The curve shown has been obtained from Figure 5.4 by integration and allowance for the greater proximity to the source of the upper parts of the slice (see Wintle and Aitken, 1977, for further discussion).

concerned. Since the natural thermoluminescence or the thermoluminescence induced by gamma irradiation will be similarly affected it would seem at first sight that optical attenuation should not matter; however, for a sample in which there is severe optical attenuation most of the thermoluminescence reaching the photomultiplier will be from the upper part of the sample and since the beta dose in this part will be lower than average due to incomplete build-up, the ratio of natural thermoluminescence to beta thermoluminescence will be higher than if there was no optical attenuation. The situation is further complicated if there is a significant alpha contribution, as has been discussed in respect of flint slices (Aitken and Wintle, 1977; Wintle and Aitken, 1977).

The experimental conclusion reached by Bell and Mejdahl (1981) with respect to 20 mg samples of 100-μm quartz grains (carried in shallow platinum planchettes with a floor area of 0.55 cm^2) is that the effective beta dose-rate for opaque 'frosty' grains can be as much as 30% lower than that for transparent 'shiny' grains; the categorization as 'shiny' or 'frosty' refers to the appearance of the grains after the hydrofluoric acid etching procedure and most pottery samples contain a mixture of the two with the proportions varying from sample to sample. As a consequence Bell and Mejdahl recommend that the effective calibration value for the beta source be found for each piece of pottery being dated, that is, intercomparison with a gamma source using the quartz grains concerned. A full quantitative explanation of the effect has not yet been established[12] but the experimental evidence is comprehensive and conclusive that such individual calibrations are necessary—especially if 20 mg samples are used; Benko (1983) reports that for monolayer samples there is no difference in the calibration factors for shiny and frosty grains.

5.2.7 SAMPLE NORMALIZATION

The additive dose technique is the mainstay of thermoluminescence dating and for this an essential need is to interrelate the thermoluminescence response from different portions of the same sample. For fine-grains the uniformity of deposition is often good enough to yield sample discs with a scatter in thermoluminescence response of $\pm 5\%$ or less but for coarse grains and slices some form of deliberate normalization is necessary. The simplest approach is to dispense a fixed volume of grains by means of a suitable device or to weigh out portions using a high precision balance. However there are limitations in this approach due to (i) optical attenuation and beta build-up effects resulting from non-uniformity in sample spreading and (ii) statistical fluctuations in the number of 'bright' grains present in each portion.

5.2 Beta Irradiation

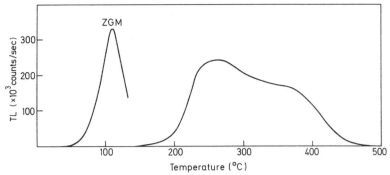

Figure 5.7 Zero glow monitoring (ZGM) for normalization between portions (Aitken and Bussell, 1979). All portions are given a small beta dose before any other thermal or radiation treatment; the thermoluminescence response to this dose in the low temperature region of the glow-curve is used as a measure of the relative thermoluminescence sensitivity of the portion concerned. The ZGM glow-curve is terminated prematurely so that the required radiation doses, differing between portions, can be administered before measurement of the high temperature thermoluminescence. Because the ZGM peak is likely to have a lifetime of less than 2 hr at room temperature, there must be strict control of time and temperature during the ZGM irradiation and during the delay until measurement, this delay being kept short.

Zero-glow monitoring (ZGM) is highly effective, usually allowing a precision of the order of ±3%, although it is not universally applicable. It requires the presence of a highly sensitive peak at a glow-curve temperature in the range 70°–200°C, see Figure 5.7; suitable peaks are usually present for feldspar, calcite, and quartz. Using this technique the initial step with each portion is to measure the thermoluminescence of this peak due to a beta dose that is small compared to the paleodose; this thermoluminescence is a measure of the overall thermoluminescence response of the portion as influenced by variations in weight, spreading, and proportion of bright grains present. In measuring the thermoluminescence from this peak the glow-curve is terminated prematurely soon after its maximum has been passed. Dating measurements on the portion then continue normally, for example, measurement of natural thermoluminescence or irradiation prior to measurement of natural-plus-artificial thermoluminescence. Because of short lifetime there is liable to be significant decay in the ZGM peak within five minutes or so; it is therefore essential to use a standardized irradiation and measurement schedule for it and to avoid variation in the temperature experienced by different portions of the same batch during ZGM irradiation and the short interval before measurement. The great strength[13] of the technique is that at the time the ZGM is measured all portions have identical radiation and thermal histories.

In the *equal pre-dose technique* the monitor measurement is made after all the dating measurements have been completed, the high temperature region of the glow-curve being used. The response to a standard dose is measured, but prior to this additional irradiations and heatings are given so that all portions are equal in the treatment they have received. Thus a portion used for measurement of natural thermoluminescence is given a dose equal to the highest additional dose involved in natural-plus-artificial evaluations, and then heated as in a glow-curve. The objective of this is to equalize transparency changes and predose effects between portions. The effort involved is substantially greater than for ZGM and it is not usually as effective.

For measurement of second-glow growth characteristics it is convenient, and effective, to use the natural thermoluminescence as monitor. This has the same advantages as ZGM. Of course it precludes the use of samples which have been used for natural-plus-artificial thermoluminescence but use of these may be undesirable anyway on account of the additional dose having affected the supralinearity intercept, see Section 5.4.

5.2.8 ALLOWANCE FOR RADIOACTIVE DECAY

The 28.0-year half-life of strontium-90 implies the dose-rate delivered by a source will decrease by 2.48% every year. Therefore after calibration against a gamma source allowance needs to be made for this. It is usual to make correction every six months.

5.3 Alpha Irradiation

Being heavy, alpha particles do not get scattered to an extent that is significant in the context of thermoluminescence; they travel in straight lines, being gradually slowed down and brought to rest at a well-defined penetration range. However, instead of complications due to buildup and backscattering there are complications in administering the irradiation, due to their short range, and in interpretation due to the dependence on particle energy of thermoluminescence effectiveness. Another important difference is that the dose delivered by an alpha particle is concentrated in a narrow cylindrical track, of the order of 0.1-μm diameter, and within most of this cylinder the ionization density is so high that all available thermoluminescence traps are filled (i.e., the thermoluminescence is in saturation); for the levels of alpha irradiation relevant to thermoluminescence dating there is not much overlapping of tracks so that the thermolu-

5.3 Alpha Irradiation

minescence observed is proportional to the total length of the tracks accumulated. This is in contrast to beta and gamma irradiation, for which the ionization is continuous throughout the solid and the thermoluminescence observed is proportional to the general level that has been reached.

There is a higher probability of creating new defects in the case of alpha particles but for the defect-ridden samples with which we are concerned in dating, this effect still remains insignificant.

5.3.1 THE IRRADIATION FACILITY

A convenient radioisotope for this purpose is americium-241, half-life 433 years; in addition to 5.5-MeV alpha particles there is also a weak emission of 0.06-MeV gamma rays but in a thin sample the dose from this is insignificant compared with the alpha dose. This radioisotope has various industrial uses (e.g., smoke detection, static elimination) and is available commercially[14] in the form of a 1-μm layer backed with 0.2-mm thick silver foil, its front face being protected by a 2-μm thick protective covering of a gold–palladium alloy. In going through this covering the alpha particles are slowed down appreciably and the emergent energy even of those that pass through in a perpendicular direction is only 4 MeV; for an oblique particle the path length in the covering is greater so that for an angle of, say, 60° to perpendicular the emergent energy is down to almost 2 MeV. The question is, what angle can be tolerated if the alpha particles are to be left with enough energy to fully penetrate the grains of the sample? If there is not full penetration then the amount of sample from which thermoluminescence is received is less than in the case of beta irradiation and hence the alpha particle effectiveness that is deduced will be erroneously low.[15]

The routine fine-grain sample has an upper grain size limit of 8 μm; assuming this is also the upper limit to the thickness of the layer (i.e., we must be careful not to deposit thick layers) we see by reference to Figure 5.8 that in order to get complete penetration it is necessary for a particle that is incident perpendicularly to have an energy of 2.5 MeV. One that is incident at 60°, for which the path length is double, needs to have an energy of 4.5 MeV. Hence it is necessary to prevent excessively oblique particles from reaching the sample either by collimation or by employing a sufficient separation between source and sample; both of these have the obvious disadvantage that the flux reaching the sample is reduced and there is also the need to employ vacuum in order to avoid energy loss in the air—at atmospheric pressure the range of a 4-MeV alpha particle is only 25 mm.

For a 12-mm diameter source and a 10-mm diameter sample in vacuum,

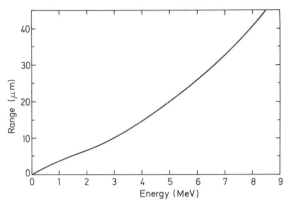

Figure 5.8 Alpha particle range, R, as a function of energy. The data shown is for pottery and derived by Bowman (1976) from the tables of Northcliffe and Schilling (1970); in the above a density, ρ, of 2.6 g/cm^3 has been assumed. The product $R\rho$ is approximately proportion to $A^{\frac{1}{3}}$, where A is the atomic weight of the absorber.

it has been concluded experimentally that the closest tolerable separation for a standard fine-grain sample is 10 mm (Singhvi and Aitken, 1978). In the six-seater irradiation facility developed (see Figure 5.9), six samples can be irradiation simultaneously by six sources, the exposure being controlled by a shutter disc having six holes in it. The samples are carried on pillars and different dose-rates can be obtained by using different pillar heights; compared to the 10-mm separation the dose-rate is successively halved (approximately) for separations of 15 mm, 20 mm, 30 mm, and 40 mm. Obviously if the samples are suspected of being thicker than standard, the maximum separation should be used—at the price of longer irradiation time—although note that even for perpendicular incidence the upper limit to thickness is only 14 μm (the range of the 4-MeV alpha particles that emerge perpendicularly from the source). Clearly, use of a radioisotope having a higher emission energy than americium-241 would be advantageous; dispensing with the protective coating would run too great a risk of contamination for routine operation.[16]

The degree of vacuum required is not high; a pressure of 1 Torr is more than adequate since even for this pressure the energy loss of a 4-MeV alpha particle in traversing the maximum source-sample distance of 40 mm is only 0.006 MeV. Evacuation can be by means of a separate facility, or by matching the housing diameter to that of the glow oven in which case the housing is placed on the latter and evacuated through a hole in its bottom.

5.3 Alpha Irradiation

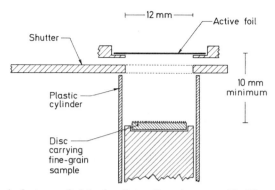

Figure 5.9 Basic features of alpha irradiator. In order to avoid oblique particles which cannot fully penetrate the layer of sample the latter must not be too close to the source; hence the device is evacuated so as to avoid energy loss in air. In the design developed by Singhvi and Aitken (1978) 6 sources are used so that 6 samples can be irradiated simultaneously; these are arranged in a circle of diameter 57 mm and to avoid crosstalk from adjacent sources there is a plastic tube surrounding each sample. Using americium-241 sources each of strength 8 MBq (0.22 mCi) the track length delivered per unit volume of a sample at a distance of 10 mm is about 0.3 μm^{-2}/min, equivalent to about 4 Gy/min in terms of absorbed dose. The dose is strongly dependent on source–sample separation, variation being about 20% per mm for 10 mm and 5% for 40 mm; the dose ratio for these two separations is 13:1. The contribution from the weak 60-keV gamma emission is negligible by comparison to alpha dose but to avoid significant irradiation of the samples if they are left in the irradiator for long periods with the shutter closed, $1\frac{1}{2}$-mm thick pieces of lead are incorporated into the shutter so that they are in front of the sources when it is closed; the dose to a sample at 10 mm separation is then only 0.1 Gy/day, and at 40-mm separation, 0.002 Gy/day.

5.3.2 CALIBRATION

As an alpha particle slows down its loss of energy per unit length of track increases sharply, see Figure 5.10. However it is found experimentally that the thermoluminescence per unit length of track shows no such increase as the particle energy is decreased, see Figure 5.11. Hence in terms of deposited dose the thermoluminescence effectiveness of alpha particles decreases sharply with decrease of energy, see Figure 5.12. Thus it is not adequate to calibrate an alpha source in terms of deposited energy; it is also necessary to specify both the energy of the alpha particles as they emerge from the protective coating and the thickness of the sample being irradiated (because this determines how low in energy the particles become during traversal). It is better to specify alpha sensitivity in terms of thermoluminescence per unit length of track and to calibrate alpha sources in terms of the length of track delivered per unit volume of

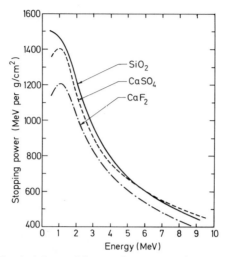

Figure 5.10 Calculated alpha particle stopping powers of quartz, CaSO$_4$, and CaF$_2$ (as derived by D. W. Zimmerman, 1971 from tables now published by Berger and Seltzer, 1982). The quantity shown is the mass stopping power and to obtain the energy loss per unit distance travelled, multiply by density; thus for $\rho = 2.6$ g/cm^3 the energy loss per μm of a 4-MeV alpha particle in quartz is 0.21 MeV. Mass stopping power is approximately proportional to $A^{\frac{1}{2}}$ for the substance concerned.

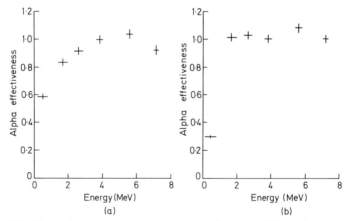

Figure 5.11 Thermoluminescence per unit length of alpha particle track as a function of particle energy; (a) for a sample of heated sand from the core of a cast bronze figure, and (b) for CaF$_2$:Mn (from Bowman, 1976). These results were obtained by irradiating very thin layers of sample with alpha particles of given energy from a Van de Graaff generator. The samples were thin enough to allow complete penetration by the particles, except possibly below 2 MeV for some samples due to coagulation of grains; this may account for the fall-off at low energy but a fall can also be expected theoretically (see Bowman, loc. cit., for discussion). The data plotted are thermoluminescence per incident particle, normalized to unity at 4.1 MeV.

5.3 Alpha Irradiation

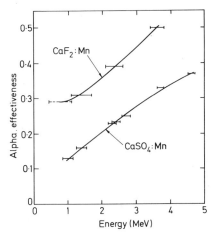

Figure 5.12 Alpha effectiveness (in terms of absorbed dose) as a function of particle energy for CaF_2:Mn and $CaSO_4$:Mn (from D. W. Zimmerman, 1970). The values shown here are the ratio of thermoluminescence per gray for an alpha particle of given energy to that for beta radiation. They are derived from experimental data such as shown in Figure 5.11 and calculated energy loss such as shown in Figure 5.10.

sample. This is the *a-value* system already mentioned in Section 4.3.1 and discussed more fully in Appendix K. Within the core of an alpha track, the thermoluminescence traps are saturated and as the alpha energy decreases, the portion of the deposited energy that is wasted becomes greater; hence a given amount of energy deposited by a low energy alpha particle is far less effective in inducing thermoluminescence than a high energy one.

There are three stages to the calibration. In the first two the thermoluminescence alpha sensitivity of a thin fine-grain sample of phosphor is determined and in the third stage this sample of known sensitivity is used to evaluate the strength of the source for the desired source-sample separation. For the first two stages a weaker source than used routinely is necessary so that the flux rate of particles from it is low enough for counting with a surface barrier detector. The fine-grain phosphor sample is irradiated, in vacuum, at a sufficient distance from the source for the incident alpha particles to be effectively perpendicular to the plane of the sample (see Figure 5.13). If n is the number of particles per unit area per unit time, d is the thickness of the sample and A its area, then the track length received by the sample is (nAd), and if G_1 is the thermoluminescence induced by unit time of this irradiation, the thermoluminescence induced per unit length of track in this phosphor is given by

$$G_1/nAd.$$

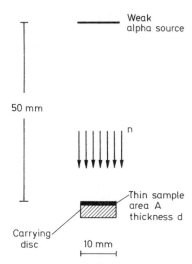

Figure 5.13 First stage in calibration of alpha source: irradiation of sample with weak source (not the one to be calibrated) prior to measurement of thermoluminescence induced. In the next stage an aperture is substituted for the sample/disc and the number of alpha particles passing through it is counted with a surface barrier detector; hence the thermoluminescence per unit length of track is evaluated for this sample. For final stage, see text. (N.B. the thickness of sample and disc have been exaggerated).

In the second stage the value of n is obtained by substituting an aperture in place of the sample and counting the number of particles going through by means of a surface barrier detector placed behind it; note that n also gives the track length per unit volume. As we shall see shortly we do not need to know A and d, although of course the latter must be small enough for complete penetration.

The third stage consists of irradiating the sample using the source that is being calibrated at the source-sample distance that is to be used. To get high enough intensity for not-too-long irradiation times in routine dating, this distance needs to be smaller than in the first stage and consequently a substantial number of the incident particles are oblique. However, if we measure the thermoluminescence induced per unit time G_2, we can obtain the track length delivered to the sample per unit time as

$$\frac{G_2}{G_1/(nAd)},$$

from which it follows that the track length delivered per unit time per unit volume of sample is simply

$$nG_2/G_1.$$

This applies notwithstanding the obliquity of the incident particles as long as the sample is thin enough for complete penetration by the most oblique ones; this condition was of course the condition that limited the source-sample distance to a minimum of 10 mm for standard fine-grain samples. For routine use it is convenient to specify source strength S in terms of micrometres of track length per cubic micrometre of sample per minute, that is μm^{-2}/min. For a source of activity 8 MBq (0.21 mCi) as commonly used in the six-seater irradiator the value to be expected for S is around 0.32 μm^{-2}/min at a source-sample distance of 10 mm, falling to 0.024 μm^{-2}/min for a distance of 40 mm. For the latter separation the particles are all close to perpendicular and so the incident flux rate of particles is also 0.024 μm^{-2}/min, that is, 40,000 per cm^2/sec.

For the purpose of rough calculations it is convenient to know the source strength in the more familiar unit of absorbed dose, the gray. From the energy loss per unit length of track (Figure 5.10) we can calculate (see Appendix K) the absorbed dose corresponding to 1 μm of track per cubic micrometre. For a particle energy of 3.7 MeV in quartz this comes to within a few per cent of 13 Gy; hence a source for which $S = 1\ \mu m^{-2}$/min at a given distance will deliver 13 Gy/min to the sample. For higher energies the dose-rate is less and for lower energies it is greater; the value for 3.7 MeV has been chosen because that was the particle energy used by D. W. Zimmerman in his basic experimental and theoretical work on this topic (1970, 1971, 1972).

5.4 Supralinearity, Sensitization, and Saturation

As illustrated in Figure 5.14, the growth of thermoluminescence with dose is not always linear; typically there is an initial supralinear portion in which the sensitivity is increasing, followed by a linear portion of constant incremental sensitivity, finally the sensitivity falls off due to the onset of saturation; the 'saturation' level is not necessarily constant, as with heavy doses there may be further increase or decrease. The initial portion of supralinearity occurs only for beta and gamma irradiation, not for alpha. It is the incremental sensitivity of the central portion that is used to define χ and to obtain the equivalent dose.

The most generally favoured explanation for supralinearity (see, for example, Chen and Bowman, 1978) is in terms of competition for electrons by traps that do not themselves give rise to thermoluminescence, these traps saturating much earlier than the thermoluminescence traps so that as the former gradually become satiated, the thermoluminescence sensitivity increases to a constant value. Alternatively, the competition

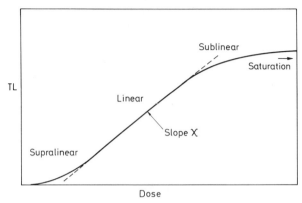

Figure 5.14 Thermoluminescence growth characteristic showing supralinear and sublinear regions. Thermoluminescence sensitivity, χ, refers to the thermoluminescence per unit dose of the linear region.

may occur during transit of electrons from the thermoluminescence traps to the luminescence centres, in which case we have 'competition during read-out' as opposed to 'competition during irradiation'.

Another type of explanation is in terms of an increase, with dose, of the number of available luminescence centres (i.e., charged with a hole and therefore ready to accept an electron). If the amount of thermoluminescence is proportional to the product of the number of traps and number of luminescence centres and both of these numbers are proportional to dose then obviously the amount of thermoluminescence will be proportional to the square of the dose. In fact when put like that it seems that the problem is more one of explaining linearity rather than supralinearity! However in the form just stated the model is oversimplistic and there are other features to be taken into account which can account for the onset of linearity, for example, a sufficiency of available centres being reached so that the probability of a released electron giving rise to a photon does not increase any further.

As noted above supralinearity does not occur with alpha particles; hence because alpha particles are much more effective in creating new defects than beta and gamma radiation, defect creation can hardly be the explanation. Defect creation is the accepted explanation for the continued rise in level of thermoluminescence at heavy doses; the decrease sometimes observed in that region is ascribed either to a subtle phenomenon termed 'radiation bleaching', or more simply, to the damage inflicted on the crystal lattice. The absence of supralinearity with alpha particles is a direct consequence of the low probability of one track overlapping an-

5.4 Supralinearity, Sensitization, and Saturation

other at the dose levels concerned—each new alpha particle finds virgin territory and the resultant thermoluminescence due to it is independent of how many other tracks have already been created; another factor is that we may regard each alpha track as a region in which the thermoluminescence versus dose curve of Figure 5.1 has been taken through to saturation.

As regards the onset of saturation with respect to beta and gamma irradiation there is a strong correlation (noted by Zimmerman, 1972; see Figure 5.15) between the dose level at which saturation begins and alpha particle effectiveness. This is as expected on the basis of the explanation given for the poor effectiveness of alpha particles—the lower the dose level at which the thermoluminescence traps are saturated the greater the wastage of ionization in the core of the alpha track.

5.4.1 SENSITIZATION (THE 'PRE-DOSE' EFFECT)

With many types of sample there is a change of thermoluminescence sensitivity when it is heated, that is, the second-glow sensitivity is different from the first-glow sensitivity; for most types the change is an increase. One reason for the change is an increase in transparency and this may occur on the first heating of the sample only. Another reason is the pre-dose effect, in which the change is always an increase, which is proportional to the total dose received previous to the heating; the 100°C peak in quartz is a prime example of it as is discussed further under pre-dose dating in Chapter 6. The magnitude of the pre-dose effect usually depends also on the temperature and the duration of the heating—see, for

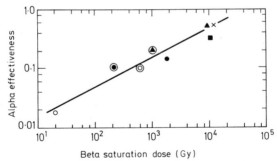

Figure 5.15 Dependence of alpha effectiveness on beta saturation dose (from D. W. Zimmerman, 1972). The saturation dose is defined as the dose at which the slope of the beta growth characteristic equals $1/e$ of its initial value. Fluorite, peak II ●, peak III ◉ ; CaF_2 : Mn X; CaF_2 : Tb, 85°C peak ▲ , 165°C peak ⓐ; $CaSO_4$: Mn ■ ; quartz, 110°C peak ○, 300°C peak ◎ .

example, the classic study by Shayes *et al.* (1967) with respect to fluorite.

The same types of explanation have been given with respect to pre-dose effect as for supralinearity, namely, competition and changes in availability of luminescence centres—the latter model having been established for the 110°C quartz peak in exhaustive studies by J. Zimmerman (1971). With respect to competition models the effect is only to be expected when the temperature reached during the heating is insufficient to empty the competing traps. A third type of explanation has been adduced for some substances; this is based on heat and radiation dependent mobility of defects giving rise to enhanced sensitivity through greater proximity between traps and centres. There is no pre-dose effect with alpha irradiation because there is no overlapping of tracks.

5.4.2 RELEVANCE TO DATING

As we have seen in earlier chapters the phenomenon of sensitization rules out the simple approach in which the paleodose is derived by comparing the natural thermoluminescence with second-glow artificial thermoluminescence, and as a consequence it is necessary to use the additive technique (see Figure 2.1). With this it can be checked whether the levels of both natural thermoluminescence and natural-plus-artificial thermoluminescence are in the linear portion of the growth characteristic, but nothing can be learnt about the supralinear portion until the sample has been heated; the critical question then arises as to whether second-glow supralinearity truly represents first-glow supralinearity.

Although an affirmative answer to this question is commonly taken for granted it must be emphasized that there is no firm basis for this assumption and we have here a weak link, as far as accurate dating is concerned. If there is no sensitivity change between first-glow and second-glow then for that sample it seems justifiable to assume that there is no change in the degree of supralinearity either. But when there is a sensitivity change, the only grounds for the assumption are either that dates of known-age samples are nearer to the truth if the assumption is made (e.g., Thompson, 1970) or that it is predicted to be correct by theory. Although theoretical models have been established for individual phosphors (usually artificial substances of known impurity content) it would hardly be a practical proposition, even if ultimately possible, to establish a model for all types of sample being dated. Likewise a test programme that establishes the validity of the assumption does so only for the types of sample involved.

An alternative possibility for resetting the thermoluminescence to zero

5.4 Supralinearity, Sensitization, and Saturation

in order to observe the low dose supralinearity is by exposure to ultraviolet light. Although this technique may be useful in the context of sediment dating (see Chapter 8) it has not so far proved useful in respect of burnt materials. Even after prolonged exposure the level is liable to be 10% or more of the natural thermoluminescence; this residual level seems partly to represent an unbleachable component and partly to be due to the conflicting process of electron transfer from other traps.

Another reason for disquiet in respect of correction for supralinearity is that for some samples the magnitude of the correction indicated by the second-glow growth characteristic is dependent on the dose received prior to the first glow, that is, it exhibits a sort of pre-dose effect (Bowman, 1975; Fleming, 1975b). For such samples the correction based on portions which have been used in the first glow for measurement of natural thermoluminescence is more than if the correction is based on portions which have been used in the first glow for measurement of natural-plus-artificial thermoluminescence.

For old samples, attention shifts from the low-dose region to problems associated with non-linearity due to the onset of saturation. The apparent sensitivity becomes lower as the dose gets higher and if this sensitivity is erroneously used as representative of the growth during antiquity, an age that is substantially too ancient will be obtained. Some authors assume that the growth characteristic has the form of a saturating exponential and determine an age on that basis. Another approach (e.g., Valladas and Gillot, 1978) is to use the shape of the second-glow growth characteristic as representative of the first-glow characteristic. As with the supralinearity correction the validity of these approaches can only be established in general terms and not for the individual sample being dated. In any case it should be noted (see Aitken, 1984) that if there is an alpha contribution to the natural thermoluminescence, as in fine-grain dating, the required growth characteristic is a composite one combining the alpha response and the beta/gamma response; in general these responses are not the same because the former remains linear to a higher dose level than the latter.

Very old samples may show a linear response to additive dose although the growth of the natural has become substantially sublinear, see Figure 5.16. This occurs when the natural thermoluminescence is in *thermal equilibrium* due to the lifetime of the relevant traps being comparable with the age of the sample, that is, the rate of trap filling due to the annual dose is equal to the rate of thermal eviction. If the linear response is taken at face value, an age that is substantially too low will be obtained; for some relevant discussion see Wintle and Aitken (1977).

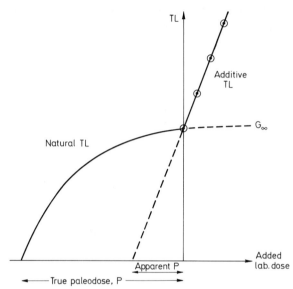

Figure 5.16 Thermal equilibrium. If the age of the sample is comparable to the lifetime of the relevant traps, then the rate of trap filling is nearly balanced by the rate of escape, and the natural thermoluminescence gradually approaches a level G_∞. For a laboratory irradiation, the rate of filling is higher by many orders of magnitude so that the rate of escape is negligible in comparison. Thus the laboratory growth characteristic is linear despite non-linearity in growth of the natural thermoluminescence. However it is unlikely that a sample for which there is non-linearity of this type will pass the plateau test because the apparent paleodose is dependent on trap lifetime and in general this is different for different glow-curve temperatures. Similarly a sample having non-linearity due to the onset of trap saturation will fail the plateau test because of the dependence of the degree of non-linearity on glow-curve temperatures.

5.5 Dependence on Dose-Rate

Figure 5.16 illustrates one way in which the rate at which the dose received may affect the thermoluminescence growth. Another possible circumstance is thermal decay of competing traps such that for low dose-rate the traps are not full (and are therefore providing competition) whereas for high dose-rates the opposite is true thereby permitting a greater rate of filling of the thermoluminescence traps[17]. As opposed to these two mechanisms which depend on decay there is the possibility of a genuine dose-rate effect, intrinsic to the trap filling process. A theoretical model for this has been given by Chen, McKeever, and Durrani (1982) who consider 'the traffic of charge carriers' between one trap and two luminescence centres, as well as between valence band and conduction band. They show that substantial changes, by a factor of 2 or more, can occur in thermoluminescence sensitivity when the dose-rate becomes high, and that depending on the parameters chosen for the model, these changes can be increases or decreases; they also show that within the same phosphor the effect may be different for peaks of different colour.

Because laboratory irradiations are given at dose-rates that are more than ten million times the dose-rates received during antiquity, the possibility of an intrinsic effect is of fundamental importance. Clearly it cannot be a strong interference as otherwise good results would not be obtained when known-age samples are dated. The work of Chen *et al.* does not give a practical prediction of the dose-rate at which dependence will begin to be significant, but experimental work[18] on quartz suggests there may be a significant effect for dose-rates as low as 10 Gy/min; this is somewhat above the dose-rate of around 1 Gy/min that is usual in pottery dating, but it does give warning against trying to save time by using higher rates. On the other hand, evidence that dose-rate dependence is not the general rule affecting all phosphors and all dose-rates has been reported by Bowman (1978) who finds that for the 200°C and 300°C peaks in natural fluorite there is no effect greater than 2% in going from 0.3 mGy/min to 3Gy/min.

5.6 Dependence on Temperature of Irradiation

As with dose-rate dependence it is to be expected that there will be temperature dependence if the traps concerned, whether directly or through competition, do not have sufficient stability. Again there is also the possibility of an intrinsic effect, for instance due to the increased velocity that ionized electrons will have at increased temperature with consequently less probability of being trapped. An experimental study[19] of several phosphors suggests that the effect is small but general; in the case of natural fluorite, a well-behaved phosphor without supralinearity, the thermoluminescence sensitivity was 3% less for irradiation at 100°C than for irradiation at 20°C. This slight dependence is hardly important except as a warning against impatience when irradiating for second-glow measurements. On the other hand, at lower-than-normal temperatures a substantial increase in thermoluminescence response has been observed[19] for quartz, sufficient to suggest that for cold regions the temperature during antiquity may have been relevant; however, an increase was observed only for total doses in excess of 100 Gy, which is beyond the useful range of quartz because of non-linearity of response.

5.7 Operator Protection

5.7.1 GENERAL

If proper care is taken, there need be no hazard to health from the radioactive sources involved in thermoluminescence dating. The starting point is the code of practice and regulations of the laboratory and country concerned; responsibility for seeing that these are followed rests with radiation protection staff at various levels, of which the most immediate will be the laboratory representative. However, in all but completely routine situations an understanding of the particular hazards involved is a necessary adjunct. In this section an introductory discussion is given of some of the hazards likely to be encountered with respect to thermoluminescence dating; it should be regarded only as a basis for discussion with the laboratory representative, which should be reinforced by reading the local regulations and by radiation surveys in the proximity of the sources in use.

There are several categories of hazard:

1. the dose received by the operator while carrying out routine irradiations;
2. the possible dose in the event of equipment failure or accident;
3. the possible dose during the initial mounting of the source in its protective housing;
4. leakage from one of the sources, giving rise to risk of ingestion;
5. handling of radioactive compounds.

The first category, the dose received routinely, is the easiest to deal with but also the one most overlooked. Initially, a survey of the dose-rate at the operator position is made using an appropriate radiation monitor. If it is not within acceptable limits then more shielding must be interposed between source and operator, or the distance between the two increased. This survey should also include the dose-rate to the fingers and so forth, while the source is being manipulated. Once a satisfactory working situation has been achieved a check is kept on it by means of a monitor badge worn by the operator. This badge, which may be based on photographic film or on a thermoluminescence phosphor such as lithium fluoride, is measured by a central radiological protection service at monthly intervals; there are also versions of such badges that can be worn on the fingers.

Although safety of the operator is of prime concern because of continual proximity to the sources, the safety of visitors must also be considered. There is a distinction here between other radiation workers and the

5.7 Operator Protection

general population, the permissible exposure of the latter being set lower on account of genetic risk.

It is also necessary to take precautions against fire and theft. When not in use, sources should be in a locked fireproof box (i.e., with asbestos and steel cladding). As far as possible, sources in use should not be left unattended unless within a fireproof enclosure in a locked room.

Permissible Limits. The International Commission on Radiological Protection recommends an annual limit of 50 mGy for radiation workers but less for the general public. For female radiation workers there are further restrictions: first, that the dose should not exceed 13 mGy in any 13 week period; and secondly, that during pregnancy the total dose should not exceed 10 mGy. The foregoing refer to whole-body doses; the limits set for doses to the hands and skin are about 10 times higher. As stated by the Commission, the limits are in terms of 'dose equivalent' but in the context of thermoluminescence dating this comes to the same thing as the absorbed doses quoted.[20]

In terms of a 40-hr week the limit for radiation workers translates to a continuous whole-body dose-rate during an 8-hr day of 0.025 mGy/hr. Obviously, higher dose-rates than this are permissible during the brief periods while sources are being handled; it is the function of the personal monitor badge to keep a check on the integrated dose received, as well as to give warning of an unsuspected increase (due, for example, to a change in working procedure); typically the radiological protection service will notify a laboratory worker if the integrated dose recorded over a month exceeds 0.2 mGy, well below the level at which there is any serious cause for alarm.

Instruments. For gamma rays and for the bremsstrahlung of beta sources (see Section 5.7.2) it is convenient to use a Geiger counter as detector, and suitable instruments are available commercially[21] with the meter scaled directly in tissue-equivalent dose-rate for gamma rays. It is essential that the detector is appropriate to the type of radiation being measured both in terms of calibration and in terms of penetration of the radiation into the sensitive volume. Thus a detector that is suitable for high energy gamma rays and X-rays, upwards of 100 kV say, may not be calibrated correctly for lower energies, and for beta radiation it may be totally insensitive. For the latter a Geiger counter with a thin window is required or alternatively a plastic scintillator (plus photomultiplier) covered with a thin light-tight foil of mylar. By incorporating a zinc sulphide screen in the latter arrangement it can be made sensitive to alpha particles also, these being electronically distinguishable because of the much bigger pulse height.[22] In general these scintillator type instruments are not calibrated to measure dose-rate. They have a much greater sensitivity than is required for detection of radiation levels that are hazardous, but they have the advantage of serving also as detectors of radioactive contamination and for the 'wipe tests' that need to be made routinely on alpha and beta sources to check the integrity of the protective coating.

5.7.2 THE HAZARDS FROM A BETA SOURCE

Whereas a gamma source is a fixed installation with massive shielding that stays unaltered once acceptable radiation levels have been achieved, a beta source may be used more flexibly and in much greater proximity to the thermoluminescence operator. Further, whereas the hazard from a gamma source once it has been loaded (which is certainly a very specialized task) is straightforwardly assessible in terms of the permissible levels mentioned above, the situation with a beta source is more complex. First, there is the evident danger from direct exposure to the beta flux; at the standard irradiation distance of 16 mm, the standard source of 1.5 GBq (40 mCi) delivers the permissible annual dose in 5 sec. Secondly, there is the sometimes unrealized danger from the scattering of beta particles.[23] Thirdly, there are the X-rays (bremsstrahlung) created when beta particles are stopped. Fourthly, there is the possibility that due to failure of the protective coating some of the active material will leak out and contaminate neighbouring surfaces, although in the author's experience the gold–palladium coatings used are of very high integrity.

Direct Beta Flux. Exposure to the direct beta flux is in the category of an accident; it is easily avoidable with care and good sense on the part of the operator. The golden rule in handling a beta source is 'never look at its front'; this applies even at a distance because beta particles have a range of several metres in air. So when inspecting the protective coating a mirror is used—a dentist's mirror on the end of a 10-cm stem is convenient. Another rule when handling a source is to wear special protective spectacles; although these protect the eyes, do not forget that the rest of the face and any other exposed or lightly clad parts of the body are also vulnerable.

In the present context the term 'handling' emphatically does not mean holding the silver disc with the fingers; as far as possible a handle 7 to 10 cm long screwed into the rear of the disc should be used with tongs and a remotely operated clamp to hold the disc itself during the initial mounting. Although in early days routine irradiations were carried out by placing the source in a mount above the sample, the end of the handle being held at arm's length, this practice is now ruled out by regulations in most countries. For weak sources, a simple shutter arrangement may be permissi-

5.7 Operator Protection

ble, but for the standard 1.5 GBq (40 mCi) source used in dating, a more complex housing incorporating safety interlocks is required if the convenience of being able to place the source above the sample, for example, while it remains in position on the oven plate, is to be retained. Two variants have been developed; the 'piston source' employing a shutter, and the 'swinging beta' type in which the source is swung away from the sample through 180° when not in use (e.g., Bailiff 1980). Of course, if the source is static and the sample is slid underneath it on a tray many of these safety problems do not arise—but the convenience of 'on-plate' irradiation is lost, and with coarse grain samples there is greater risk of disturbance. Bremsstrahlung is a serious hazard in source manipulation too, necessitating use of tongs and other remote handling devices. The dose-rate at a distance of 10 cm behind the standard 1.5 GBq source is about 250 μGy/hr. Bremsstrahlung also need to be considered with respect to any shutter devices; the shutter needs to be thick enough to ensure that the dose-rate to the sample in the shutter-closed position is negligible.

Wipe Tests. Although routinely there is no reason for anything to come in contact with the face of the source, there is obviously less risk of contamination if there has been no diffusion of strontium-90 through to the surface of the anti-corrosion coating of gold or palladium. This is checked by means of the wipe test in which a swab of cotton wool, moistened with ethanol, is lightly wiped over the surface (both source and swab being held remotely, of course). The activity of the swab is then measured, a limit of 200 Bq (0.005 μCi) being considered acceptable; it is usual practice to make this wipe test annually. In twenty years experience involving a dozen or so sources the author has only encountered one which showed a positive wipe test.

Another reason for wiping the face of a source is to remove the occassional adhering grain of sample which gets there during irradiation at close separation. Such a grain becomes heavily dosed, and if it drops off into another sample at some subsequent time, the operator is suddenly faced with a glow-curve of alarmingly large magnitude.

5.7.3 THE HAZARDS FROM AN ALPHA SOURCE

The hazards in this case are rather different. Alpha particles are stopped by a small thickness of air, approximately 5 cm for a 5 MeV particle; consequently it is safe to view an alpha source directly as long as an appropriate minimum distance is maintained. Also, the outer layer of skin is effectively dead and just about thick enough to shield the inner living tissue (see ICRP 23); direct handling of an alpha source is permissible with the additional obvious protection of rubber gloves being worn in any case to avoid risk of contamination. It is this latter, with the consequent possibility of ingestion, that can be a serious hazard with an alpha source.

The possibility of source leakage arises for three reasons not applicable in the case of beta sources. First, alpha emitters tend to move along and across surfaces because of the recoil of the nucleus when an alpha particle is emitted; this seems to happen even when the daughter nucleus is stable, presumably because atoms that have not yet decayed get carried along with the recoiling one. Secondly, any coating on the active layer must be very thin in order to allow the alpha particles themselves to get out. Thirdly, the passage of the alpha particles through this coating causes structural damage in it; an associated reason for deterioration is connected with the combined effect of high ionization density and moisture so that storage in a dessicator or under vacuum is recommended.

An alternative to the integral foil type of source is one having a discrete 'window'. The radioisotope is electroplated onto a suitable substrate and then to avoid the leakage dangers inherent in a naked source a thin foil (e.g., 1 μm of nickel) is positioned across it as a window (this word is often used also in respect of the coating on the integral foil source). Such an arrangement has the advantage that the window can be changed if it becomes defective, but on the other hand if it does fracture there is the possibility of finding a scrap of it lying on the sample after irradiation—and of course it is likely to be contaminated on its inner surface. There is no doubt that the integral foil source is the more robust (being subjected to similar tests of temperature, impact, abrasion, etc., as the beta sources) although with the disadvantage of emergent energy being a little lower, as mentioned earlier; a discrete window source must not be put under vacuum unless it is known to be suitable for this.

Alpha emission is often accompanied by some form of gamma emission; in the case of americium-241 this occurs with 36% of the alpha emissions and has an energy of 0.06 MeV. Although insignificant by comparison with what the sample receives from the alpha emission, the doserate associated with this gamma emission is enough to give rise to a small hazard unless simple precaution is taken; for gamma rays of this energy ten-fold attenuation is obtained with about 1 mm of lead.

Wipe Tests; Permissible Surface Contamination. For integral foil americium sources the wipe test is the same as for beta sources, the permissible activity[24] on the swab being 200 Bq (0.005 μCi) as before. Swabbing is quite out of the question for a discrete window type of source due to its fragility and the alternative used is to place a piece of filter paper under the source during storage; this is monitored for alpha activity before each day's usage.

With due precaution there should be no possibility of working surfaces becoming contaminated, but in case it happens it is useful to note that according to International Commission recommendations the permissible level[25] is 0.4 Bq/cm^2 (10 pC/cm^2), with averaging over an area of up to 300 cm^2 being allowed; the same level is permitted for the body except that the averaging limit is then 100 cm^2, or over the whole area of the hand in the case of the hands.

5.7.4 HAZARD DUE TO MONAZITE AND PITCHBLENDE

In routine operation there is no occasion to handle radioactive compounds, and consequently hazards associated with radiochemistry will not normally be encountered; if an unusual research project does require radiochemistry, then of course, a specialist's advice should be sought. The nearest a thermoluminescence practitioner is likely to come to hazard from unsealed radioactivity is through risk of ingestion and inhalation in the handling of finely powdered monazite or pitchblende during assembly of a calibration source for alpha counting, but it is more a perceived risk than a real one. According to International Commission recommendations (see ICRP 30) the annual limit of inhalation is on the order of a gram in the case of either monazite (assuming 10% thorium) or pitchblende (assuming 50% uranium). For ingestion the limits are substantially higher. Intake of such quantities is hardly a practical proposition even if self-inflicted injury was attempted, but even so the use of a face mask and subsequent washing of hands are reasonable precautions.

Technical Notes

[1] This is to be expected since gamma radiation induces thermoluminescence only through the secondary electrons it creates. Although the energy spectrum of the latter will not necessarily be the same as relevant beta spectra the thermoluminescence per unit dose is unlikely to be dependent on electron energy until the ionization density becomes comparable to that in an alpha track. Relevant experimental observations have been reported by Almond, McCray, Espejo and Watanabe (1968) and by Tochlin and Goldstein (1968). As emphasized by the latter authors it is important to distinguish between dose per unit radiation flux, which

does vary below 0.1 MeV, and thermoluminescence per unit dose. For alpha particles both of these quantities are strongly dependent on particle energy.

[2] A convenient container is a glass tube of wall thickness 3 mm, internal diameter 3 mm, and length 15 mm; according to Murray (1981) the effective gamma flux (photon energy 0.66 MeV) as seen by the phosphor is attenuated by about 7% for a completely filled tube.

[3] The *absorption* coefficient is to be distinguished from the *attenuation* coefficient, which quantifies the decrease in the flux of primary photons. Some of the energy lost from the primary flux is carried by degraded photons produced in Compton interactions and in narrow beam geometry these scattered photons will not contribute significantly to the absorbed dose. Whereas the absorption coefficient is to be used for calculating dose, the attenuation coefficient is appropriate for estimating the decrease of flux due to the container wall (and to some extent to the sample also). For 0.66-MeV photons in aluminium the absorption coefficient is 0.0077 per mm and the attenuation coefficient 0.0202 per mm.

[4] For fine grains deposited on disc it is to be expected that the dose will be determined by the absorption coefficient of the disc material because the thickness of the grain layer is negligible compared to the path length of the secondary electrons. However Murray and Wintle (1973) found that for fine grains of fluorite on an aluminium disc the dose received was 6% too low, and in the case of obsidian about 6% too high. Difference in electron stopping power between grains and disc is a possible explanation in the case of obsidian.

[5] There is also beta emission from strontium-90 itself. However E_{max} is only 0.54 MeV for this spectrum and as a consequence there is severe attenuation by the 0.1-mm thick silver screen on the front face of the source.

[6] The source shown in Figure 5.2, formerly designated by code SIP, is obtainable on special order from Amersham Internation (U.K.). Codes SIA and SIQ are routinely available and consist essentially of the same active silver foil; in code SIF the strontium-90 is incorporated in ceramic thereby reducing the health hazard due to bremsstrahlung. There are various other suppliers of suitable sources, for example, Isotope Products Laboratories, Burbank, California.

[7] When a nucleus emits a beta particle it also emits a neutrino, but the latter is so highly penetrating that its contribution to thermoluminescence is totally insignificant. For a given beta spectrum, the energy emitted by the nucleus is the same for each event, that is, the neutrino makes up the difference between E_{max} and the beta energy.

[8] German for 'braking radiation'; so called because the emission occurs when an electron is brought to an abrupt halt through collision with a nucleus.

[9] If the half-value thickness is expressed in terms of g/cm^2, it is nearly independent of the absorber concerned, rising only slightly with increasing Z. For precise calculations reference should be made to published values for stopping powers and ranges (Berger and Seltzer, 1982).

[10] Uniformity of dose across the sample is important primarily in avoidance of error due to non-uniformity in sample spreading; it is also relevant when there is

non-linear dependence of thermoluminescence on dose, for example, in studying the onset of saturation.

[11] It is important to check whether the particular spray used gives rise to any parasitic thermoluminescence. One satisfactory spray is Silkospray, obtainable from Willy Rusch, 7053 Rommelshausen, Stuttgart, West Germany; however, since manufacturers change their recipes from time to time intermittent checks of spurious-free performance should be made. In any case it should be noted that whereas this spray was satisfactory in combination with stainless steel discs, use with aluminium discs may give rise to a weak parasitic thermoluminescence, not enough to upset phosphor measurements but unacceptable for quartz measurements in the dose region below 10 Gy.

[12] Although the build-up effect gives a semi-quantitative explanation, it is difficult to accept (from Figure 5.4) that that effect is strong enough to account for the ratio of 1.5 between beta source calibration factors for shiny and frosty grains that these authors find. They also eliminated an explanation in terms of the backscattered dose being stronger in deeper parts of the sample—the same ratio was observed for planchettes of silver, copper, and aluminium. Another possibility is that the effect is connected with backscattering when the beta rays enter the sample; because of the rough surface of the frosty grains an appreciable percentage of the incident beta flux may never enter the sample, according to Kalefezra and Horowitz (1979) the order of 40% can be rejected in this way. However, an explanation in terms of backscattering does not fit with the observation by Benko (1983) that there is no difference between the calibration factors if monolayer samples are used. Another possibility, mentioned by the authors, is that there is a difference in thermoluminescence effectiveness between beta and gamma radiation.

[13] It can be argued that if the wavelength of the ZGM peak is different from that of the dating thermoluminescence there may be portion-to-portion variations in the ratio between the two because of differing degrees of optical attenuation. In practice such interference does not seem to be important, and the same is true with respect to polymineral fine-grain samples when the ZGM peak is due to one particular mineral. For coarse grains there is limitation due to statistical variations in the number of bright grains present coupled with some grains being bright in the ZGM region but dim at high temperature, and vica-versa.

[14] The 20-mm wide silver strip supplied under code AMM 3 by Amerham International (U.K.) has an active width of 12.5 mm and a loading of 7 MBq/cm^2 (0.2 mCi/cm^2). Thus for the 12-mm diameter active area used in the six-seater irradiator of Figure 5.8, the effective activity is 8 MBq (0.22 mCi). An alternative isotope is curium-244, half-life 18 years; this has the advantage of a slightly higher alpha emission energy, 5.8 MeV instead of 5.5. Curium-242 has a higher alpha energy still (6.1 MeV) but the disadvantage of a half-life of only 163 days. Both the curium isotopes have the disadvantage of accompanying beta emission ($E_{max} \approx$ 0.04 MeV) with about 25% of the alpha emissions; in the case of americium-241 there is a 0.06-MeV gamma photon with about 36% of the alpha emissions.

[15] Another approach to alpha irradiation is to use a sample that is thick enough

for all the particles to be stopped in it and evaluate the effectiveness in terms of total energy deposited The difficulty then is that unless the sample is completely transparent the thermoluminescence from beta irradiation (and also the natural thermoluminescence) will be affected to a different degree from alpha attenuation. Correction for this is complex—see for example calculations by Aitken and Wintle (1977) for a flint slice.

[16] Curium sources are available for which a discrete 'window' substitutes for the protective coating. These are usually thin foils of nickel (~ 1 mg/cm^2) or tantalum (~ 2 mg/cm^2), and so the energy loss is somewhat less than in the coating; however there is risk of fracture, *particularly on evacuation*.

[17] This mechanism has been considered with respect to the supralinearity of the 370°C peak in quartz, the 110°C peak being hypothesized as the competitor; experimental investigation by Fleming (1969) did not give support to this model. On the other hand, observations made by Smith (1983) on another type of quartz suggested the possibility of such interaction; the important practical finding with this type was that delay of a few days between irradiation and measurement removed the substantial scatter of points in the growth characteristic that occurred when there was no delay.

[18] Using Brazilian quartz rock crystal Groom *et al.* (1978) report a decrease in thermoluminescence sensitivity for the high temperature peak for dose-rates in excess of 100 Gy/min if the total dose exceeds 300 Gy. Using detrital quartz and quartz extracted from granite, G. Valladas and H. Valladas (1982), extending the work of Valladas and Ferreira (1980), report increases of up to 10% and decreases of up to 20% when the dose-rate is changed from 0.01 to 10 Gy/min depending on type of quartz and spectral response of the detection system, the total dose being 30 Gy; the changes quoted are for blue and ultraviolet filters, respectively, the changes for a green filter are within experimental error.

[19] As reported by Aitken, Huxtable, Wintle, and Bowman (1974) the sensitivity for irradiation at 100°C relative to that at 20°C for other phosphors was: fluorite, decrease of 30%; CaSO$_4$:Tm, decrease of 5 to 10% depending on dose; CaF$_2$:Mn, $<\frac{1}{2}$% decrease; LiF (TLD 100), increase of 13%. For the 325°C peak of quartz there was a decrease of 17%. King and Johnson (1983) found no effect in CaF$_2$:Mn, CaF$_2$:Dy, CaF$_2$:Tm, or Li$_2$B$_4$O$_7$:Mn but a 20% increase with respect to LiF:Mg; these authors mention enhanced trap migration and aggregation in respect of the latter. As reported by Durrani *et al.* (1977) the sensitivity of quartz for irradiation at 113K is higher than at room temperature by a factor of 5 or more for an X-ray dose of 600 Gy and above, but there is no difference for a dose of 100 Gy; for heavy gamma doses the variation is stronger.

[20] The relevant report is ICRP No. 26, published in the *Annals of the ICRP*, Vol. 1, No. 3 (1977, Pergamon Press). The unit used is the now obsolete rem, equal to the rad in the case of beta and gamma radiation; in the case of more heavily ionizing radiation a given dose in rads is assessed as having a greater radiobiological equivalence than 1; for neutrons, for instance, 1 rad corresponds 10 rem. In the SI system the unit for dose equivalent is the sievert (Sv), equal to

Technical Notes

the gray (= 100 rad) in the case of lightly ionizing radiation. It may be useful to mention here the now obsolete unit of exposure, the roentgen, which does not have a named unit in the SI system; for a radiation field in which the radiation flux has an exposure value of 1 röentgen the absorbed dose is roughly 1 rad (i.e., 10 mGy) in soft tissue; more precisely, exposure to 1 roentgen of 1-MeV gamma rays will produce an absorbed dose of 8.7 mGy in air or quartz, and 9.7 mGy in human tissue. For gamma energies below 0.1 MeV, the corresponding absorbed dose become substantially different (see Appendix M, Figure M.2).

[21] Such as the Mini-Monitor type 5.10R made by Mini-Instruments, Burnham-on-Crouch, Essex, U.K.; the detector is separate from the electronics, and having a diameter of only 16 mm it is useful for surveying in congested regions. The response extends down to 50-KeV photons; below this, their counter type 5.40 can be used.

[22] Such as the Portable Contamination Meter type PCM5, made by Nuclear Enterprises, Edinburgh.

[23] Experiments by the author, using LiF (TLD 100) chips, indicate that alongside a 1.5-MBq (40-mCi) source 5 cm away from its centre, the dose-rate due to backscattering from a lead block placed 2.5 cm in front of the source is about 0.5 Gy/hr (i.e., between 1 and 2% of the direct beta dose to a sample 16 mm from the source); this is excluding the contribution from bremsstrahlung. The attenuation coefficient for absorption of the backscattered flux appears to be not much greater than that of the primary flux, namely, the half-value thickness for aluminium is about 0.3 mm.

[24] This is the limit specified by Amersham International. Using a scintillation contamination meter of the type mentioned in note 22, this limit corresponds to a count-rate of about 5 counts/sec.

[25] This level, which is for high toxicity (Class I) radionuclides, corresponds to a count-rate of about 5 counts/sec for the above mentioned monitor (detector area 50 cm^2), the efficiency of detection being ten times greater than for radioactivity on a swab.

CHAPTER 6

SPECIAL METHODS

In the preceding chapters the basic framework of thermoluminescence dating has been established. Now, in the three remaining chapters of the book we build on that framework, discussing first the use of certain properties as special methods (e.g., pre-dose dating, feldspar dating, zircon dating) and secondly, the extension into paleolithic and geologic applications. Whereas the first five chapters are essential reading for any thermoluminescence practitioner, some sections of these final three are rather specialized; depending on their particular interests, readers may want to exercise some degree of selectivity.

6.1 Pre-Dose Dating

As discussed in Chapter 5, it is not uncommon for the thermoluminescence sensitivity of a sample to be different after taking a glow-curve from what it was before. One cause is the pre-dose effect, and with this the change is dependent on the previous irradiation received by the sample. In the particular case of the 110°C peak in quartz this 'memory' can be used for evaluation of paleodose, as was realised by Fleming (1969), who established it as a dating tool in the early 1970s (Fleming, 1973). Properly used it is a powerful technique, though limited in use to samples of the last 1000–2000 yr because of early onset of saturation—usually at around several gray.

The 110°C peak in quartz has a half-life at room temperature of only 1–2 hr and consequently it is not present in the natural thermoluminescence (unless the sample has been inadvertently exposed to light during preparation—see Section 6.2). Because it is a highly sensitive peak, exposure to quite a small dose (around 0.01 to 0.1 Gy) is usually sufficient to induce an accurately measurable thermoluminescence. It is the change in response to this *test-dose* by which the paleodose is evaluated. The important feature discovered by Fleming was that for the memory to be unlocked it is necessary to heat the sample to around 500°C; this *thermal activation* causes a strong increase in subsequent response to the test-dose, as illustrated in Figure 6.1. The increase is proportional to the total dose received since the pottery was fired, that is, to the paleodose.

The basic sequence of measurements is as follows:

1. Give test-dose and measure 110°C peak response, S_0;
2. Heat to the activating temperature (this is usually a brief heating to 500°C as in the course of a normal glow-curve);
3. Give test-dose and measure 110°C response again, S_N;
4. Give laboratory calibrating dose, β (usually several gray);

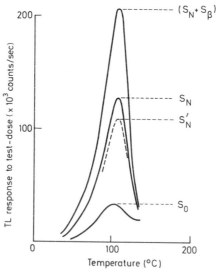

Figure 6.1 The pre-dose effect in the 110°C peak of quartz. The glow-curves, all for the same portion, show the response to the same beta test-dose of 0.01 Gy at various stages: S_0 before any heating, S_N after heating to 500°C, S'_N after giving a calibrating dose of 1 Gy and heating to 150° and ($S_N + S_\beta$) after a second heating to 500°. The reason why S'_N is different from S_N is discussed in Section 6.1.2.

6.1 Pre-Dose Dating

5. Repeat step 2;
6. Give test-dose and measure response, $(S_N + S_\beta)$.

Since $(S_N - S_0)$ is proportional to the paleodose P and the further increase between steps 3 and 6 is proportional to the calibrating dose, we have

$$P = \frac{S_N - S_0}{(S_N + S_\beta) - S_N} \beta. \tag{6.1}$$

This assumes of course that for the further increase between steps 3 and 6 the constant of proportionality is unchanged, that is, the effect is linear with dose; this needs to be checked. Other important points discussed later are the need to make a *quenching correction* and the need to check that the pre-dose characteristics have not been changed by the heating in step 2.

This pre-dose effect can usually be observed in any sample, fine-grain or coarse-grain, that contains quartz, but for highest accuracy there is advantage in extracting the quartz as in the inclusion technique.[1] For a 5-mg sample of quartz, the typical increase in sensitivity is by a factor of between 2 and 10 for a pre-dose of 1 Gy; hence a paleodose of as little as 0.1 Gy (sometimes even less) can be measured. Thus it is highly advantageous for young samples in which the conventional method is liable to be inapplicable because of too low a thermoluminescence intensity in the high temperature region of the glow-curve; a notable non-archaeological application is for evaluation of the dose recorded in roof tiles at Hiroshima and of the fall-out dose associated with Nevada nuclear bomb testing as recorded in the bricks of houses (e.g., Bailiff and Haskell, 1984; Haskell, Kaipa, and Wrenn, 1985; Ichikawa and Nagatomo, 1983; Maruyama *et al.*, 1983). It is also an important tool in the authenticity testing of types of ware for which the true thermoluminescence is masked by spurious thermoluminescence or invalidated by anomalous fading; use on porcelain is a particularly important application (Stoneham, 1983). It can also be used to give an indication of whether the lack of thermoluminescence in an object is due to partial reheating rather than to the object being a modern forgery; this is because the temperature needed for erasure of conventional thermoluminescence is not always sufficient for full activation of the pre-dose (see Section 6.1.3); this was done in the authenticity investigation of Hacilar ware to discount the hypothesis that the thermoluminescence had been erased from many of the objects because of heating in domestic ovens for drying after recovery from marshy ground (Aitken, Moorey and Ucko, 1971).

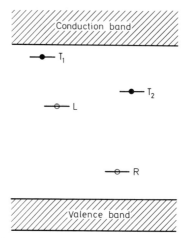

Figure 6.2 Energy-level model for the pre-dose effect of the 110°C peak in quartz (after J. Zimmerman, 1971).

6.1.1 THE MECHANISM RESPONSIBLE

Studies made by J. Zimmerman (1971) led her to conclude that the sensitivity enhancement is due to an increased availability of charged luminescence centres. The model is shown in Figure 6.2. In this T_1 and T_2 are electron traps and R and L are hole traps; if charged with a hole, L traps become luminescence centres, that is, subsequent capture of an electron gives rise to emission of a photon as the electron and hole mutually annihilate. T_2 is presumed to be deep enough not to be emptied by heating to 500°C, and it is introduced into the model so that charge balance can be maintained. The thermoluminescence process consists of the thermal release of electrons from the traps T_1 and the capture of some of these into the luminescent centres L. Such capture only occurs for centres which are charged with a hole and so the thermoluminescence sensitivity is proportional to the number of centres which are so charged.

Firing of the pottery in antiquity is hypothesized to empty nearly all traps, and consequently the sensitivity after firing, S_0, is low. Subsequently, the natural radiation dose-rate produces electrons in the conduction band and holes in the valence band; the electrons are trapped in T_2 (because T_1 is too shallow to retain them for more than about an hour) and the holes are captured at R (rather than L because of the former's presumed higher capture cross-section). Thus during antiquity the hole population in R gradually builds up, and it is usual to refer to R as the *reservoir* traps.

6.1 Pre-Dose Dating

In step 1 of the pre-dose procedure, the test-dose charges the traps T_1 with a small number of electrons and puts a small number of holes into R. On heating through 150° the electrons are released and those that find a charged luminescence centre give rise to thermoluminescence. Because there has been no change in the number of holes trapped at the centres of type L the observed sensitivity S_0 is the same as if the measurement had been made immediately after firing.

The heating to 500°C in step 2 causes thermal release of holes from R, and these are then captured at L (because any recaptured at R are immediately re-ejected). Hence the sensitivity measured in step 3, S_N, is proportional to the number of holes that had been accumulated in R during antiquity.[2]

UV Reversal. If an already activated portion is illuminated with ultraviolet (in the wavelength region of 240 nm) there is reduction of S; this is interpreted as due to photoexcitation of electrons from the valence band into activated L centres. If after this illumination the portion is activated again, then S is substantially restored. It is presumed that the holes left in the valence band by the photoexcited electrons are captured at R traps so that effectively the illumination causes transfer of holes from L centres to R traps; hence, when a further activation is carried out they are put back into the L centres with consequent restoration of S. In the quartz studied by Zimmerman (1971) this transfer cycle from L to R and back to L could be accomplished with a loss of less than 10%. Using a 10-min illumination by an 80-W medium-pressure mercury lamp, with the portion at a distance of about 10 cm, there was a reduction in S by 80% in the L to R step; using a 100-W lamp Bailiff (1979) reports 90% reduction for a 2-hr illumination. Bailiff also studied the thermoluminescence emission spectra (which peaks at 360 nm), finding a tendency for the residual S after ultraviolet illumination to be biased towards longer wavelength; this was interpreted as due to the presence of some additional centres, termed L_0, which are not associated with the pre-dose mechanism.

6.1.2 QUENCHING

If the sensitivity is measured after administration of the calibrating dose, step 4, but before activation in step 5, it will usually be found that it is less than the value of S_N as measured in step 3. In order to make correction for this, two extra steps are inserted in the sequence:

4a. Heat to 150°C in order to empty the 110°C traps of the electrons acquired from the calibrating dose;

4b. Give test-dose and measure response; call this S'_N.

When the calibrating dose is activated, step 5, the resultant increase in sensitivity is the difference between the responses in steps 4b and 6; hence to make allowance for quenching, the expression for the paleodose needs rewriting as

$$P = \frac{S_N - S_0}{(S_N + S_\beta) - S'_N} \beta. \tag{6.2}$$

This quenching of sensitivity is readily explicable in terms of the model. During activation of the natural dose, step 2, a certain number of L centres have been charged with a hole, giving an overall sensitivity S_N. Electrons reaching these charged L centres, either directly during step 3 or via traps during step 4a, annihilate with the hole thereby leaving the L centre uncharged and so no longer contributing to the sensitivity. The amount by which the sensitivity is decreased is of the order of 10% per gray of calibrating dose.

With some types of quartz, enhancement is observed rather than quenching, that is, S'_N is greater than S_N. Presumably the explanation is that during the clearing of the 110° traps by heating to 150°, step 4a, there has been sufficient partial activation of the calibrating dose, that is, transfer of holes from R traps to L centres, to overcompensate for the quenching effect as described above. Such partial activation is liable to occur for samples having activation characteristics of type (a) in Figure 6.3, discussed below.

The procedure as so far discussed, in which more than one activation is applied to the same portion, is usually referred to as the *multiple activation* technique. This is to distinguish it from the *additive dose* method, discussed shortly, in which more than one portion is used and the calibrating dose added on top of the natural one before any activation. This parallels the additive dose method in conventional thermoluminescence and it avoids upset due to possible changes in pre-dose characteristics that are engendered by heating; it also avoids the need for a quenching correction. On the other hand, there has to be normalization between portions, and in general this means poorer precision.

6.1.3 ACTIVATION TEMPERATURE

So far the activation temperature has been quoted as 500°C, following the practice in early work. Depending on the type of quartz and the temperature and duration of firing in antiquity, there are variations in the *thermal activation characteristic* (TAC) that is observed by heating the portion to successively increasing temperatures from 200°C upwards, the

6.1 Pre-Dose Dating

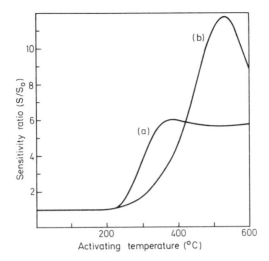

Figure 6.3 Two examples of thermal activation characteristic (TAC) for the 110°C peak in quartz (from Fleming and Thompson, 1970). S_0 is the sensitivity before activation and S is the sensitivity after rapid heating to the temperature indicated. Curve (a), which shows 'early' activation, is for geological quartz fired for 18 hours at 700°C and then given a pre-dose of 0.5 Gy; curve (b), which shows 'late' activation followed by thermal de-activation, is quartz fired for 18 hours at 1000°C and given a pre-dose of 10 Gy.

sensitivity being measured after each heating. Figure 6.3 gives two examples of TAC, curve (a) showing 'early activation' and curve (b) showing 'late activation'. It is evident that in the latter case 500°C is not sufficient to achieve full activation. The form of the TAC is taken to reflect a distribution in activation energies of the R traps; traps close to the valence band have their holes transferred to L centres at a lower temperature than traps that are not as close.

Another feature of Figure 6.3 is that the sensitivity reaches a maximum and falls off when higher temperatures are used; this is referred to as *thermal de-activation* and presumed to be due to thermal eviction of holes from L centres (through the mechanism of thermal excitation of electrons from the valence band).

Of course the temperature at which the sensitivity maximum of the TAC is reached depends on the time spent at high temperature. If the heating rate is slow, and more particularly if the peak temperature is held for a minute (say) before being cut off, then the maximum will shift downwards in temperature. Obviously it is essential that exactly the same thermal treatment is maintained for all activations when dating is being done.

Ambient Activation; High S_0. For a sample exhibiting early activation, an immediate question is whether there has been partial activation during antiquity, that is, whether during prolonged storage at ambient temperature there has been a significant amount of transfer of holes from the reservoir traps R into the luminescence centres L. If so, then S_0 will be higher than its value immediately after firing, and consequently $(S_N - S_0)$ will be low, giving rise to too young an age estimate.

If the measured value of S_0 is only a few percent of S_N, then there is direct evidence that the amount of ambient activation is unimportant; a low value for S_0 also rules out activation through inadvertent heating at some time subsequent to firing (unless the heating was so severe as to cause complete thermal de-activation). If S_0 is not negligible then the question arises as whether some part of it arises from ambient activation (or inadvertent heating). If it is suspected that the major part of S_0 arises in this way then S_N should be used in the numerator of Equation 6.2 instead of $(S_N - S_0)$, and the age so derived regarded as an upper limit.

One way in which evidence about the extent to which ambient activation contributes to S_0 can be obtained is by comparing the shape of the TAC for the natural dose with that for an artificial dose. This is illustrated in Figure 6.4; the early activation apparent in the TAC for the latter is largely absent in the former, indicating that the shallow R traps have suffered ambient activation.

In the absence of ambient activation it might be thought that S_0 should be zero. That this is not the case can be shown experimentally by measuring the pre-dose sensitivity of freshly fired quartz (fired say at 1000°C—enough to erase the pre-dose memory); the value obtained for S_0 is finite and for some types it can be substantial, depending on the thermal treatment.[3] In terms of the model this initial value for S_0 is explained (see Section 6.1.1) by the presence of an additional type of luminescence centre, L_0, which although associated with 110°C traps, does not participate in the pre-dose mechanism.

In the case of fine-grain samples a high value for S_0 may be due to the thermoluminescence from other minerals; for instance from feldspar there is a peak at about 150°C and the lower part of this may be making an appreciable contribution in the region of 110°C. Some discrimination in favour of quartz thermoluminescence may be obtainable by using an appropriate colour filter.[4] Another approach is to utilize the UV reversal effect mentioned in Section 6.1.1. By illumination with UV before measurement of S_0 the holes that reached L centres through ambient activation are transferred back to the R traps; the residual sensitivity should then be that due to other minerals or to non-participating centres in

6.1 Pre-Dose Dating

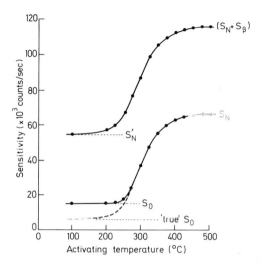

Figure 6.4 Thermal activation characteristics (TAC): effect of ambient activation. After measuring the TAC, $S_0 \to S_N$, the calibrating beta dose, β, is administered. This causes suppression of S_N to S'_N. The TAC, $S'_N \to (S_N + S_\beta)$, is then measured. In the first TAC, activation does not start until 250°C because the shallow R traps have already been activated in antiquity. In the absence of ambient activation the two TACs would match; however for some samples there is an intrinsic change of shape between first and second TAC (see Figure 6.5). When measuring TAC or using the multiple activation technique allowance needs to be made for the cumulative addition of test doses on to the effective pre-dose (unless the test dose is negligible by comparison).

quartz, but it appears that some of the sensitivity contributed by the latter can be eroded also.

Change of TAC due to Heating. For some samples the TAC observed for an artificial dose applied to a portion for which the natural dose has already been activated is different in shape from the natural TAC even though there has been no ambient activation; this was first noted by Bailiff (1983) and is illustrated in Figure 6.5. It is interpreted as being due to a change in the energy distribution of the R traps due to the heating that has intervened.

If there is change of TAC on heating, then there is risk that a different fraction of R traps are activated for the calibrating dose (step 5) than for the natural dose (step 2). To avoid this it is necessary that the activating temperature should be high enough to achieve the maximum sensitivity possible; it should not be too high as there is then deactivation. Thus when the multiple activation technique is used it is first of all necessary to

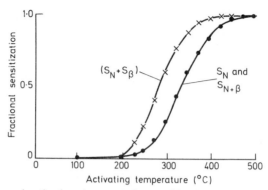

Figure 6.5 Thermal activation characteristics: modification of R trap distribution due to heating. For some samples the shape of the first TAC carried out is different to subsequent ones on that portion (Bailiff, 1983). The difference is not associated with one being for natural dose and other for artificial because the TAC for a second portion to which a calibrating dose has been added before any activation has the same shape as the TAC $S_0 \rightarrow S_N$.

In this figure the sensitivity is expressed as fraction of the total sensitivity change during the TAC.

measure the TAC and decide on the appropriate activating temperature for the sample.

It appears that by so doing satisfactory results are obtained, as evidenced by tests on samples that have been given a known dose (Haskell, Kaipa and Wrenn, 1985). However it cannot be ruled out that for some samples heating may cause a change in the total number of R traps (or L traps for that matter) in addition to altering the R-trap energy distribution, that is, the overall pre-dose capacity is changed.

6.1.4 THE ADDITIVE DOSE METHOD

Error from such alteration can be avoided by using more than one portion and making comparison of S_N with $S_{N+\beta}$, the latter form of nomenclature being used to denote that the beta dose has been administered before activation of the natural. With the first portion of sample steps 1–3 of the multiple activation technique are implemented in order to evaluate S_N; with the second portion the following four steps are implemented:

(a) Give test-dose and measure S_0;
(b) Give laboratory calibrating dose β;
(c) Heat to the activating temperature;
(d) Give test-dose and measure $S_{N+\beta}$.

6.1 Pre-Dose Dating

The expression for the paleodose is now

$$\frac{S_N - S_0}{S_{N+\beta} - S_N} \beta. \tag{6.3}$$

Besides avoiding upset due to any change in pre-dose characteristics on heating this method avoids interference by quenching. On the other hand, there is the need to normalize between portions. In practice normalization by weight does not give good precision, and better results are sometimes obtained by using each portion's value of S_0 for normalization; however this cannot be done for a sample having the otherwise desirable characteristic of low S_0 because the peak will be too small.

Equations 6.2 and 6.3 both assume linearity of predose response and this needs to be checked, as discussed in Section 6.1.7 and illustrated in Figure 6.6. In the additive dose method several portions are used for this, all needing normalization.

The Spike Technique (Bailiff, 1985; Haskell and Bailiff, 1985; Haskell, Kaipa and Wrenn, 1984). This is essentially an additive dose method which incorporates the multiple activation technique in order to avoid the need for explicit normalization. Steps 1–6 are implemented with the first portion as in the multiple activation technique. The same steps are imple-

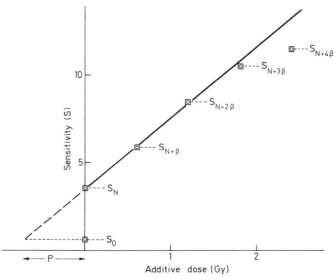

Figure 6.6 Linearity check using additive dose method. The paleodose P is the intercept on a horizontal line through the S_0 data point.

mented with a second portion except that it is first 'spiked' with a dose β_{sp}. From the first portion we obtain an evaluation of the apparent paleodose, P', and from the second portion $(P' + \beta'_{sp})$, where β'_{sp} denotes the apparent spike dose. 'Apparent' dose is used in order to allow the possibility that in the multiple activation technique there has been a change in pre-dose capacity due to the heating that intervened between measurement of S_N and measurement of $(S_N + S_\beta)$. As long as the apparent dose is proportional to the true dose, with a constant of proportionality that it is independent of dose, the true paleodose can be obtained as in Figure 6.7.

6.1.5 ERASURE OF THE PRE-DOSE MEMORY

Heating to the activating temperature unlocks the memory but it does not erase it. Illumination with ultraviolet reduces the activated sensitivity by transferring holes from the L centres back to the R traps, but this is *reversal* rather than erasure; if the sample is thermally activated again the sensitivity reappears. For erasure quartz usually needs to be heated to a temperature in excess of 600°C, and for some types of porcelain 1000°C is necessary. Erasure is presumed to be due to thermal eviction of holes from L centres—the thermal deactivation referred to in connection with Figure 6.3. It might be expected that erasure would be associated with the α–β transformation that occurs in quartz at 575°C, or the conversion to tridymite at 800°C, but this has not so far been noted.

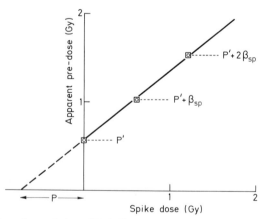

Figure 6.7 The spike technique (Haskell and Bailiff, 1985). The multiple activation technique has been used to evaluate the apparent pre-dose in three portions; to two of these a known spike dose has been administered before measurement. The true paleodose P is obtained as the intercept on the dose axis.

6.1 Pre-Dose Dating

Besides erasure, subjection to high temperatures causes change in the subsequent pre-dose capability. It has been suggested (Sunta and David, 1983) that changes in capacity observed in laboratory re-firing experiments could be used to determine the temperature of the firing in antiquity—on the basis that changes in capacity should not begin until the ancient firing temperature is exceeded. However, successful application has not so far been achieved (e.g., Watson and Aitken, 1985).

6.1.6 EFFECTIVENESS OF ALPHA IRRADIATION

As discussed shortly, the onset of saturation occurs at a few gray of beta pre-dose. Thus in the high ionization density core of an alpha track, the proportion of holes that go to waste because all available R traps are full is even greater than the wastage of electrons and holes in the case of conventional thermoluminescence. Hence the effectiveness of alpha particles relative to beta is very low. It is denoted by κ; values are typically around 0.01 whereas for conventional thermoluminescence in quartz the k-value is around 0.1. Such a low κ-value means that the alpha contribution is rarely more than 5% for fine-grains and negligible for coarse grains.

6.1.7 NON-LINEARITY IN PRE-DOSE RESPONSE

The samples for which pre-dose dating is most appropriate are those having a paleodose in the region below a few gray; this is on account of the early onset of saturation effects. However, when there are problems and uncertainties in the use of conventional thermoluminescence for samples having a higher paleodose, inevitably the attempt is made to use pre-dose dating there too. Also, there are some samples for which the onset of saturation causes non-linearity of pre-dose response at a level somewhat less than a few gray. Utilization in the non-linear region compounds the complication of an already complicated technique; let the practitioner beware!

Non-linearity can arise either from the R traps or from the L centers, or from both. Which has dominance may depend on whether multiple activation or additive dose is being used. R trap saturation is more restrictive in the latter because of the more holes to be stored before transfer. In respect of the T traps we are not reliant on linear response since the test-dose is kept the same throughout the measurement of any given sample; in any case the test-dose should be small compared to the calibrating dose, that is, 0.1 Gy or less, and this is negligible compared to the saturation level.

Whichever method is being used the first step in testing for saturation effects is to make two determinations of the paleodose P using a different value of the calibrating dose β for each; these doses should differ by a factor of about two and the higher should not be less than the approximate magnitude of the paleodose. Unless the two values of P agree to within 5%, and in any case if the highest accuracy is being sought, further investigation should be made.

Non-Linearity Effects: Multiple Activation. L centre linearity can be tested by repeated dosing and activation using the same dose each time. The latter condition avoids any interference from R trap non-linearity since the same degree of saturation will exist on each occasion. By plotting S against the accumulated dose, the level of artificial dose at which L centre non-linearity begins can be determined.

To test R trap linearity the sensitivity obtained from activation of a single dose is compared with that obtained by individual activations of a number of smaller doses, which in total equal the single dose. By repeating this for different values of the single dose, the level of dose is found at which the R trap response becomes non-linear. Obviously the L centre test can be incorporated into the incremental activation part of this procedure.

It should be appreciated that unless proper correction is made for quenching there will be a pseudo non-linearity when multiple activations are employed. The correction is made by adding to each observed sensi-

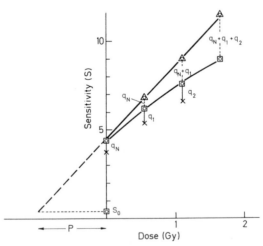

Figure 6.8 Correction for quenching in multiple activation technique. The ⊚ are experimental points obtained in step 5 after activation; X are experimental points obtained in step 4b after administration of calibrating dose but before its activation. The quenching at each stage is indicated by q, q_1, q_2. The △ are corrected points.

6.1 Pre-Dose Dating

tivity the cumulative loss due to quenching, as illustrated in Figure 6.8. Note however that if this procedure is carried on until there is saturation, a pseudo linear region can result.

Another distortion that can arise is associated with incomplete activation; this too can cause apparent linearity when in reality there is non-linearity. When activation is incomplete, a second activation carried out without any additional calibrating dose being given gives rise to an increase in S (over and above what is to be expected from activation of the test dose). If, for such a sample, successive calibrating doses are being activated there is more 'hung over' contribution in subsequent activations than in early ones; hence what is truly linear may appear to be supralinear and what is sublinear may appear to be linear. The remedy is of course to ensure that activation is complete by appropriate choice of temperature. In any case it is necessary in all this work to use a temperature that achieves maximum activation so as to avoid distortion due to TAC change on heating.

There is a practical problem that becomes more acute when there are many successive activations of the same portion; this is the difficulty of avoiding loss of grains. The high temperatures required for activation preclude the use of oil as an adhesive and so it is necessary to have a stiff enough heater plate to avoid grain movement. An alternative is to use slices, though of course quartz separation is not then possible; slices are necessary for porcelain in any case in order to avoid the risk of falsely high values of S_0 that are liable to be engendered in crushing porcelain.

Non-Linearity Effects: Additive Dose. If the multiple activation technique is applied using only one value of calibrating dose without making the specific test for R trap non-linearity mentioned above, there is the risk that R trap saturation will go undetected even though the growth of S with successive calibrating doses (all of the one value) is plotted out; this only tests L centre linearity. With the additive dose technique linearity of S for increasing values of additive dose does test both L centres and R traps for linearity. On the other hand, as mentioned earlier, in this method the R traps are required to be linear to a higher dose level than with the multiple activation technique.

Supralinearity. No evidence has been reported for non-linearity in growth of S at lower doses.

Evaluation in Non-Linear Cases. With a sample for which the paleodose is sufficient to be in the region of non-linearity for R traps (but not for L centres) evaluation is possible in the multiple activation technique by using calibrating doses which give sensitivity increases that bracket

$(S_N - S_0)$, and then interpolating. Note however that if the doses are sufficient to saturate the R traps then the same sensitivity increase will be obtained irrespective of dose.

Otherwise, an analytical approach such as proposed by Chen (1979) is necessary. In this the non-linearity is taken to be exponential in form, and Chen gives a procedure for evaluating the growth parameters from a plot of S against dose.

The accuracy obtainable for a sample in which non-linearity intervenes is inevitably degraded even though the analytical approach may be comprehensive; also, allowance for the complicating effects mentioned in earlier sections becomes that much more difficult and uncertain.

6.2 Phototransfer

The 110°C peak is not present in the natural glow-curve of a quartz sample because of its short lifetime. However if the sample is subjected to ultraviolet illumination before measurement the 110°C peak will appear; also, the height of the 370°C peak will be somewhat less than for an equal sample which has not been illuminated, see Figure 6.9. The effect of the ultraviolet has been to evict electrons from the deep traps ('donors') and some of these were re-trapped in shallow traps ('acceptors'). The 110° peak observed in the subsequent glow-curve is termed *Phototransferred*

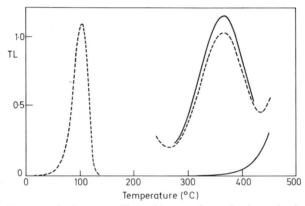

Figure 6.9 Phototransfer in quartz. The solid curve shows the thermoluminescence peak from a portion of quartz extracted from ancient pottery. The dashed curve shows the glow-curve obtained from a second portion which has been exposed to ultraviolet illumination before measurement. The 110°C peak appears because there has been transfer of electrons from deep traps to shallow traps.

Thermoluminescence (PTTL). As might be expected the transferred thermoluminescence is proportional to the electron population of the 370° traps. Hence, in principle at any rate, the transferred thermoluminescence can be used to evaluate the paleodose and determine the age. Phototransfer from deep traps to shallow traps is exhibited by most thermoluminescent minerals and by artificial phosphors; in the latter it is used as a method of measuring the amount of ultraviolet exposure. Although its use for dating has been investigated (Bowman, 1979), as outlined below, it is not at present a method of general use. Nevertheless some discussion of it is appropriate not only because of potential advantages for age determination but also because of its relevance to sediment dating (Chapter 8).

The basic advantage of PTTL is in giving access to electrons stored in deep traps without the need to heat the sample to the high temperature necessary to empty those traps. The most obvious advantage is in the avoidance of the thermal signal that is the inevitable accompaniment of direct observation of high temperature thermoluminescence; this thermal signal is likely to mask the thermoluminescence unless suitable colour filters (blue, violet, or ultraviolet) are interposed so that in conventional thermoluminescence we are to a large extent restricted to minerals which have substantial short wavelength emission. There is also avoidance of spurious thermoluminescence and another fundamental advantage is in allowing the possibility of dating bone and aragonitic shell; these are out of bounds to conventional thermoluminescence because of decomposition or phase change on heating.

Associated with the first advantage is the possibility of access to deep traps from which the direct thermoluminescence is dim due to *thermal quenching* as mentioned in Section 3.3—at elevated temperatures there is a decrease in the intrinsic efficiency of the luminescence process due to an increasing probability that the luminescence centre will de-excite by emission of phonons (i.e., transfer of heat to the crystal lattice) rather than photons. Thus there may be traps which empty at higher temperatures, say 500°C, which do not give rise to much thermoluminescence and which consequently would be difficult to detect even if there was no thermal signal. Such very deep traps are important in minerals such as zircon for which the conventional thermoluminescence exhibits anomalous fading; there is less likelihood, on theoretical grounds, that the very deep traps are affected in this way and the phototransfer work of Bailiff (1976) gives experimental support to this.

Finally there is the possibility of a 'single-portion technique' in which all of the necessary thermoluminescence measurements can be made on a few milligrams of sample—in contrast to the additive dose method of conventional thermoluminescence which requires at least a dozen portions. This possibility arises because of avoidance of sensitivity changes associated with heating (causing second-glow response to be different to first-glow response) and because repeated measurements on the same portion are possible since only a small percentage of the donor population needs to be transferred in order to observe the PTTL.

6.2.1 QUARTZ

Of various procedures tried, Bowman (1979) found the 'total bleach' method the most encouraging. In this procedure the steps were as follows:

1. Thermal 'wash' for 1 min at 240°C in order to empty donor traps in that region of the glow curve;
2. Illuminate with 260 nm ultraviolet from a monochromator (using a high-pressure xenon lamp as source), and measure 110°C peak;
3. Repeat with 320 nm ultraviolet;
4. Bleach sample using an unfiltered medium pressure mercury lamp (usually the PTTL was reduced to less than 5% in about 17 hr);
5. Administer calibrating beta dose and repeat steps 1–4 to give four or five measurements of PTTL for different values of the calibrating dose. From the growth characteristic so obtained the paleodose corresponding to the 'natural PTTL' could be read off.

In six out of the eight archaeological samples tried, the paleodose assessed in this way agreed with the value obtained using conventional thermoluminescence; the average value for the ratio between PTTL paleodose and conventional paleodose was 1.00 with a standard deviation of 0.09. In the other two samples the value obtained with 320 nm illumination was in reasonable agreement with the conventional paleodose but for 260 nm illumination the value was low by a factor of about 2. Bowman suggests that agreement between the two wavelengths would be an important reliability criterion. The measurements were made on 100-μm quartz grains which had been subjected to standard hydrofluoric acid etching. Careful correction had to be made for pre-dose sensitivity changes in the 110°C peak. One observation made during this work was that whereas the PTTL is reduced to low level by prolonged bleaching (step 4), the level of conventional thermoluminescence remains comparatively high; this suggests that it might be advantageous to use PTTL for dating sediment for which the residual level of conventional thermoluminescence from quartz is a drawback (see Chapter 8). Another observation was that for some samples and the range of transfer wavelengths used (260–400 nm) there is negligible contribution from the traps associated with the 370°C peak; the 'natural' PTTL signal being associated with the 320°C peak (Bailiff, Bowman, Mobbs and Aitken, 1977).

Encouraging results have also been reported by Sasidharan, Sunta and Nambi (1979). However with respect to quartz there are no strong advantages for pottery, except possibly with respect to less sample being required, and these investigations should be regarded as a base from which extension to minerals not datable by conventional thermoluminescence can be made.

There is another dating possibility associated with photoeviction of electrons from their traps. This is by observation of the prompt luminescence due to electrons that go directly to luminescence centres. Although studies by Kristianpoller (1983) were not encouraging, more recent work has shown 'optical dating' to be feasible (see Section 8.2.6.).

6.2.2 OTHER MINERALS

Whereas in the quartz studies mentioned above, the residual PTTL observed after prolonged bleaching was small, usually equivalent to less than 0.5 Gy, in other minerals that have been investigated so far the residual level is much higher, sometimes so high as to make the PTTL relatively insensitive to dose; even after erasure of conventional thermoluminescence by heating, there is likely to be a substantial PTTL signal.

As mentioned earlier, one potential role of PTTL is in circumventing anomalous fading. In fluorapatite and zircon, both seriously prone to this affliction, it is possible to obtain a PTTL signal from traps which empty in the range 500–700°C (Bailiff, 1976; Bailiff *et al.*, 1977); in the case of zircon stored for 6 months there was no detectable fading compared to 15% in one hour for conventional thermoluminescence. Zircon has also been studied by Zimmerman (1979b) and by Mobbs (1978). Despite extensive work, the former found it impossible to reduce the residual PTTL to a level equivalent to less than several hundred Gy, concluding that transfer dating of zircon extracted from pottery was unlikely to be a useful technique.

If the sample is held at low temperature, for example, at −200°C by cooling with liquid nitrogen, then observation of PTTL from acceptor traps which empty below room temperature becomes possible. The PTTL from aragonitic shell and from bone has been observed in this way (Mobbs, 1978), but the size of the residual signal was discouragingly high. Low temperature PTTL has also been used (Mobbs, 1979) with respect to deep sea sediment using a peak at −85°C; the paleodoses evaluated were in good agreement with the results of conventional thermoluminescence measurement.

6.3 Zircon Dating

Zircon (zirconium orthosilicate $ZrSiO_4$) is a minor but ubiquitous mineral in pottery, bricks, tiles, and many types of rock. It has particular interest for thermoluminescence because of its high content of uranium and thorium, ranging from 10 to 100 times the content of typical clay. As a result, the internal dose-rate within a zircon grain is high enough for the gamma contribution from the soil to be negligible; this eliminates the need for knowledge of burial environment and makes possible the dating of samples from museum shelves. The beta contribution from the pottery matrix is not important either. Hence there is elimination of the need for any water content correction with its attendant uncertainty due to changes in climate during antiquity; zircon itself is almost impervious to water because of its compact structure. For the latter reason it seems likely that escape of radon will be unimportant but so far this has not been investigated.

In view of the foregoing it might be expected that use of zircons would be widespread in thermoluminescence dating. However there is another side to the coin. One fundamental difficulty is that a substantial number of zircon grains are afflicted by anomalous fading. Another is that grains are often zoned both in radioactivity and thermoluminescence sensitivity, the two being anticorrelated; such zoned grains have an erroneously low paleodose because the dose has been greatest in regions of the grain where the thermoluminescence sensitivity is lowest.[5] Lastly, there are variations by orders of magnitude in radioactivity and thermoluminescence sensitivity from grain to grain as a whole. These difficulties dictate that measurements are made on single grains and the consequent technical problems are considerable. Nevertheless satisfactory zircon dating of pottery has been developed by Sutton and Zimmerman (1976); however their procedure is extremely time consuming and no general use of the technique has been made.

A particularly powerful application is in authenticity testing when there is suspicion that clandestine irradiation has been used to fake the thermoluminescence signal, that is, the object has been irradiated with gamma rays or high energy X-rays so as to make a modern forgery appear to be old. Such irradiation delivers the same dose to all grains whereas in a genuinely old object, the paleodose carried by zircon grains is very much higher than in other grains (quartz in particular). The same considerations apply with respect to objects which have been legitimately exposed to an uncertain dose of X-rays for the purpose of radiography. It was because of this latter problem that zircon dating was applied to the sandy clay core from the Bronze Horse of the New York Metropolitan Museum of Art

6.3 Zircon Dating

(Zimmerman, Yuhas and Meyers, 1974), to confirm its ancient Greek or Roman origin; whereas the paleodoses for quartz grains and fine grains were the order of 10 Gy or less, the paleodoses for several zircon grains were in the region of 1000 Gy.

6.3.1 THE BASIC TECHNIQUE OF SUTTON AND ZIMMERMAN (1976)

Extraction of zircon grains from pottery sherds is similar to the procedure outlined in Chapter 2 for quartz and feldspar inclusions. After treatment which mainly leaves zircon and quartz, the latter is removed by the heavy liquid separation technique: in methylene iodide, of specific gravity 3.33, the quartz (sg = 2.6) rises whereas the heavy minerals, including zircon (sg = 4.6), sink. Typically, for a sherd of 10 to 40 g, the high density fraction is about 100 μg and of this about half is composed of zircon grains, giving a dozen or more crystals greater than about 60 μm in diameter and hundreds of smaller ones. Using a binocular microscope[6] sizeable zircon grains are picked out with a vacuum tweezer device.

For measurement of thermoluminescence, individual grains are carried on an aluminium 'stub disc' that is placed on the standard heater plate as shown in Figure 6.10. By shielding the photomultiplier from all hot parts except the grain and the 200-μm diameter tip of the stub, it is possible to measure thermoluminescence glowcurves up to 475°C without the use of restrictive colour filters to discriminate against the thermal signal; this is done by using a photomultiplier with a bialkali photocathode (e.g., the EMI 9635QA in common use for dating).[7] After measurement of the natu-

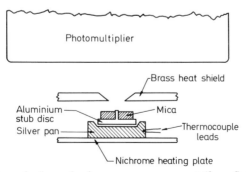

Figure 6.10 Single grain thermoluminescence measurement (from Sutton and Zimmerman, 1976). The grain is placed on the 0.2-mm diameter tip of the stub, which is slightly recessed to help in retention. The diagram is to scale except for the photomultiplier, which should be much larger; the diameter of the aperture in the thermal radiation mask is 2 mm.

ral thermoluminescence, the grain is irradiated with alpha particles by supporting it on thin plastic foil between a matched pair of americium sources. Following a 24-hr delay to allow the decay of short-term anomalous fading, the artificial thermoluminescence is measured; typical glow-curves are shown in Figure 6.11.

The paleodose is evaluated from the ratio between the natural thermoluminescence and the artificial thermoluminescence; this presumes that there is no change in sensitivity between first glow and second glow and also that the growth of thermoluminescence with dose is linear; as discussed in Section 5.4 both of these assumptions are valid for alpha irradiation, as long as, with respect to the first, there is no change in transparency. Because the grain is large compared to the range of the alpha particles the average dose received in the irradiation is affected both by the weight and by the shape of the grain. Therefore the grain must be weighed; this is done with a quartz fibre balance to an accuracy of ±2% in the range of 0.2 to 10 μg; cubical shape has to be assumed. Another assumption is that the surface layer of about 10 μm reached by the alpha particles has a thermoluminescence sensitivity that is representative of the whole grain.

The count-rate from a single grain being far too low for alpha counting, the thorium and uranium contents of the grain are determined by induced fission track techniques. For this, the grain is mounted in a high-temperature epoxy which is then ground to expose the central plane of the grain and it is the radioactivity only in the region of the central plane that is obtained. The average dose-rate to the grain is calculated from the con-

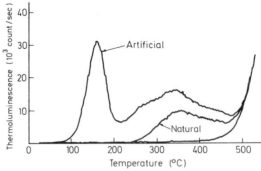

Figure 6.11 Glow-curves from a relatively bright zircon grain from Bronze Age Cypriot pottery (from Sutton and Zimmerman, 1976). The artificial dosage is 10 kGy of alpha irradiation; there was a 24-hr delay before measurement to allow decay of short-term anomalous fading.

6.3 Zircon Dating

centrations deduced, making allowance for escape of alpha particles from the surface of the grain.

In some cases it may be necessary to make a small correction for the beta dose-rate from the pottery and the gamma dose-rate from the soil; this is done by evaluating the paleodose carried by quartz grains in the pottery and making appropriate subtraction. As far as the dose-rate due to the thorium and uranium in the grain itself is concerned, the beta plus gamma component is insignificant. This is because the grain is small compared to the range of a beta particle, and it is only for much larger grains that this component would become significant (see Appendix C).

Finally, indication of the spatial uniformity of thermoluminescence sensitivity is obtained by examining the cathodoluminescence, from the polished section of the grain, using a scanning electron microscope. This assumes that the luminescence centres involved in the prompt luminescence from electron bombardment are representative of the centres involved in the thermoluminescence. Of the grains studied by Sutton and Zimmerman about 75% showed significant zoning of cathodoluminescence, and frequently these also showed inhomogeneous uranium distribution. Comparison of the cathodoluminescence photographs with the fission track maps showed a strong anticorrelation: regions bright in cathodoluminescence tended to be low in uranium and vice versa. Hence for zoned grains much of the natural dosage is being delivered where it has least effect, giving rise to low ages as was indeed found—by up to a factor of four.

The ages calculated for the 15 grains (from four sherds) which were not zoned lay within 20% of the correct archaeological age and the average values for four sherds, each based on about four grains, were on average about 7% low with a scatter about that average of 7%. The sherds were from Roman Britain and from the Bronze Age in Cyprus.

Thus the evidence of the test programme is encouraging, suggesting that as long as there is a 24-hr delay before measurement of the artificial thermoluminescence, upset by anomalous fading is not too serious. However the amount of effort involved in getting a date is formidable. Mention is now made of some less tedious techniques which may be feasible. The use of phototransfer has been discussed in the previous section; although it may circumvent anomalous fading it is unsatisfactory because of the high residual signal, and in any case it does not help with the problem of zoning. The need to reject 75% of the grains at the final stage of measurement is a particularly discouraging aspect of the basic technique just described; this stage cannot be done before measurement of the thermoluminescence because, of course, the thermoluminescence due to electron bombardment would completely swamp the natural signal.

Spatial Mapping of Thermoluminescence Using an Image Intensifier. Besides having to be done late in the sequence of measurements, mapping by cathodoluminescence has the additional disadvantage that the centres responsible for prompt luminescence are not necessarily the same as those responsible for the thermoluminescence. Direct photography of the thermoluminescence is generally out of the question because of the extremely low light levels involved. It is therefore necessary to use a photomultiplicative device which has spatial resolution; the image intensifier tube (IIT) is one such device.

Figure 6.12 shows the basic operation of a high gain IIT. Photoelectrons produced at the input photocathode by a single photon are accelerated across to the first phosphor by an electrostatic field. An axial magnetic field maintains the spatial position of the photoelectron as it strikes the phosphor. The adjacent photocathode converts the photons produced in the phosphor back to photoelectrons. The process is repeated until the last stage when the photoelectrons strike the output phosphor, which is effectively the anode. The gain at each stage is such as to produce between one and ten million photons for every photoelectron leaving the photocathode. Hence each input photon which provides a photoelectron (the efficiency of the photocathode is such that about 25% do so) gives a visible spot of light on the output phosphor, which is photographed.

In the system developed by Templer and Walton (1983) for the thermoluminescence of zircon grains (see Figure 6.13) images of thermoluminescence indicate a resolution of approximately 10 μm in the transverse plane. However, light is collected from a layer corresponding to the finite depth of focus of the optical system, and this is greater than 10 μm (Templer *et al.*, 1985; Walton, Templer and Reynolds, 1984). Neverthe-

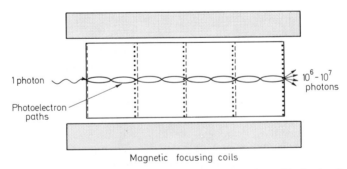

Figure 6.12 Schematic diagram showing operation of image intensifier having four stages of multiplication by phosphor–photocathode sandwiches, accelerating voltage approximately 10 kV per stage: $\cdots$ phosphor, --- photocathode (from Templer and Walton, 1983).

6.3 Zircon Dating

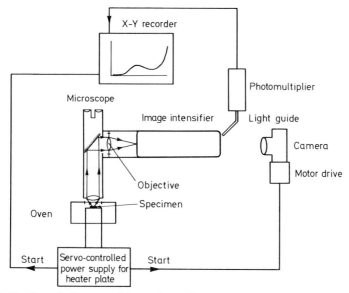

Figure 6.13 Overall diagram of image intensifier system (from Templer and Walton, 1983).

less it does appear possible to detect zoning on a scale small enough to be relevant to alpha particles. Of course when using cathodoluminescence the light comes only from a very thin surface layer, determined by the short range of the low energy bombarding electrons.

In use for dating, the natural thermoluminescence of a group of grains would be photographed and quantitative evaluation of individual intensities would be made. This would then be done for the artificial thermoluminescence from the same group, hence obtaining the paleodose for each individual grain. The photograph of the artificial thermoluminescence would be used to reject grains showing zoning, and only the remainder would go forward for fission track evaluation.

6.3.2 AUTOREGENERATION

In this proposed technique, zoning would not matter because the age is determined by comparison of the natural thermoluminescence with the thermoluminescence that grows again in a given storage time; this 're-grown thermoluminescence' refers to the same peak as used for the natural thermoluminescence and is induced by the grain's internal radioactivity. The need for artificial irradiation and radioactivity measurements is avoided, the age equation being simply

$$\text{Age} = \frac{\text{(Natural TL)}}{\text{(Regrown TL)}} \times \text{(Storage time)}. \tag{6.4}$$

This is, of course, strictly true only for those grains in which the dose-rate contributions during antiquity from the pottery matrix and from the burial soil are negligible. In a trial of the technique (Sutton and Zimmerman, 1979), the storage period was six months, but even so, the level of regrown thermoluminescence was very low—as expected. Measurements were on a grain-by-grain basis because in a group of grains the whole measurement would have been invalidated if within the group there was a grain afflicted by anomalous fading. Ages obtained for grains from a two thousand year old baked soil ranged from being correct to being low by a factor of six, there being rough correlation with the amount of anomalous fading observed in one day.

Even with the special low-level detection system built, measurement of the regrown thermoluminescence was at the limit of feasibility; apart from severe demands with respect to suppression of spurious thermoluminescence there was the basic limitation due to the small number of photons available. Another possibility is to use the low temperature peak at $-120°C$ which is at least an order of magnitude more sensitive. The proposed method (Mobbs, 1978; Templer, 1985a) is to allow regrowth of that peak while cooled by liquid nitrogen ($-196°C$) and to infer the regrowth of the high temperature peak by measuring the relative sensitivities. The age equation would be

$$\text{Age} = \frac{\text{(Natural TL)}}{\chi_{HT}} \times \frac{\chi_{LT}}{\text{(Regrown TL)}} \times \text{(Storage time)}, \tag{6.5}$$

where χ_{HT} and χ_{LT} are the sensitivities of the high temperature and low temperature peaks. This assumes the degree of zoning is the same for both peaks, which may not be the case.

It has also been suggested that for quartz the 110° peak could be used in this way (McKerrell and Mejdahl, 1980), the grains being remixed into the pottery matrix for storage (Liritzis and Galloway, 1982). The 140°C peak in potassium feldspar is another possibility, more attractive because of its quite high internal radioactivity.

6.4 Kinetic Methods

These are based on subtle interpretation of the natural glow-curve using information contained in the temperature region below that at which the associated traps have sufficient stability for linear accumulation during

6.4 Kinetic Methods

antiquity. As with natural regrowth the need for explicit evaluation of dose-rate is avoided.

In a shallow trap which at burial temperature has a short electron lifetime τ the electron population reaches an equilibrium level such that the rate of escape is equal to the rate of filling by ionizing radiation. The level reached is proportional to the rate of filling, which is of course determined by the dose-rate. Hence the level reached can be used as a measure of this. Since the thermoluminescence peak height G is proportional to the electron population in the trap and τ^{-1} is the probability of escape, the equation determining equilibrium is

$$\frac{1}{\tau} G_{eq} = \chi \dot{D}, \qquad (6.6)$$

where χ is the thermoluminescence per unit dose and $\dot{D}$ is the effective dose-rate. Hence the latter can be evaluated by measuring G, χ and τ. Since τ is strongly dependent on temperature, changing on the order of 10 to 20% per °C, a good estimate of burial temperature is essential; the age accuracy obtainable is dependent on the reliability with which such estimate can be made, and this varies with circumstances.

In practice the level of thermoluminescence is usually too small to be measurable in the glow-curve region where trap lifetimes are short enough for equilibrium and an intermediate peak is used instead. For such a peak the thermoluminescence is not yet in equilibrium, but assuming first-order kinetics, it is building up according to

$$G = G_{eq}(1 - \exp(-t/\tau)), \qquad (6.7)$$

where G_{eq} represents the peak height that would be reached for $t \gg \tau$ so that its value is as given by equation (6.6). If G_{HT} is the height of the stable thermoluminescence peak used for dating, then since

$$G_{HT} = \chi_{HT} \dot{D} t,$$

we have from (6.6) and (6.7)

$$\frac{G}{G_{HT}} = \frac{\chi \tau (1 - \exp(-t/\tau))}{\chi_{HT} t}. \qquad (6.8)$$

As used in quartz inclusion dating (Langouet et al., 1979; Langouet, Roman, and Gonzales, 1980) τ was evaluated empirically using known-age samples, a value of 4500 yr being obtained; hence by measuring the ratio of peak heights and the ratio of sensitivities, the age t can be obtained from Equation 6.8 for samples of unknown age. It is of course essential that known and unknown samples experienced the same burial temperature.

Another way of utilizing a peak of intermediate lifetime is by using elevated temperature to simulate prolonged burial (Charalambous and Michael, 1976; Charalambous et al., 1982). Successive trials of irradiation at elevated temperature, employing different times and temperatures, are made until the same peak ratio as in the natural glow-curve is obtained. Suppose the combination that gives the natural ratio is a temperature of T' for a time t'. Then it follows from Equation 6.8 that

$$t'/\tau' = t/\tau \qquad (6.9)$$

where τ' is the lifetime of the intermediate peak at T'. For first-order kinetics the expression for lifetime is $s^{-1}\exp(E/kT)$, so that (6.9) may be written

$$\frac{t}{t'} = \frac{\exp(E/kT)}{\exp(E/kT')}, \qquad (6.10)$$

where T is the burial temperature, both T and T_1 being in degrees absolute. Hence by determining the trap depth E for the intermediate peak and finding the paired values t' and T', the age can be evaluated from Equation 6.10. As with the previous method there is strong dependence on burial temperature T.

Both these methods are highly attractive in their avoidance of dose-rate determinations. However they do demand that the sample is well-behaved in following precisely the predictions of theory; in the author's experience this is often not the case with archaeological samples and it is advisable to get experimental confirmation of good behavior before proceeding with any particular sample.

6.5 Feldspar Dating

It is through historical circumstances that the use of feldspars, widely present as crystalline inclusions in pottery and rocks, is included in the chapter entitled Special Methods; it seems that in the future it will be regarded as a standard procedure along with quartz dating and fine-grain dating. Its development was retarded because of the strong anomalous fading reported for feldspars extracted from lava. Subsequently, there were indications that this did not apply to feldspars of granitic origin,[8] and through work at the Nordic Laboratory for Thermoluminescence Dating in Denmark, it has been found that for feldspar grains extracted from Scandinavian pottery and burnt stones, the fading is not serious (Mejdahl 1983); also, by using a high temperature peak, at 650°C, fading difficulties can be avoided even in feldspars from lava (Guérin and Valladas, 1980).

6.5 Feldspar Dating

There are also the indications from results of sediment dating discussed in Chapter 8 that fading of feldspar thermoluminescence is not a serious source of error in that application, at least for samples of the last 100,000 yr.

So it may be that feldspar is really the Cinderella of the standard techniques. Its thermoluminescence is 10–50 times brighter than that of quartz, and its dose range is substantially greater—extending to around 2000 Gy. Furthermore, in the case of potassium-rich feldspars the annual dose in a large grain originates predominately from within the grain itself thereby substantially reducing dependence on the gamma dose from the soil and water content, although not to the same degree as zircon; on the other hand, there is the advantage over zircon that when the potassium dose is dominant, any possible effect due to radon escape is minimized. In thermoluminescence dating, discussion is usually only in terms of the major types—*alkali feldspars* (potassium-rich or sodium-rich) and *plagioclase feldspars* (calcium-rich); this rather simplistic approach arises because these are what it is practical to obtain as separate fractions from the sample of pottery (or stone). In reality there is a wide spectrum of compositions principally based on varying proportions of *orthoclase* ($KAlSi_3O_8$), *albite* ($NaAlSi_3O_8$) and *anorthite* ($CaAl_2Si_2O_8$); the two former are predominant in alkali feldspars and the latter in plagioclase feldspars; other elements, such as magnesium and iron may also substitute for the K, Na, or Ca. The upper limit to the potassium content is determined by the composition of orthoclase, which has a content of 14% K; the content of the alkali feldspars extracted from samples is more commonly in the range 5–10%.

The use of the high temperature peak in plagioclase feldspar for dating lava flows is discussed in Chapter 7.

6.5.1 TECHNIQUE (MEJDAHL AND WINTHER-NIELSEN, 1982; MEJDAHL, 1983)

The present procedure employed at the Nordic Laboratory is as follows. After crushing, most of the clay matrix (in the case of pottery) is removed by wet grinding and repeated washings in water and dilute hydrochloric acid. This leaves a residue of some remaining clay particles and a mixture of quartz, feldspars, and heavier minerals. As for zircon dating, the next step is separation of these minerals by successive suspension in heavy liquids[9] of different densities; the procedure employed is indicated in Figure 6.14. It is not possible to separate quartz (sg 2.65 g/cm³) and plagioclase feldspar (sg 2.62–2.72) in this way but the thermoluminescence of the fraction containing them is dominated by feldspar be-

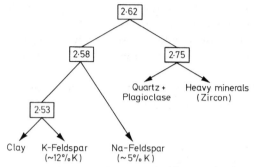

Figure 6.14 Feldspar dating: flow chart for heavy liquid separation (as used in the Nordic Laboratory, 1984). The numbers shown indicate specific gravity (g/cm^3); the left branch at each step represents the fraction which floats and the right branch the fraction which sinks. The specific gravity required at each step is obtained by mixing tetrabromoethane (sg 2.96) and dipropyleneglycol (sg 1.03) in appropriate proportions.

cause of its greater brightness unless the proportion of quartz is high. A quartz-only separate can be obtained by dissolving away the feldspar with hydrofluoric acid, as in the quartz inclusion technique.

It is important that the alkali feldspars should be separated into potassium-rich (10–13% K) and sodium-rich (4–6% K) fractions. This is because the thermoluminescence sensitivity of the former is often as much as four times that of the latter, that is, there is correlation between potassium content and thermoluminescence sensitivity. If the alkali feldspars are not split into these fractions but the paleodose and the potassium activity are measured for the alkali feldspars as a whole, then the age obtained will tend to be too high.[5]

The separates are next sieved into a number of grain size fractions within the range 100 μm to a few millimetres. This is necessary because the correction for attenuation of the beta dose-rate coming from the matrix depends on grain size and because in the case of alkali feldspars there is the internal dose-rate from within the grain and this is dependent on size too, see Figure C.2. Grains greater than 0.5 mm are crushed down before measurement; all sizes are washed in 10% hydrofluoric acid and then in 10% hydrochloric acid. This etches away the outer alpha-irradiated layer of the grains, the hydrochloric acid removes any fluorides formed by the first washing. These washings also eliminate any spurious thermoluminescence induced by the grinding.

The paleodose for each size fraction is measured by the standard additive dose method with correction for supralinearity by means of second-

6.5 Feldspar Dating

glow measurements; preheating for 10 sec at about 300°C is employed before all thermoluminescence measurement so as to remove all but the stable part of the glow-curve. Subsequently the samples are tested for fading by irradiation and storage over a period of several weeks.

Thermoluminescence dosimetry methods are used for the soil gamma dose-rate and the matrix beta dose-rate. The potassium content of the alkali feldspar grains is measured by beta counting, being simpler than flame photometry and at least as precise (Bøtter-Jensen and Mejdahl, 1985).

6.5.2 INTERNAL BETA DOSE

As an illustration let us consider a 1-mm diameter grain of potassium feldspar for which the K content is 10% embedded in a sherd for which the contents of K, Th, and U are, respectively, 1%, 10 ppm, and 3 ppm, the sherd being buried in soil for which the combined gamma plus cosmic annual dose is 1 mGy/a. From Table 4.4 we see that the beta doses in an infinite matrix will be 8.3 and 1.55 mGy/a for grain and sherd, respectively. According to Figure C.2, the average dose-rate in the grain due to its inherent K content is 34% of the infinite matrix value, that is, 2.8 mGy/a. As far as the potassium beta dose-rate is concerned, the average value in the grain due to the potassium in the sherd will be $(100 - 34)\%$ (i.e. 66%); that this is so can be seen by considering the case of a grain having the same K content as the sherd. The attenuation for the beta contributions from thorium and uranium is a little stronger because there is a greater proportion of low energy particles and 62% is appropriate as the attenuation of the overall sherd beta dose for the concentrations given; thus the average beta dose-rate received by the grain from external beta particles is $1.55 \times 0.62 = 0.96$ mGy/a. Hence the total dose-rate is $(2.8 + 0.96 + 1.00) = 4.8$ mGy/a, so that the dose rate from the grain's inherent K content is 58% of the whole.

The validity of this approach has been demonstrated on a number of sites in Scandinavia (Mejdahl, 1983); Table 6.1 gives results obtained with quartz and feldspar grains from a pottery fragment found in a Viking Age settlement in Denmark archaeologically dated as A.D. 950–1050. We see from the table that the paleodose is indeed higher for the large alkali feldspar grains and that the six ages obtained from the different fractions are in satisfactory agreement, the standard deviation of the scatter being only 4%. One other pottery sample and two stones were dated from the same site giving ages of 949, 887, and 995 years leading to a mean date for the site of A.D. 1040.

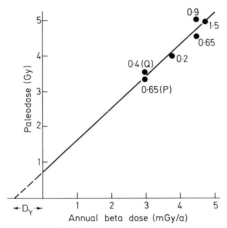

Figure 6.15 Isochron dating of Viking pottery making use of the difference in potassium content between grains of potassium feldspar and grains of plagioclase and quartz (redrawn from Mejdahl, 1983). The paleodose and beta activity are measured for the various fractions of grains extracted from the pottery: the numbers indicate the grain size (in mm); Q is quartz; P, plagioclase; the remainder are potassium feldspar. The age is derived from the slope of the line; hence it is not necessary to know the soil gamma dose.

6.6 Subtraction Dating

Figure 6.15 shows, for the Viking pottery of Table 6.1, a plot of paleodose versus the beta dose-rate calculated for the different grain fractions, that is, the sum of sherd and grain contributions. The straight line fitted to these points has a slope which corresponds to a paleodose increase of 0.92 Gy for each dose-rate increment of 1 mGy/a. We have here an incremental form of the basic age equation, and the implied age is $(0.92/0.001) = 920$ years, in excellent agreement with the average age of Table 6.1. However in plotting the figure we did not use the measured value of the gamma dose-rate and so by this technique, which is essentially on the same principle as the subtraction technique of Section 2.3, the sample can be dated without knowledge of burial circumstances. The soil gamma dose-rate can incidentally be derived from the figure; it is the intercept of the line on the dose axis and its value is 0.77 mGy/a.

In Section 2.3 the comment was made that the accuracy attainable with a subtraction technique was inevitably rather poor because it was a matter of the difference between two quantities of comparable size each with significant uncertainties. In the method of Mejdahl just described—which incidentally is similar to the isochron technique used in geological dating—more than two quantities are involved and consequently there is

6.6 Subtraction Dating

TABLE 6.1
Dating of Mineral Fraction from Viking Pottery Fragment[a]

Mineral and grain size (mm)	Annual dose[b] (mGy/a)			Paleodose (mGy)	Age (yr)
	Sample (beta)	Grain (beta)	Total		
Quartz					
0.3–0.5	2.98	0	3.76	3360	894
Plagioclase					
0.5–0.8	2.69	0.28	3.75	3570	952
Alkali-feldspar					
0.1–0.3	3.26	0.51	4.55	4050	890
0.5–0.8	2.69	1.76	5.23	4600	880
0.8–1.0	2.42	2.03	5.23	5080	971
1.0–2.0	1.93	2.77	5.48	5000	912
				average	917
				(SD	4.0%)

[a] From Mejdahl, 1983.
[b] The gamma-plus-cosmic annual dose was 0.78 mGy/a.

improvement in accuracy. Clearly, samples that contain large grains of potassium-rich feldspar are a very attractive proposition for thermoluminescence dating, at least in Scandinavia, for it remains to be seen to what extent anomalous fading intervenes elsewhere. Apart from coarse pottery, burnt stones are good candidates for this technique.

The fine-grain–quartz inclusion combination has already been discussed in Chapter 2. Another category of subtraction dating which has been found promising (Plachy and Sutton, 1982) is subtraction of the paleodose found in quartz grains in a low radioactivity burnt rock, such as quartzite or sandstone, from that found in quartz grains from a nearby high radioactive burnt rock, such as granite; unfortunately evaluation of the dose-rate to the quartz grains from the radioactive components in the granite turns out to be complex on account of inhomogeneity; another drawback is that the method is upset if the gamma dose-rate is different at the two locations.

Another possibility for subtraction is between large grains, of equal size, of alkali feldspar and quartz (McKerrell and Mejdahl, 1980). Another is between the paleodose evaluated for fine-grains evaluated by conventional thermoluminescence and that evaluated by the 110°C predose technique; because of very low pre-dose effectiveness of alpha particles the difference in the two paleodoses is predominantly due to the alpha activity.

6.7 Gamma-Thermoluminescence

The emphasis in the previous three Sections has been on elimination of the need to know the gamma dose-rate. We now come to a method (Schvoerer *et al.*, 1982) which has the opposite philosophy—that the gamma dose-rate is the simplest component to measure (using a thermoluminescence capsule), and that within limits, the total dose-rate can be derived from it thereby dispensing with the need for any other radioactivity measurements. The method relies on the radioactivity of the sample being the same as that of the surroundings, as for example in the case of a pottery sherd from a large deposit of similar pottery, a brick from a kiln, or looking ahead to Chapter 8, a sample of sediment.

Referring to Table 4.5, which gives the components of the annual dose for 'typical' pottery and soil, we see that for fine-grain dating, the ratio of total dose to gamma dose is 4.2 for potassium, 3.7 for thorium, and 5.6 for uranium; the average is 4.5 and unless thorium and uranium are dominant in abnormal proportions, or the alpha effectiveness differs strongly from that assumed, the value for actual samples will not be far different. Schvoerer *et al.* developed an analysis defining the limits of composition within which it is to be expected that the total dose-rate can be reliably evaluated from the gamma dose-rate and the alpha effectiveness k. Within these limits (1–3% for K, 10–20 ppm for Th, and 1–5 ppm for U) the value of the ratio Γ between total dose and gamma dose is given by the relation

$$\Gamma = 11.7k + 2.7. \tag{6.11}$$

The maximum deviation of the actual value from the value given by this relation for extremes of composition within the limits quoted depends on k, being about 4%, 11%, and 16% for k-values of 0.1, 0.2, and 0.3, respectively; below $k = 0.8$, the possible deviation increases and the authors suggest restriction to samples having k-values in the range 0.08–0.15. No correction is made for water content, it being assumed that there is no difference between sample and soil; similarly there is no serious upset due to radon escape as long as it is assumed that the degree of escape is the same for both. Application has been made to a variety of sites, comparison being made with conventionally derived thermoluminescence dates or archaeological indications; within the restrictions noted, basically that the sample and surroundings have the same radioactive composition, the results are encouraging.

Of course the dose-rate recorded by a capsule includes a contribution from cosmic rays and this needs to be subtracted in order to obtain the gamma dose-rate, although the authors do not explicitly insist on this. For a burial depth of one or two metres at low altitude and cosmic dose-rate is around 0.15 mGy/a (see Section 4.2.2) and for soil of 'typical' composition (1% K, 10 ppm Th, and 3 ppm U) this amounts to 14% of the gamma dose-rate; soils having a gamma dose-rate lower by a factor of two are not uncommon and the cosmic rays then contribute nearly a quarter of the capsule dose.

Another approach that avoids any laboratory evaluation of radioactivity is through on-site measurements with a portable gamma spectrometer (Section 4.3.4); this has the advantage that the individual concentrations of the radioelements are obtained, as well as the cosmic ray dose-rate. However there is of course the same limitation that sample and surroundings must have identical radioactivity and water content.

Technical Notes

[1] Initially there were reports that pre-dose dating was upset by hydrofluoric acid etching but this does not now seem to be generally accepted. Of course etching is not so important in pre-dose dating because the alpha effectiveness is substantially lower than in conventional thermoluminescence.

[2] As pointed out by Chen (1979) the model as it stands is not compatible with the experimental evidence that the thermoluminescence is proportional both to the pre-dose and to the size of the test-dose. It is necessary to assume the existence of an additional trap which competes for electrons after their thermal eviction from T.

[3] In the nomenclature introduced by Chen, S_{00} is used for the sensitivity immediately after firing in distinction to S_0, the value as measured in the laboratory. For an archaeological sample S_0 may be greater than S_{00} due to ambient activation or it may be less due to quenching by the natural dose.

[4] Although Bailiff (1979) obtained no worthwhile improvement in the ratio (S_N/S_0) by using various colour filters for a quartz sample (for which S_N/S_0 was only about 3) this does not rule out possible advantage for polymineral samples; note however that feldspar is likely to have a substantial ultraviolet component. For the same sample he notes that S_0^{rev}, the residual value of S_0 after ultraviolet reversal, is lower than corresponds to the transfer out of L of holes associated with the archaeological dose which have undergone ambient activation. Thus if $(S_N - S_0^{rev})$ is used for the numerator of Equation 6.2, the age obtained may be erroneously high; however this sets a tighter upper limit to the age than is obtained by ignoring S_0 altogether. For some samples S_0^{rev} is not much less than S_0 and there is then firm evidence that the amount of ambient activation has been unimportant.

[5] This can be seen by considering a sample having two parts, in one of which the thermoluminescence sensitivity is χ_1 with an internal dose-rate D_1, and in the other χ_2 and D_2. In the case of zircon grains these are zones of the same grain; in the case of alkali feldspars the two parts are the potassium-rich and sodium-rich

fractions; in the case of calcite they are different regions in the sample from which the grains are obtained.

If the two parts are equal in size then the paleodose evaluated for the sample will be

$$P = \frac{\chi_1 P_1 + \chi_2 P_2}{\chi_1 + \chi_2},$$

where P_1 and P_2 are the paleodoses for the individual parts (which cannot be evaluated without separation). Assuming for the purpose of argument that there is no other dose-rate than the internal one the age obtained for the sample will be

$$A' = \frac{P}{\frac{1}{2}(D_1 + D_2)},$$

whereas the true age, which would be obtained from measurement on the individual parts, is

$$A = \frac{P_1}{D_1} = \frac{P_2}{D_2}.$$

If for zoned zircons we take as example $D_1 = 10D_2$, that is, the uranium and thorium contents of one zone is ten times that of the other, then $P_1 = 10P_2$. Suppose there is anti-correlation with thermoluminescence sensitivity and that $\chi_1 = 0.3\chi_2$. Then the paleodose found experimentally will be

$$P = \frac{(0.3 \times 10 + 1)\chi_2 P_2}{(0.3 + 1)\chi_2} = 3.1P_2,$$

and so

$$A' = \frac{3.1P_2}{5.5D_2} = 0.56A.$$

Thus for the values given the apparent age obtained will be approximately half the true age. Whenever there is anti-correlation erroneously low apparent ages will be obtained, though of course the extent of the error depends on the degree of anti-correlation; calcite is another material in which this can occur (see Section 7.5).

Alkali feldspar exhibits positive correlation. Since the potassium-rich fraction has about twice the potassium content of the sodium-rich fraction, we have $D_1 = 2D_2$, and hence $P_1 = 2P_2$. According to Mejdahl (1985) the sensitivity ratio of the two fractions is typically around 4, and hence $\chi_1 = 4\chi_2$. Thus we obtain

$$P = \frac{(4 \times 2 + 1)\chi_2 P_2}{(4 + 1)\chi_2} = \frac{9}{5}P_2,$$

and so

$$A' = \frac{1.8P_2}{1.5D_2} = 1.2A,$$

that is, the apparent age is 20% higher than the true age.

⁶ Recent work (Templer, 1985b) suggests that there is risk of bleaching at this stage, even though subdued amber light is used.

⁷ Spectral studies of zircon thermoluminescence have been made by various workers (e.g., Jain, 1978; Iacconi, 1980; Templer, 1985b). In respect of dating the important emissions are centred on 490 and 550 nm (the bialkali photocathode quantum efficiency falls to just below 10% for the latter wavelength). In rejection of the thermal signal there is advantage to be gained by using a light blue colour filter (e.g., Corning 4-96).

⁸ The implication is that the anomalous fading observed in feldspars of volcanic origin is the result of the comparatively rapid cooling they experienced. However, given the wide range of feldspar types it seems likely that the composition could also be relevant, as well as the impurity concentration (which will be lower in slowly-cooled crystals). The main slowly-crystallized forms are all alkali feldspars: microcline (high K), perthite (more K than Na), antiperthite (more Na than K) and albite (high Na). The main rapidly crystallized forms of potassium feldspars are orthoclase (14% K) and sanadine, of sodic feldspars are analbite and monalbite, and of plagioclase feldspars are oligoclase, andesine, labradorite, bytownite, and anorthite (these five being listed in increasing order of content of pure $CaAl_2Si_2O_8$, ranging from 10 to 30% for oligoclase to 90 to 100% for anorthite). The types in which anomalous fading has been reported are sanidine, andesine, labradorite, and bytownite, all by Wintle (1973).

⁹ An alternative to the rather toxic liquids traditionally used is sodium polytungstate powder, obtainable from: Metawo, Falkenried 4, D-1000 Berlin 33, West Germany.

CHAPTER 7

BEYOND POTTERY

Pottery was not made much before ten thousand years ago and it is earlier, in the paleolithic, beyond the 40,000-yr range of radiocarbon, that thermoluminescence makes its most important contribution. In this penultimate chapter we deal with burnt flint, burnt stones, volcanic lava, and unburnt stalagmitic calcite, and in the final chapter with unburnt sediment; with these two latter, and sediment particularly, the question arises as to how well the thermoluminescence clock was set to zero at $t = 0$. Given the importance of paleolithic application it might seem inappropriate that types of material relevant to it are reached only in this chapter; however, although the foregoing chapters were concerned with pottery explicitly, nearly all of their contents has relevance to non-pottery materials, and in any case pottery can be regarded as the proving ground on which thermoluminescence was established before setting off into the unknown. Of course too, although there was no pottery in the paleolithic there was baked clay in the form of hearths and ovenstones, and occasionally figurines.

7.1 Burnt Flint

Flint[1] is the specific name given to the nodules and bands of *chalcedony* that occur in chalk but as is common it should here be taken to refer to other occurrences of chalcedony, that is, to *chert* in general. Chalcedony is a form of quartz and it is therefore to be expected from the experience with the quartz inclusion technique in pottery that it will be reliable for thermoluminescence dating. Flint implements (blades, scrapers, arrow heads, etc.) were in widespread use by paleolithic man, and in addition to the implements themselves, there remain some of the flakes, chips, and cores associated with manufacture. Some of these artefacts were heated accidentally by falling into the fire, some were heated deliberately in order to improve the 'knapping' properties. Besides use for dating, thermoluminescence is an effective technique for determining whether a given piece has in fact been heated, simply by determining the plateau and paleodose (Melcher and Zimmerman, 1977); it seems that visual appearance (reddening, vitreous luster, potlids, crackling, etc.) is not always reliable. With more complication a thermoluminescence estimate can be made of the temperature that was reached (Valladas, 1983), as is also the case for the electron spin resonance technique (e.g., Robins *et al.*, 1981).

7.1.1 SAMPLE PREPARATION

For the first thermoluminescence dating of flint (Göksu *et al.*, 1974), the measurements were made on thin slices; this was done to avoid spurious thermoluminescence. However with slices there are severe experimental complications associated with the need to make allowance for transparency effects[2] (except when, as was the case in this first application, the alpha contribution to the annual dose is unimportant). There was little further work until it was found that the spurious thermoluminescence associated with crushing could be suppressed by washing in dilute hydrochloric acid (Valladas, 1978). Such washing also removes the 'regeneration thermoluminescence' which is otherwise liable to occur as a result of storage (Göksu and Fremlin, 1972); it seems likely that this is due to the presence of residual carbonate in the flint.

Crushing a flint is a violent business and it would not be surprising if there were other adverse effects too. The shock might cause eviction of electrons from traps or even an increase or decrease in the number of traps; the absence of such effects can be tested by administering a known gamma dose to an uncrushed flint from which the natural thermoluminescence has been erased by heating and then making thermoluminescence evaluation of the administered dose, employing the same procedure as used for dating. Excellent agreement has been reported (Huxtable, 1982)[3] using the same technique for crushing as used for pottery, that is, squeezing in the jaws of a vice as gently as possible; use of a steel percussion mortar is also satisfactory (Bowman and Sieveking, 1983). The resulting

7.1 Burnt Flint

grains being washed in dilute hydrochloric acid or acetic acid; it is of course necessary, as for pottery, to do this crushing after sawing off the beta transition zone (2 mm of the surface) in red light. For adequate suppression of spurious thermoluminescence the usual measures are still needed in addition to the acid washing, that is, evacuation of the oven before admission of continuously flowing dry argon. Elimination of spurious thermoluminescence can be checked in three ways: (i) a good plateau in the ratio between natural thermoluminescence and natural-plus-artificial thermoluminescence; (ii) zero thermoluminescence in a flint from which the natural thermoluminescence has been erased before crushing; and (iii) agreement between the paleodose evaluated from coarse grains with that evaluated from fine grains. This latter is a good test because fine grains are much more prone to spurious thermoluminescence because of their greater surface-to-volume ratio. In pottery the fine grains carry a higher paleodose than quartz inclusions because alpha particles only penetrate the outer surface of the latter; in the case of flint the observed agreement[4] between the two paleodoses indicates there is no radioactive differentiation between grains sizes. This is as might be expected since the rods and platelets of quartz comprising the flint[1] are 'alpha-thin' and in any case the fine-grains are to some extent produced from coarse grains during the crushing. Besides being important in testing for suppression of spurious thermoluminescence fine-grain discs are of course necessary for measurement of alpha effectiveness in the same way as for pottery; the difference here is that the alpha particles contribute to the coarse-grain annual dose too.

7.1.2 STABILITY

The natural thermoluminescence glow-curve from flint (Figure 7.1) is a broad peak centered on about 370°C, as for some quartz. In any given sample assurance of adequate stability is obtained from the plateau test but in considering how far back in time flint dating might be used it is useful to know the storage lifetime of the associated trapped electrons, assuming this to be the same for all flints. Estimates indicate a lifetime in excess of 50 million years (Wintle and Aitken, 1977; Bowman, 1982b), even if the burial temperature was as high as 20°C; storage of irradiated samples over periods of some months gives no evidence of any anomalous fading (see Section 3.4). In practice the time range of flint dating is limited by the onset of saturation; this varies from type to type but commencement of non-linear response may be expected somewhere in the region of 100 to 500 Gy. For an annual dose of 1 mGy/a the corresponding age limits are 100–500 thousand years but of course there are substantial

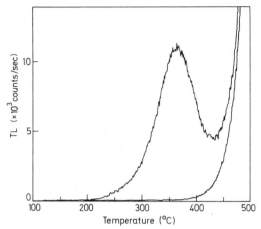

Figure 7.1 Natural glow-curve for burned flint from the Lower Paleolithic site at Belvédère, Holland (measurement by J. Huxtable).

variations in the values of the annual dose from site to site; actual dates, archaeologically acceptable, have now been determined for several sites in the region of 200 thousand years ago (see, for example, Huxtable and Aitken, 1985).

Contrary to early indication there is evidence that significant bleaching of thermoluminescence by sunlight can occur even in whole flints (Huxtable, 1981; see also Bowman and Sieveking, 1983); it had been expected that the material would be insufficiently transparent for this to occur beyond the depth of the surface layer that is discarded in sample preparation. Hence flints should be transferred to a black bag as soon as excavated; for flints that have not been kept in the dark there is the possibility of making an experimental check as to whether bleaching has occurred but it is highly undesirable for the laboratory to have this extra complication. Of course, flints will have been exposed to light during their utilization by ancient man but at that time there is no accumulated thermoluminescence to be bleached; in principle there is the possibility of light-induced thermoluminescence being acquired at that stage but from studies of the phototransfer process in quartz it seems an unlikely one.

7.1.3 RADIOACTIVITY

As is the case for quartz inclusions in pottery, the radioactivity of flint is low but by no means always negligible. Typically[5] the alpha count-rate for a 42-mm diameter screen is in the range 0.5–3 counts/ksec (corresponding to 0.3 to 1½ ppm of uranium, assuming the thorium content to be much lower)[5] and the potassium content in the range 0.03–0.1%. Taking the range of alpha effectiveness to be 0.05–0.15 these limits give internal dose-rates in the range 0.05–1 mGy/a. Since typical values of soil gamma dose-rate lie in the range 0.5–1.5, we see there is a range of circumstances extending from the internal dose being negligible to it being dominant.

For flints in which the internal dose-rate is significant, consideration needs to be given to possible effects due to non-uniform distribution of radioactivity. As in the case of zircons and calcite, if high radioactivity correlates with low thermoluminescence sensitivity then the age obtained will be erroneously low. As yet there is no indication that this is an actual source of error[6]; in any case such error would only be serious when the internal dose is dominant. It is more common for the soil gamma dose to be the main contribution, and the primary uncertainty is then the degree to which a reliable estimate can be made of the average value of soil water content over the flint's burial period keeping in mind possible changes in drainage conditions and climate; the possibility of radon escape and other forms of radioactive disequilibrium need to be considered and high-resolution gamma spectrometry measurements on soil samples in the laboratory are desirable in addition to on-site measurements by thermoluminescence capsule or portable gamma spectrometer. In the flint itself there is negligible uptake of water and because of its compact structure there is little scope for radon escape and so forth; so long as checks made for uniformity (by means of fission track mapping or image intensifier observation of the thermoluminescence) are positive, best accuracy is to be expected when the internal dose is dominant.

7.1.4 SAMPLE SELECTION

The primary necessity of sample selection is that at the time of its involvement with ancient man the flint was sufficiently heated to reset the thermoluminescence to zero. Visual appearance gives some guide but the final arbiter is the thermoluminescence plateau obtained in the course of dating. Inevitably it is small fragments that are most likely to be best burnt but for them it is difficult to remove the 2-mm surface layer and still be left with enough material for thermoluminescence and radioactivity measurements; for the latter one needs a piece approximately 5-mm thick by 30-mm across so that the initial thickness has to be[7] at least 10 mm.

Six well-burnt flints of adequate size are desirable for a reliable date but there are many sites for which the laboratory has to make do with less. Despite this limitation the technique is a dating tool of major importance for paleolithic research.

7.1.5 APPLICATION

Published results at present extend from flints burned only 2000 yr ago at the Alice Boer site in Brazil (Beltrao et al., 1982) to flints burned nearly 300,000 yr ago at the Lower Paleolithic site of Belvédère in Holland (Huxtable and Aitken, 1985). Other dating reports for flint include the following sites: Pech de l'Aze, France (Bowman et al., 1982); Combe Grenal, France (Bowman and Sieveking, 1983); Cariguela Cave, Spain (Göksu et al., 1974); Hengistbury Head and Longmoor in England, Abri Vaufrey, Biache and Seclin in France, Sclayn and Mesvin in Belgium, Pontnewydd in Wales, La Cotte in Guernsey (Aitken, Huxtable and Debenham, 1985; Huxtable and Jacobi, 1982; Huxtable and Barton, 1983); Etiolle, France (Valladas, 1978); Terra Amata, France (Wintle and Aitken, 1977b).

7.2 Burnt Stones

In pre-pottery cultures it was common practice to use 'pot-boilers' for cooking. Stones, or balls of clay if stones were not plentiful, were heated in a fire and then placed in the food container, which might be a skin bag or a stone trough; of course, when pottery vessels became available these could be placed directly on the fire and pot-boilers became obsolete. Because the stones tended to shatter when placed in the water re-use was not convenient; consequently burnt stones are plentiful on sites where this practice had been employed and often there are large mounds of these fragments, for instance, on sites of the first millennium B.C. in the Orkney Islands (thermoluminescence dates by Huxtable et al., 1976). Besides this culinary source there are stones which were used in the construction of fireplaces or found their way into the fire accidentally.

Obviously there are a very wide variety of rock types that may be encountered and also a corresponding variation in component minerals. Large grains are much more likely than in pottery and problems of radioactive inhomogeneity become more acute. However, this has been put to good use in the dating of granitic stones from Scandinavian sites using the feldspar technique (Section 6.5); the dose-rate in large grains of potassium-rich feldspar has a substantial component from the inherent radioactivity of the grain thereby weakening dependence on external radioactivity, with the advantages already discussed; in a grain of several millimetres the inherent potassium may contribute as much as 90% of the total dose. At the other end of the scale are quartz grains from sandstones. For these the dose-rate is dominated by radioactivity external to the grain and in some cases radioactive mapping of adjacent grains is necessary (Plachy and Sutton, 1982); the fine grain technique may also be used. Other reports on the dating of burnt stones have been given by

Huxtable *et al.* (1972), Wintle and Oakley (1972), Schvoerer *et al.* (1974), Poupeau *et al.* (1976), Ichikawa and Nagatomo (1978), Bechtel *et al.* (1979), Valladas (1979), and Mejdahl (1983).

7.3 Volcanic Eruptions

Use of thermoluminescence for volcanic products, ash, and lava, was a natural extension from pottery dating, but early trials did not succeed because of excessive spurious thermoluminescence and anomalous fading, respectively (Aitken *et al.*, 1968a, Aitken and Fleming, 1971; Wintle, 1973). Although primarily concerned with geology such use is also relevant to archaeology. Ordinarily this is rather indirectly, e.g., when layers of ash or lumps of lava occur on archaeological sites. However there has been one volcanic eruption of very direct archaeological relevance. This was the cataclismic explosion of the Thera volcano on the Aegean island of Santorini which destroyed the highly developed Minoan civilization on that island around 1500 B.C., perhaps causing destruction on Crete also—although this is still being argued. Up to the present, thermoluminescence dating of the Theran volcanic products remains a challenge; early attempts were not successful, but with the encouraging results noted below for volcanic glass there are now better prospects. Also mentioned below is a successful technique for the dating of feldspathic lava (but there was no emission of lava during the Thera explosion). Most lava is deficient in quartz but this mineral is sometimes present in sufficient quantity for dating in other volcanic emissions (e.g., Miallier *et al.*, 1983, 1985).

7.3.1 VOLCANIC GLASS
(BERGER AND HUNTLEY, 1983)

Volcanic glass is one component of airborne ash (*tephra*), the others being rock fragments and mineral crystals. The authors chose glass because of the notoriety of volcanic feldspars for showing anomalous fading and because there could be no doubt that the glass fraction had been heated sufficiently for complete erasure of thermoluminescence. The fine-grain (2–11 μm) technique was used, deposition being made after cleaning the sample of unwanted components.[8] With one exception good agreement with known ages was obtained, five samples from the Seattle region in the age range 450–13000 years being dated. The authors do not report any interference from spurious thermoluminescence as had been experienced in the earlier attempts to date volcanic ash (which had not been given the same pretreatment as here). Whether or not this successful dating of volcanic glass presages success with archaeological glass (see Section 7.4) remains to be seen; an important difference is of course that the former occurs naturally in fine-grain form whereas either slicing or crushing is necessary for the latter, with attendant likelihood of spurious thermoluminescence being induced.

Linearity of thermoluminescence response up to at least 20 Gy is reported, although for the oldest sample, having a paleodose of 500 Gy, the natural thermoluminescence was already in the non-linear region necessitating a modification of the additive dose procedure; this used regeneration after optical bleaching as for sediments, see Chapter 8.

7.3.2 LAVA (VALLADAS, GILLOT AND GUÉRIN, 1979; GUÉRIN AND VALLADAS, 1980)[9]

The lava in which anomalous fading had first been detected (Wintle, 1973) had come from the Chaîne des Puys in the Massif Central of France, the thermoluminescence ages obtained being low by an order of magnitude; absence of quartz from these lavas means that dating has to be based on the feldspar component. When samples from this region were studied using thermoluminescence in the 600°C region of the glow curve, instead of 300°C as formerly, it was found first that a plateau could be obtained, second that any fading in a gamma-irradiated sample stored for a year was less than a few percent, and third that the ages obtained were in agreement with other techniques. These latter were: thermoluminescence using quartz from a granitic inclusion and from pebbles heated by the lava; radiocarbon; uranium-series; and potassium-argon. This is in line with the experimental finding that anomalous fading could be circumvented by means of phototransfer from deep traps (Section 6.2); theoretically it is to be expected that the probability of escape of electrons by tunnelling should be lower for deep traps because the energy barrier to be penetrated is higher.

In order to observe thermoluminescence in this high-temperature region it is necessary to restrict the wavelengths reaching the photomultiplier to the ultraviolet (in the region of 325 nm) in order to avoid it being swamped by the thermal signal. This is done by means of a colour filter,[10] and Figure 7.2 shows the glow-curve for a mineral extract (obtained using similar techniques to those outlined for zircon in Section 6.3). It will be noted that the peak at around 300°C which had been used in the earlier studies barely show in the natural thermoluminescence. In order to obtain a good plateau it is necessary to eliminate this lower-temperature thermoluminescence by pre-heating the sample for a minute before the glow-curve proper is taken; this is done for all glow-curves. The onset of

7.3 Volcanic Eruptions

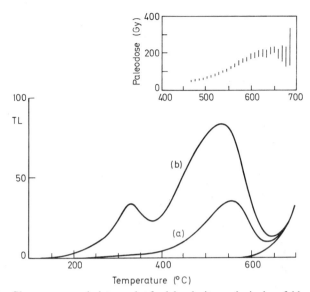

Figure 7.2 Glow-curves and plateau plot for labradorite, a plagioclase feldspar, from lava of the Chaîne des Puys (from Guérin and Valladas, 1980): (a) natural thermoluminescence; (b) thermoluminescence from another portion to which an artificial dose of 160 Gy has been added to the natural dose. The plot above is the paleodose deduced from the ratio of (a) to (b); there is a tendency to level off above 580°C. By pre-heating for 1 min at 560°C before measurement (both for the natural thermoluminescence and for the natural-plus-artificial thermoluminescence) a satisfactory plateau extending down to about 520°C was obtained.

saturation for the high-temperature peak is in the region of 3000 Gy and the useful age range suggested for the technique is 3,000–300,000 yr.

The dosimetry situation of feldspar grains in lava is similar to that of quartz inclusions in pottery as far as alpha activity is concerned, that is, the thorium and uranium contents of the grains are almost negligible by comparison with that in the matrix in which they are embedded. In the comparatively fine-grained lavas used by Guérin and Valladas (1980) the dating was by means of 100-μm grains of plagioclase feldspar, larger grains not being plentiful. The alpha contribution to these grains is substantially more important than with quartz grains because the alpha effectiveness in feldspar is about five times that in quartz; as a result the alpha contribution to the thermoluminescence is about one-third of the total whereas in the case of quartz grains it is typically 10–15%. Rather than etch away the alpha-irradiated layer, the alternative of making an evaluation of the alpha contribution is preferred. This is because of uncertainties in the etching process, as discussed in Appendix C. There are two parts to this evaluation. First the alpha flux incident on the grains while they were embedded in the lava is found by means of fission track mapping. Second the response of these 100-μm grains to a known alpha flux is measured by exposing them to a laboratory source,[11] the grains being vibrated so as to simulate the omnidirectional flux incident from the lava.

7.3.3 CONTACT-BAKED SEDIMENT

An indirect way of dating volcanic eruptions is by means of the soil or rock over which the molten lava had flowed and solidified; for soil, standard pottery techniques can then be used (e.g., Huxtable and Aitken, 1978). Pebbles within such soil can also be used, and in the case of the Chaîne-des-Puys work quartz grains were extracted from such pebbles for measurement (Valladas and Gillot, 1978). Another indirect method employed in this work was the use of large quartz grains from a granite enclave that had been carried along in the lava; special evaluation of the internal dose profile was necessary (Gillot *et al.*, 1978). All of these rely on the chance finding of suitable material heated by the lava; the two latter were from lava flows of particular interest because of their reversed magnetization (see Section 7.7) and many days of painstaking search were needed to find the samples.

7.4 Meteorites and Craters, Metallurgical Slag, and Glass

7.4.1 METEORITES

While in space meteorites are subjected to a substantial cosmic-ray dose, approaching 50 mGy/a; hence, unless modified by some event the thermoluminescence will be in saturation, at least for those peaks associated with traps deep enough to have long electron lifetimes. Heating during the rapid transit through the earth's atmosphere causes a fusion crust to form and obviously that causes complete drainage of thermoluminescence. There is some penetration of heat into the interior too with partial drainage resulting; study of the thermoluminescence in different regions of the meteorite allow estimates to be made of various parameters such as the rate of ablation (e.g., Sears and Mills, 1973); other studies have allowed estimates to be made of the pre-atmospheric mass (e.g., Sears, 1975). Another cause of partial drainage is proximity of the meteor's orbit to the sun and information on this too has been obtained through study of thermoluminescence (e.g., Melcher, 1981b). Despite these pre-arrival effects the study of the ratio between natural thermoluminescence peaks at 'low' temperature, circa 200°C, and 'high' temperature, circa

400°C, has been used as a rough guide to the date of the meteorite's fall to earth; the ratio should be higher for a recent fall than for an earlier one (for a review, see McKeever, 1983; see also Sears and Mills, 1974; Sears and Durrani, 1980; Melcher, 1981a). The rate of decay does of course depend on the storage temperature but the majority of available meteorites are those that have been preserved in the Antarctic ice sheet. More precise dating, using the regrowth of thermoluminescence in the drained fusion crust, is also a possibility as long as the difficulties inherent in dealing with such thin samples can be overcome.

7.4.2 METEOR CRATERS

Shock-heating of some rocks in a meteor crater is sufficient to reset the thermoluminescence to zero, thus allowing the date of impact to be determined in the same way as in pottery dating. This has been demonstrated recently by Sutton (1985a, 1985b) with respect to Meteor Crater, Arizona, an age of circa 50,000 years being obtained. The measurements were made on quartz grains; fission track mapping was necessary for evaluation of the annual dose. A shock of the order of 100 kbar is necessary for resetting.

7.4.3 METALLURGICAL SLAG

Turning back to archaeology, mention is now made of an adjunct of smelting that is available in prolific quantities; slag relates directly to the archaeological event itself and the contemporaneity or otherwise of developments in metal technology in different parts of the world is a matter of considerable interest. Slags consist essentially of silicates and since there is no doubt of the temperature having been high enough for erasure, thermoluminescence should be applicable. However in the first trials (Carriveau, 1974) there was great variability in thermoluminescence signal and the ages obtained were much too high in some cases. Subsequently in application to some Aegean slags, it was found (Elitzsch et al., 1983) that while the glassy type of slag did indeed give poorly reproducible thermoluminescence signals, fine-grain samples from porous macrocrystalline slag had TL characteristics that were much more encouraging. In this type of slag there were radioactive-free inclusions of quartz and iron silicate embedded in a glassy matrix, with the uranium (and presumably the thorium also) being carried in the glassy matrix; the thermoluminescence signal was emitted predominantly by the inclusions giving a similar situation to that in pottery. However in the case of slags it is not feasible, in preparation of fine-grain samples, to avoid crushing down the inclusions and so allowance has to be made for a rather complex dosimetry situation; nevertheless, with some exceptions, thermoluminescence ages obtained were in reasonable agreement with expectation. Use of the inclusion technique, which would have simpler dosimetry, was impeded by the difficulty of making separation of intergrown phases.

7.4.4 GLASS

As regards archaeological glass (vessels, windows, etc.) a basic question is whether continued exposure to light before burial would give rise to a significant zero-age thermoluminescence. In a study made on a range of English samples of the first millennium A.D. (Sanderson *et al.*, 1983) such generation was indeed observed when drained samples were exposed to ultraviolet, as well as bleaching in the case of undrained samples, but other difficulties prevented an assessment of how serious an impediment to dating this particular effect might be. Foremost among these difficulties was a very low thermoluminescence sensitivity, two or three orders of magnitude below that of quartz in some cases; coupled with rather low thorium and uranium contents the expected level of natural thermoluminescence was below the limits of detection except in one or two cases. However, it was established that archaeological glasses can give reproducible thermoluminescence signals after irradiation and perhaps other types (e.g., uranium-rich glasses) might be more appropriate for dating.

There are a wide range of volcanic glasses, and with these there is less likelihood of a light-induced zero-age signal; successful dating of the glass component of airborne ash has been noted in Section 7.3. A type of volcanic glass used in prehistoric times for knives and arrowheads and so forth is obsidian. Some examples of this, for example, from Obsidian Creek in British Columbia, have very high thermoluminescence sensitivity with consequent usefulness as a quartz-like substance for laboratory dosimetry experiments (e.g., Murray and Wintle, 1979) as long as correction is made for short-term anomalous fading. Determination of the geological source of obsidian by means of trace element analysis has been a powerful tool in elucidating ancient trade routes and there is prospect that thermoluminescence glow-curve shape could be diagnostic in this respect (Huntley and Bailey, 1978; Göksu and Turetken, 1979).

7.5 Stalagmitic Calcite (Unburnt)

Paleolithic caves in limestone usually contain *stalactites* (from the roof), *stalagmites* (on the floor) as well as *flowstone sheets* across the

7.5 Stalagmitic Calcite (Unburnt)

floor.[12] These deposits of calcium carbonate are formed by spontaneous precipitation from water which has passed through limestone, the precipitation occurring either through outgassing of carbon dioxide as the water comes into contact with the atmosphere or through evaporation of the water. Calcite is the most common form of these deposits, aragonite and dolomite also being encountered. A flowstone sheet may be the lateral extension of the base of a stalagmite, which like a stalactite, forms as water drips from the roof, or it may be a *spring deposit*—the result of water seeping across the floor of the cave from a spring.

These deposits are important in various aspects of Quaternary research, one reason being that prolific deposition is indicative of certain climatic conditions. Their dating is important for archaeology because of involvement with human occupation. Artefacts and skeletal remains may be contained in the sediment layers above or below a flow or even within the flow itself; a stalagmite may be stratigraphically related to an occupation level; within sediment layers there may be fragments of broken stalagmites and of stalactites which have been cracked off the roof by frost action. The first method[13] established for dating these deposits was by means of uranium-series disequilibrium. At formation a calcite crystal takes in uranium atoms but not daughter atoms; subsequently thorium-230 (formerly called *ionium*) is formed by radioactive decay, building up with its 75,400-yr half-life; the useful age range is from 10,000 yr to around 300,000 yr when the daughter has accumulated to within a few per cent of equilibrium. The reason for considering thermoluminescence as an alternative are first that it might extend the age range, and second that a confirmatory method having a different basis is desirable on account of possible interferences due to recrystallization, leaching, and high sediment content. Although these may affect thermoluminescence too, they are unlikely to give an error in age that is quantitatively the same.

The most critical question in dating unburnt material, calcite or other, is whether at crystal formation the latent thermoluminescence was zero. A second question is whether, given that the dominant peak in the glow-curve (see Figure 7.3) is at only 280°C, the traps are deep enough for the stored electrons to have a long enough lifetime for useful dating of paleolithic sites—quite apart from the possibility of anomalous fading. The answers to both of these questions are positive, and calcite proves to be a reliable material in these respects; there is of course the possibility of error due to recrystallization but so far this does not appear to be common in samples which are acceptable on the basis of visual inspection. Like flint its radioactivity tends to be low so that the main contributor to the thermoluminescence is the gamma dose from the surroundings. These tend to be less homogeneous than in the case of flint but it is evident

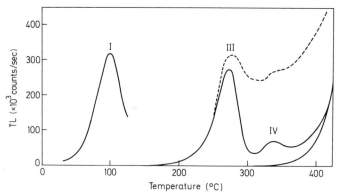

Figure 7.3 Glow-curves for calcite. Peaks III and IV are natural thermoluminescence and peak I the response to a small beta dose (typically ~5 Gy) used to normalize between portions by the ZGM technique discussed in Section 5.2. Peak III is the peak used for dating, though its lifetime may be as low as a million years; although peak IV has better stability it is unsatisfactory because of earlier saturation. The dashed line shows the glow-curve likely to be obtained unless strict precautions are taken against spurious TL.

nevertheless that stalagmitic calcite can provide valuable dating evidence (e.g., Debenham and Aitken, 1984). The main peak, III, has a linear response to at least 1000 Gy but probably an age limit of around a million years is imposed by its lifetime. Peak IV, though more stable, has a response which is appreciably non-linear from 100 Gy onwards; nevertheless it has been used in confirmation of Peak III results for old samples (Debenham, 1983), as discussed later. Calcite can also be dated by means of electron spin resonance (ESR); like thermoluminescence this measures the number of trapped electrons that have accumulated and has the same limitations in respect of strong dependence on environmental gamma dose. ESR will be discussed further in Section 7.6.

There is a strong thermoluminescence signal in limestone also. Initially it had been hoped that this could be used for geological dating but unfortunately there is strong interference by the effects of pressure due to overlying strata (Zeller *et al.*, 1955); on the other hand useful geological results have been obtained from contact-baked limestone (Johnson, 1963).

7.5.1 THERMOLUMINESCENCE

As with flint, thin slices were used initially for measurement in order to avoid spurious thermoluminescence (see Figure 7.3) but subsequently it was found[14] that, following crushing by squeezing in a vice, the simple expedient of washing grains in dilute acetic acid makes them usable as

7.5 Stalagmitic Calcite (Unburnt)

long as standard anti-spurious procedures are rigorously followed. This applies too with respect to fine grains, as required for evaluation of alpha particle effectiveness; an a-value of 0.3 is typical.

The usual additive dose method is used for evaluation of paleodose, with correction for supralinearity by means of second-glow growth characteristic. Coarse-grain (100 μm) portions are prepared as monolayers by sprinkling the grains on a disc coated with silicone oil and normalization between grains is by means of Zero Glow Monitoring (see Section 5.2).

For reasons given in the next paragraph it is important to use a colour filter which does not transmit wavelengths longer than about 500 nm; thus a blue-pass filter such as the Corning 5-60 should be used rather than the blue-green Corning 4-96 with which much early work was done.

7.5.2 ZERO-AGE THERMOLUMINESCENCE

Using the Corning 4-96 filter the paleodose evaluated for young calcite is liable to be substantially too high, (e.g., a paleodose of 50 Gy for a 4000-yr old calcite for which a paleodose of about 5 Gy was expected; Debenham *et al.*, 1982). Photographs of the thermoluminescence emission using an image intensifier, such as shown in Figure 7.4, showed that most of the natural thermoluminescence originates from small isolated 'hot spots' which are not present in artificial thermoluminescence. Examination of the spectra of the thermoluminescence by these authors, using a spectrograph coupled to the image intensifier (Walton, 1982), indicated that whereas the artificial thermoluminescence lies in a broad band centred on 520 nm the natural thermoluminescence occurs at longer wavelengths, as shown in Figure 7.5. Also shown in the figure is the spectrum of a limestone and it will be seen that this is similar to that of the natural thermoluminescence of the young calcite. Hence a possible explanation is that during formation of the young calcite, specks of limestone were incorporated in it; these specks would carry a strong natural thermoluminescence acquired over geological times. There may also be a contribution to the zero-age thermoluminescence from specks of sediment which, like the specks of limestone, are liable to be blowing about in a cave; it is common to find, after dissolution in hydrochloric acid, an insoluble residue[15] amounting to a few percent by weight.

As will be seen from Figure 7.5 the remedy is to discriminate against wavelengths longer than 500 nm. Using the Corning 5-60 filter correct ages have been obtained for fairly young samples; it appears that any zero-age thermoluminescence observed is equivalent to a paleodose of not more than the order of 5 Gy (Debenham and Aitken, 1984).

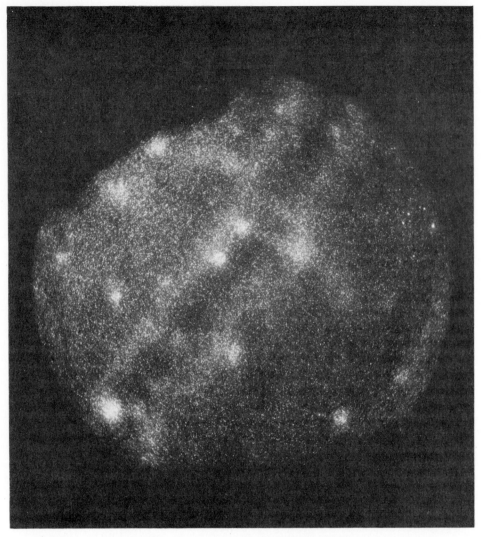

Figure 7.4 Image intensifier photograph of natural thermoluminescence (250–300°C) from a 'young' calcite (Debenham *et al.*, 1982). Diameter of slice is 10 mm.

7.5.3 ZONING

Image intensifier studies (Walton and Debenham, 1982) have also shown, in some samples, a broader type of spatial variation in the thermoluminescence emission; this takes the form of bright and dark bands of the order of a millimetre wide as illustrated in Figure 7.6. Comparison with

7.5 Stalagmitic Calcite (Unburnt)

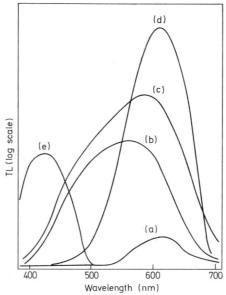

Figure 7.5 Spectra of thermoluminescence emissions, 250–300°C: (a) natural thermoluminescence of a 'young' stalagmite; (b) second glow following gamma irradiation of 400 Gy; (c) natural thermoluminescence of an old stalagmite (having a paleodose of 400 Gy); (d) natural thermoluminescence of limestone. The intensities are on a logarithmic scale and not corrected for spectrometer response; this is approximately constant between 430 and 590 nm but decreases rapidly below 400 nm and above 600 nm. Curve (e), on a linear scale, shows the transmission characteristics of the Corning 5-60 (blue) filter (from Debenham and Aitken, 1984).

the spatial distribution of uranium as given by induced fission track mapping indicates anti-correlation, that is, the bright bands correspond to low uranium content and vice versa; comparisons with optical transparency maps indicate that optical attenuation is an important factor affecting the brightness of thermoluminescence emission, though sensitivity variations probably play a part also. Whatever the cause, samples showing this zoning are liable to give an erroneously low age because certainly alpha particles from the high uranium regions will not be able to reach the regions of bright thermoluminescence and to some extent the same is true also for beta particles. Note that the variation observed for natural thermoluminescence will be less marked than for artificial thermoluminescence.

By no means all samples show this zoning and in any case the effect is very frequently unimportant because of dominance by the external gamma dose. This is particularly likely for fragments buried in sediment

Figure 7.6 Image intensifier photograph of natural thermoluminescence (250–300°C) from lowest stalagmitic floor of the Lower Paleolithic site in Caune de l'Arago, France. Diameter of slice is 10 mm; blue colour filter (Corning 5-60).

for regions in which the calcite has low radioactivity (e.g., in the Dordoyne region of France the uranium content is less than 0.1 ppm—see Aitken and Bussell, 1982).

A related problem to zoning stems from gross variations in uranium content between adjacent growth layers; this gives risk that the material used for uranium determination is not representative of the material used for paleodose measurement (Wintle, 1978b).

7.5.4 RADIOACTIVITY; GROW-IN OF THORIUM-230

The internal radioactivity varies with time also. We must distinguish here between that part of it which is carried in with the specks of limestone and sediment and that part which is intrinsic to the calcite. As mentioned earlier it is only uranium that is chemically acceptable to the calcite lattice, that is, uranium-238 and uranium-234 in the case of the main series.[16] Being short lived, the first two daughters of uranium-238 grow to equilibrium immediately, but the first daughter of uranium-234 is thorium-230 which has a half-life of 75,400 yr, and so it grows towards equilibrium only slowly, the activity being given by

$$C = C_\infty\{1 - \exp(-\lambda t)\}, \tag{7.1}$$

where C_∞ is the equilibrium activity, t is the age, and λ the decay constant is equal to (0.693/75400). All the subsequent daughters are short lived by comparison and hence they grow so as to remain in equilibrium with the thorium-230.

Thus the annual dose[17] provided by thorium-230 and subsequent daughters increases exponentially according to Equation 7.1. This necessitates a trial and error approach for calculating the age; essentially this means plotting the integral[18] of annual dose against t and seeing at what value of t the integral is equal to the paleodose.

7.5.5 STABILITY

Contrary to the situation in feldspars we are not now concerned with anomalous fading—tests on calcite for storage over months have shown none (Wintle, 1978a; Debenham, 1983)—but instead with thermal fading as determined by trap depth and frequency factor. These parameters are determined by laboratory measurements (see Appendix E) and from them the lifetime of electrons in the trap can be calculated from Equation 3.4. Despite the low temperature of the peak, such studies indicate a remarkably deep trap: 1.75 ± 0.03 eV according to Wintle and 1.56 ± 0.08 eV according to Debenham, the former using a blue-green filter (Corning 4-96) and the latter a blue filter (Corning 5-60); the corresponding lifetimes for storage at 15°C are around 30 million years and $0.3 - 3$ million years respectively; the occurrence of the peak at such a low glow-curve temperature is due to a high frequency factor, 5×10^{15}/sec according to Wintle.

Prediction of million-year-long lifetimes from these kinetic parameters places a very heavy reliance on the correctness of theory, that is, on the validity of Equation 3.4. While there is no reason to doubt theory, in view of the very long extrapolation it is sensible with old samples to look for more direct evidence about stability. Routinely this evidence is provided by the plateau test but with a sharp single peak there is the possibility that a plateau will be obtained despite thermal decay having occurred; it is only if there is a continuum of trap depths that the shape of the peak is changed by thermal decay, the maximum moving upwards in temperature because of less decay occurring for the deeper traps. Debenham (1983) has investigated this for 350,000 yr old calcite from the early hominid site of Caune de l'Arago in southern France; by holding artificially irradiated samples at elevated temperature it was confirmed that thermal decay did upset the plateau, and from the quality of the plateau observed for the natural thermoluminescence an upper limit of 15% could be set for the fading. This corresponds to a lower limit of 1 million years for the lifetime.

Further direct evidence about stability is available from peak IV. Although this becomes non-linear in response to doses of above 100 Gy, it can be used to evaluate paleodose by means of curve fitting techniques, although with rather poor precision. For the calcite just mentioned it was concluded that if fading had occurred in peak III the amount could not have exceeded 30%. Thus the plateau evidence gives tighter limits and Debenham suggests an upward correction to the age of $8 \pm 8\%$. The lifetime is of course strongly dependent on storage temperature and this needs to be taken into account for sites having an average temperature during antiquity higher or lower than the 15°C assumed for Caune de l'Arago.

7.6 Bone and Shell; Use of Electron Spin Resonance

Because of their organic content, unburnt bone and shell tend to give chemiluminescence on heating and so be out of bounds to thermoluminescence; use of electron spin resonance (ESR) seems much more appropriate because heating is not then involved.

7.6.1 BONE

Despite difficulties attempts have been made to use thermoluminescence to date bones. In very old bone the organic content has largely disappeared and measurements are feasible on mammoth samples (Jasinka and Niewiadomski, 1970). Otherwise it is a matter of using techniques for removal of organic material which do not upset the natural thermoluminescence (Christodoulides and Fremlin, 1971; Driver, 1979) or alternatively avoiding heating by use of the phototransfer technique (Mobbs, 1978); results so far have not been encouraging.

7.6.2 SHELL

The same problem of organic content occurs with shells (Driver, 1979), and the phototransfer technique is again a possibility (Mobbs, 1978) that would be worth pursuing were it not for the availability of ESR.

There is the additional problem that if the calcium carbonate is in the form of aragonite, there is interference from light emitted during the glow-curve when, at around 400°C, this transforms to calcite (Johnson, 1960; Johnson and Blanchard, 1967). In old enough shell most of the aragonite will have transformed to calcite naturally although it then seems likely that the event being dated is this transformation (but see discussion in Gallois et al., 1979).

As far as archaeology is concerned most of the shells involved seem to be aragonitic. However there are exceptions which might be worth pursuing; for instance in slugs there is a calcite secretion (*Arion granules*) which gives a satisfactory thermoluminescence signal. Because these are only a millimetre or so across, a special technique would be necessary to assess the annual beta dose received from the soil (see note 7).

7.6.3 MOVEMENT OF RADIOACTIVITY

Apart from problems of thermoluminescence measurement, another adverse factor in respect of bone and shell is the tendency for the radio-elements, by leaching action, to have moved in and out of the sample so that the annual dose has been varying.

It is well known from uranium-series studies that there are both 'closed systems' and 'open systems', and indeed the uptake of uranium is a rough relative method of dating bone. In some soil environments it seems that once uptake has occurred the uranium does not subsequently move out and the question then is how soon after burial did the uptake occur.

It is known from uranium series dating that the foregoing is not a serious problem with aragonitic coral; the initial uranium content is high and the mobile component unimportant by comparison. Of course it needs to be checked that no appreciable conversion to calcite has taken place.

7.6.4 ELECTRON SPIN RESONANCE

Electron spin resonance, which gives a spectrum of the microwave power absorbed by the sample as the strong magnetic field in which it is placed is varied, is not really within the territory of this book, but brief mention will be made because it gives an alternative way of measuring the trapped electron population and is increasingly being used in long range dating. A comprehensive review has been given by Hennig and Grun (1984). It is a much more diagnostic tool in understanding the mechanisms of what is going on in a crystal, but for dating there is the disadvantage that because of the complexity of the spectrum the age-dependent signal may be difficult to isolate. Another disadvantage is that it is not as sensitive as thermoluminescence, the minimum detectable paleodose being substantially higher.

On the other hand, there are a number of aspects in which it is highly advantageous compared to thermoluminescence. The most important is that it does not require the sample to be heated as far as actual measurement is concerned (though storage at elevated temperature may be necessary for stability investigations, or for erasure prior to low-dose linearity checks). Another is that the electrons are not evicted from their traps during measurement, so the natural signal can be measured many times using only a single portion; this means too that only one portion is needed for the additive dose method. Third, since luminescence is not involved there are no transparency problems, nor are there any effects associated with the involvement of luminescence centres. In principle one would expect ESR investigation to be a valuable complement to thermoluminescence in giving better understanding of, say, supralinearity and the extent to which luminescence centres are responsible, but so far direct intercomparisons between the two techniques have been few.

Extensive application has been made to stalagmitic calcite and now attention is being given increasingly to bone and shell (e.g., Ikeya, 1978, 1980). In some of this work there has been a tendency to assume typical values for annual dose so that it is only the paleodose for which an accurate value is obtained; in part this is due to the uncertainties about uranium mobility, as noted above, together with the experimental difficulty of measuring alpha effectiveness. For the latter a thin deposition is required and to get enough material, say 50 mg, for ESR measurement it is necessary to irradiate about 20 fine-grain discs and then scrape off the grains to form a composite sample; there is also the alternative possibility of irradiating large grains as discussed in Appendix K. Some workers assume that the alpha effectiveness for ESR will be the same as measured for thermoluminescence; since luminescence centres are additionally involved in the latter this assumption must remain questionable until comprehensive comparison has been made. Another assumption often made is that the growth with dose is linear in the region below the natural level, a region well known for its supralinearity in thermoluminescence studies; in a pre-heating technique used for calcite it has been shown that this assumption can lead to substantial overestimate of age, that is, there is sublinearity rather than supralinearity (Skinner, 1983; Yokoyama *et al.*, 1983).

However these shortcomings are no more than is to be expected of a technique that is in the early stages of development and ESR is now well established as a powerful dating tool. Its use in dating the human skull found in Petralona Cave, Greece is one illustration, albeit controversial. Measurements were made on bone fragments and on calcite encrustations attached to the skull, a total of not more than a gram of each being sufficient (Hennig *et al.*, 1981). An application that holds high promise is tooth enamel, because of its compactness.

Geological application has been made to unburnt flint with respect to a signal attributed to lattice damage caused by the recoil of nuclei after emission of alpha particles (Garrison *et al.*, 1981). This signal is enhanced by fast neutrons, which are used for artificial irradiation, but not by gamma radiation; hence there is no dependence on soil radioactivity. This would have strong advantages for archaeological burnt flint; such extension depends on being able to distinguish the much weaker recoil signal from interfering background.

7.7 Thermoluminescence Dating of the Lake Mungo and Laschamp Geomagnetic Excursions

It is now well established from paleomagnetic studies that during certain periods of the geologic past the direction of the earth's magnetic field was reversed by 180° with respect to its present direction, a polarity *epoch* lasting on the order of a million years; the present epoch is termed the *Brunhes* and it began 0.7 million years ago. Within an epoch there are occasional much shorter occurrences of abnormal polarity termed *excursions* or *events* depending on the degree of reversal achieved. All this geomagnetic history is obtained from the remanent magnetization acquired by volcanic rocks as they cool, and by lake and ocean sediments on settling; exceptionally, as at *Lake Mungo* in Australia the remanent magnetization in a prehistoric fireplace may indicate reversed polarity.

The Mungo excursion (Barbetti and McElhinney, 1972) has been dated by radiocarbon measurements on charcoal as being in progress from 28,000 to 32,000 yr ago on the uncorrected radiocarbon timescale but using the revised half-life. This timescale is distorted by variations in atmospheric radiocarbon content and its calibration into calendar years is only well known, using dendrochronology, for the last eight or nine thousand years, though comparisons against uranium-series dating suggest that in earlier periods it is not out by more than around a thousand years. Thus it was of interest to attempt dating this fireplace by thermoluminescence. Using quartz inclusions extracted from the baked clay, an average age of 33,500 ± 4300 yr was obtained (Huxtable and Aitken, 1977) for the first phase of the excursion, and using the same technique on heated underlying sediment, ages for the individual five fireplaces were from 31,400 ± 2500 to 36,400 ± 2,500 yr (Bell, 1978). The latter author suggests that the radiocarbon ages may be too young by several thousand years; however, in the light of the discussion given in Appendix B the possibility that the difference may be due to systematic error in the thermoluminescence ages cannot be ruled out.

A question of particular interest for geophysicists is whether a given excursion represents a reversal of the earth's field as a whole or whether it arises from a strong irregularity in the motion of the fluid near to the mantle boundary which affects only a localised region. Because of this the dating of excursions recorded in different parts of the earth's surface is critical. Another record of reversed magnetization is carried in recent volcanic lava of the Chaîne des Puys, France—the *Laschamp* event (Bonhommet and Babkine, 1967). At first the age for this had been set as 8000–20,000 yr, the former limit being from radiocarbon dating of an overlying paleosoil and the latter being a limiting age from potassium–argon measurements. On this basis there seemed to be no possibility of contemporaneity with the Mungo excursion but a likelihood that it was associated with the Gothenburg event; this latter had been inferred from measurements on 12,400-yr old sediment in Sweden but its validity was discounted subsequently (Thompson and Berglund, 1976) leaving Laschamp as the most recent example of geomagnetic irregularity. Application of thermoluminescence has shown that the Laschamp event was by no means as recent as had been supposed and furthermore that it may have been contemporary with the Mungo excursion, giving the possibility of a worldwide occurrence.

The difficulty in dating the Laschamp event was due to the absence of any wood remnants within the soil fossilized by the flow as far as radiocarbon was concerned, and the absence of quartz inclusions in the lava as far as thermoluminescence was concerned; as has been mentioned earlier use of feldspars from this lava had given grossly low ages on account of anomalous fading. As far as potassium–argon and uranium series were concerned the lava was considered as below the limiting age of those techniques; subsequently they have given results concordant with the thermoluminescence.

Although there was no quartz intrinsic to the lava, diligent searching located some quartz pebbles which had been heated by it and also, within the lava, a granitic enclave containing quartz inclusions. The former were under the associated *Olby* flow, also reversed, and the latter was from the Laschamp flow itself. The ages obtained were 42,000 ± 3,000 and 34,300 ± 3000 yr, respectively (Gillot et al., 1978; Valladas and Gillot, 1978). At about the same time further indication that the Laschamp event was less recent than supposed was obtained by fine-grain dating of sediment baked by a lava flow which, while not reversed, had a sufficiently abnormal direction to suggest that it was close in time to the event; this was the Royat flow and the age obtained was 25,800 ± 2,200 yr (Huxtable, et al., 1978), somewhat lower than the age of 43,500 ± 3,900 yr subsequently obtained for the lava itself (Guérin, 1981) perhaps on account of anomalous fading having been more serious than indicated by short-term tests.

Finally, with the development of the high temperature technique for feldspars, dating of the lava itself became possible. The ages obtained were 39,200 ± 2,800 yr for Olby and 30,200 ± 2,900 yr for Laschamp (Guérin and Valladas, 1980; Guérin, 1982b, 1983) thus allowing the possibility that the Laschamp and Mungo recordings are indeed manifestations of the same event. On the other hand, as pointed out by Gillot et al. (1979), the patterns of magnetic directions recorded in the two localities are not consistent with this interpretation.

Technical Notes

[1] A concise account of the geology of flint will be found in an article by Sieveking et al. (1972). It is there noted that the quartz (SiO_2) of which it consists is in its alpha form and present as small plates and rods 0.2–30 μm long and 0.6 μm thick, with pores or spaces between the plates and the rods ranging in size from 0.02 to 0.3 μm, some of these spaces being filled with aqueous solution. Some flints are the result of silicification of calcium carbonate.

[2] The complications arise in measurement of the effectiveness of alpha particle irradiation relative to beta. In the former only a thin surface layer of the slice is reached whereas in the latter it is the whole thickness. However it is not just a matter of making allowance for this according to the ratio between the alpha layer thickness and the slice thickness because there is significant optical attenuation of the beta-induced thermoluminescence emitted in the deeper part of the slice. Hence it is necessary to evaluate the attenuation coefficient for each slice dated. Fortunately this laborious technique has been made obsolete by the finding (Valladas, 1978) that adequate suppression of spurious thermoluminescence can be achieved for crushed samples of flint, even for fine-grains (Huxtable, 1982), hence allowing measurement of alpha effectiveness in the usual way. Another problem with the slice technique is that of normalization between slices (see Bowman and Seeley, 1978).

[3] Measurement conditions used (see also Huxtable and Jacobi, 1982): heating rate 10°C/sec; bialkali photocathode (EMI 9635 QA) with a blue or violet filter (e.g. Corning 5-60 or 7-59) plus a Chance Pilkington HA3 for infrared rejection; coarse grains (size range 90–150 μm) directly on the heater plate (using silicone oil) with normalization by weighing, average sample weight 2 mg; no normalization for fine-grain discs (aluminium), the disc to disc scatter being no more than 3% (although the difficulty of preparing sufficient well-covered fine grain discs from such a hard material as flint is to be noted).

[4] Although Huxtable (1982) reports one case for which the fine-grain paleodose is *less* than the coarse grain value by about 30%, this is exceptional. Similarly Bowman and Sieveking (1983) report two cases where the fine-grain evaluation of a known gamma dose is substantially low.

[5] The ranges quoted are based on values reported for archaeological flints; these include neutron activation results for two sites in the Dordogne, France, which indicate the thorium to uranium concentration ratio to be only 0.1–0.2 (Bowman *et al.*, 1982), although for one flint the ratio was 1.4 (Bowman and Sieveking, 1983). The latter authors report agreement between neutron activation and alpha counting, marginal in two cases.

For five non-archaeological flints analysed in connection with ESR dating, concentrations as high as 3 ppm were observed both for thorium and for uranium (Garrison *et al.*, 1981). In that work distinction is made between 'binder' and silica microcrystals, the former being removed by etching in hot aqua regia; the bulk of the radioactivity is in the binder, the highest uranium concentration for the microcrystals being 0.5 ppm.

[6] Fission track analysis of ten slices from four flints from a single site (Combe Grenal, France) indicated the uranium distribution within the slices to be essentially uniform although there were slice-to-slice variations within two of the flints (Bowman and Seeley, 1978); neutron activation of 6 portions from one flint ruled out gross inhomogeneity (Bowman and Sieveking, 1983).

Image intensifier photographs of the thermoluminescence itself have given indications of uniformity for the three or four slices so far measured (Walton, Templer, private communications) with one exception; for this sample the glow-curve shape was abnormal.

If all of the radioactivity is carried in the lattice then uniformity is to be expected. It is when it is along grain boundaries then there will be non-uniformity. It seems likely that the fine-grain paleodose would then be greater than the coarse grain paleodose; agreement between the two, as is usually the case, is a criterion for reliability.

[7] If there are no flints thick enough and there is no other dating evidence for the site, so that a rough date is worthwhile, there is the possibility of proceeding without sawing off the surface layer, as long as the flints have not been exposed to light. An estimate is then made of the beta dose contribution from the soil (See Appendix H). Another possibility would be to make a sodium silicate replica loaded with phosphor grains and find the dose acquired through immersion in pitchblende for a period (Aitken *et al.*, 1985).

⁸ There was successive removal of (i) carbonate by dilute HCl, (ii) organics by 6-day treatment in 10% H_2O_2, (iii) iron oxide coatings by CBD treatment (Na citrate–Na bicarbonate–Na dithionite), and (iv) feldspars and hydration rinds by a few minutes in 10% HF at 20°C. For measurement of thermoluminescence a standard bialkali photocathode was used with a blue-pass Corning 5-58 filter.

⁹ Relative methods for dating a given lava site have been reported for Hawaii (May, 1977).

¹⁰ Specivex type 325 is obtainable from MTO, 11 rue ampère, BP6, 91302 Massy Cedex, France. The oven used has been described by Brou and Valladas (1975). The trap depth of the deep traps concerned was estimated to be 2.75 eV with a frequency factor of about 10^{15} sec^{-1}, giving a half-life of about 4×10^{23} yr (Guérin, 1983).

¹¹ This measurement is repeated for different incident energies so as to be able to evaluate the average response to the spectrum of particle energies received from the thorium and uranium of the matrix. The word 'response' is being used here rather than 'effectiveness' because the measurement takes into account the fact that penetration of the grains is only partial. For the small amount of internal alpha activity, about one-tenth of the matrix value, it is the effectiveness that is required and this is measured on fine-grains in the usual way; this component amounts to about one quarter of the total alpha contribution. The potassium content was no more than 2%, substantially the same as in the groundmass in which the grains were embedded.

¹² In North American usage *travertine* includes all these three whereas in Europe that word tends to mean open air deposits, much less pure, *speleothem* being used for cave deposits.

¹³ For archaeological application the reader should refer to a review by Schwarcz (1980), to whom a debt in respect of the text paragraphs is acknowledged.

¹⁴ The use of acetic acid was introduced in the first place to avoid effects in samples obtained by means of a power drill that gave rise to an erroneously low paleodose (Wintle, 1975a). The standard anti-spurious recipe is a continuous flow of dry, high purity argon admitted after evacuation of the oven to 20 mTorr (as measured *in the oven*); besides achieving removal of residual air this evacuation is probably important also in removal of adsorbed water vapour from surfaces in the oven and from the grains themselves. It has also been found (Aitken and Bussell, 1982) advantageous to 'immerse' the grains in a patch of high viscosity silicone oil (e.g., General Electric Viscasil 60,000).

¹⁵ A difficulty with this explanation is that the glow curve shape of the natural thermoluminescence of the residue is not a good match to that of calcite; the same applies to a lesser extent in respect of limestone. The effects of sediment being present have been reported by Bangert and Henning (1979).

¹⁶ In respect of the minor series it is uranium-235 that is incorporated and except for the short-lived first daughter thorium-231 the growth of the rest of the series is controlled by protoactinium-231, which has a half-life of 32,800 years.

[17] The dose rates corresponding to separate sections of the series can be found from Appendix G. For a fuller treatment see Wintle (1978b); this includes allowance for the small disequilibrium between uranium-238 and uranium-234 that is found in groundwater, the latter typically being in excess of the former by around 10% due to selective leaching from rocks during incipient weathering.

[18] If we write the dose-rate at any time t' since formation as $D = D_1 + D_2(1 - \exp(-\lambda t'))$ then the integral of D over the time t elapsing between formation and measurement is given by

$$D_1 t + D_2 \{t - \frac{1}{\lambda}(1 - \exp(-\lambda t))\}.$$

CHAPTER 8

SEDIMENT DATING—SOLAR RESETTING

8.1 Introduction

In Figure 8.1 glow-curves are shown for windblown (*aeolian*) sediment from a paleolithic site of NW Europe; curve (a) is the natural thermoluminescence, curve (b) is the same except that this portion had been put out in the February sunlight for an hour, and in the case of curve (c) there had been 24 hr illumination with a sunlight simulator. Besides illustrating why thermoluminescence work has to be done in subdued red light, these glow curves show the most likely basis for being able to use thermoluminescence to obtain the deposition date of unheated sediment; exposure to sunlight prior to deposition causes 'bleaching' to near zero of previously acquired thermoluminescence and when the sunlight is cut off by later sediment being deposited on top the thermoluminescence re-accumulates in the same way as in pottery after firing.

The sediment illustrated was from a deposit of *loess,* the name given to layers, sometimes several meters thick, of silt-sized sediment (2–60 μm) which extend from NW Europe across northern Asia to China as well as

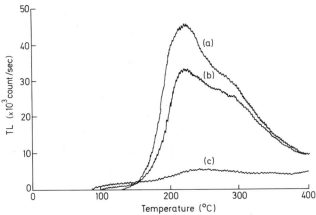

Figure 8.1 'Bleaching' of TL due to exposure to light. Curve (a) is natural thermoluminescence; curve (b) after exposure to sunlight for 1 hr; curve (c) after exposure to simulated sunlight for 24 hr.

occurring in the Americas and many other parts of the world. According to the majority of geologists[1] it originates from the grinding action of glaciers but in dry windy conditions it may have been carried up to several hundred kilometers from its source. During this transportation, perhaps high up in the atmosphere, there is ample opportunity for thermoluminescence to be bleached quite apart from the time it lay on the surface before being blown by the wind or after deposition prior to being covered. The involvement of glacial action suggests there may be correlation with climate and for this reason its dating is of considerable interest in Quaternary research; equally it is important in the dating of paleolithic sites either when remains of human occupation are found in the loess itself or in establishing the chronology of a stratigraphic sequence.

Sunlight is not the only agent that has been suggested for erasure of previously-acquired thermoluminescence. Dating of loess was initiated in the USSR around 1965 at the Kiev Institute of Geology and in this work[2] weathering and glacial grinding are put forward as additional possibilities (e.g., Morozov, 1968; Shelkoplyas, 1971; Morozov and Shelkoplyas, 1980). However it is currently taken for granted in nearly all laboratories that the effect of sunlight is at any rate dominant.

Dating now extends to many other types of sediment, for example, 'desert loess', sand dunes, ocean sediment, sediment in lakes filled by melting glaciers, and even the dust incorporated in the ice of a glacier itself. This explosion of interest was triggered by work on ocean sediment in Canada at Simon Fraser University in the late 1970s; this research led to the realization[3] by Wintle and Huntley (1979, 1980) that solar resetting was the mechanism that explained the observed increase with depth of paleodose in the ocean cores being studied and that it might be relevant in many other situations. In hindsight, given the insistance in thermoluminescence laboratories on the need for subdued red light, it is surprising that this widespread extension of thermoluminescence dating was so long delayed.

Despite different regions of origin, glow-curves from sediment are fairly similar. Because of its brightness the thermoluminescence from feldspar dominates over that from the quartz, which in terms of quantity is usually more plentiful. The former is more rapidly bleached and is therefore likely to be better set to zero; on the other hand there is the possibility of anomalous fading. It may therefore be desirable to measure the quartz signal separately—after mineral separation; however with quartz from old samples there are complications due to its non-linear response.

8.2 Methods for Evaluation of Paleodose

Evaluation of paleodose is complicated by the need to allow for the fact that the natural thermoluminescence is composed of two components: the thermoluminescence acquired since deposition and the zero-age thermoluminescence that the sample had when it was deposited. If this latter is substantial, as is often the case, failure to make allowance for it will lead to a thermoluminescence age that is much too old. Although weathering and glacial grinding may play a part in resetting the thermoluminescence, because of the difficulty of making laboratory studies of those effects the methodology evolved is based on the presumption that sunlight is the agent. From the success achieved it seems that this is a satisfactory working hypothesis.

8.2.1 ZERO-AGE THERMOLUMINESCENCE: THE UNBLEACHABLE RESIDUAL

For a sample exposed to the sun there is a certain probability per incident photon that an electron will be evicted from its trap and so the rate of eviction will be proportional both to this probability and to the number of trapped electrons, as well as to the intensity of the light. This simple model leads to the expectation that the number of electrons remaining trapped at time t will be exponentially dependent on t, similar to the case of radioactive decay. However experiment indicates that this is not the case (see Figure 8.2) and that as the 'bleaching' proceeds the

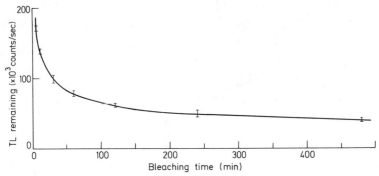

Figure 8.2 Thermoluminescence remaining after bleaching for various time. Data are for the level of natural thermoluminescence (at a glow-curve temperature of 280°C) of fine grains of loess from the paleolithic site at Biache, France. Bleaching was by means of a 300-W Solar Simulator set at half power, giving an irradiance that was approximately half that of natural sunlight. The unbleached level, not shown, was about 4 times the thermoluminescence remaining after 5 min of bleaching (the highest point shown). (Measurements by V. Griffiths).

probability of eviction decreases.[4] Thus we have the concept of the 'easy-to-bleach' thermoluminescence being removed first followed in turn by thermoluminescence which is harder and harder to bleach until eventually an unbleachable residual component is left.

The time required to remove all but this latter depends of course on the intensity and spectrum of the bleaching illumination, as well as the susceptibility to bleaching of the minerals present in the sample. The practical question in evaluating paleodose is whether, at deposition, the bleaching had been sufficient to remove all but this residual component and it is only the third method outlined below, the *partial bleach method*, that satisfactorily deals with a sample that, at deposition, had retained some of its bleachable thermoluminescence. However the two methods described first are simpler and are applicable where the presumed circumstances of deposition were such as to ensure full removal of the bleachable component; the more the thermoluminescence is dominated by feldspar, the more is this likely to be the case.

8.2.2 THE ADDITIVE DOSE METHOD

The additional dose method is basically the same procedure as used for pottery except that instead of reading off the equivalent dose Q from the intercept on the dose axis (see Figure 8.3) it is read off from the intercept on a horizontal line through G_0, the latter being the level of thermolumi-

8.2 Methods for Evaluation of Paleodose

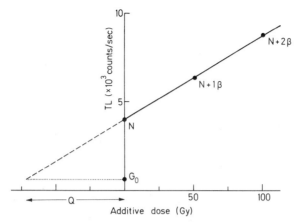

Figure 8.3 Additive dose method as used in sediment dating. Several separate portions are averaged for each of the data points; G_0 is the residual thermoluminescence after a prolonged laboratory bleaching, that is, the unbleachable component; Q is the equivalent dose.

nescence remaining after a long laboratory bleaching. Although it can be established that the laboratory bleaching is long enough for only the unbleachable component to remain, there is the risk, as just discussed, that the exposure to sunlight prior to deposition was not sufficient to reach this level. A check of this is given by the plateau test in which the paleodose is plotted against glow-curve temperature; as is illustrated in Figure 8.1 the degree of bleaching is dependent on glow-curve temperature and if the bleachable component had not been fully removed at deposition there will not be a plateau.

First use of this method was by Singhvi *et al.* (1982) in respect of sand dunes of the Indian desert.

8.2.3 THE REGENERATION METHOD

Except for measurement of natural thermoluminescence all portions are subjected to long bleaching and then artificial irradiations (beta or gamma) are administered, thereby regenerating the thermoluminescence growth characteristic; of course a new portion must be used for each data point. The paleodose is read off from a horizontal line (see Figure 8.4) drawn at the level of the natural thermoluminescence. An advantage is that if there is no alpha contribution to the thermoluminescence, separate evaluation of any correction for non-linear growth is unnecessary; if there is an alpha contribution then as in other methods allowance needs to be

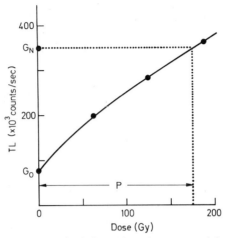

Figure 8.4 The regeneration method. Several portions are used for measurement of the natural thermoluminescence, G_N; the remaining portions are given a long bleach and the level of residual thermoluminescence, G_0, checked with one of them; the rest are irradiated to various doses and the growth characteristic regenerated. The value of the paleodose, P, is read off as indicated.

made for the fact that the alpha growth characteristic is likely to reach sublinearity at a higher dose level than the beta/gamma characteristic (Aitken, 1984).

A practical drawback is that each portion needs to be given long bleaching whereas in the previous method this was only required for evaluation of G_0; a short cut is to use partial bleaching for the portions used for regeneration and make allowance for the bleachable thermoluminescence then remaining, as in 'method a' of Wintle and Huntley (1980). A more serious disadvantage is the possibility that bleaching has altered the sample's sensitivity so that the regenerated growth does not truly represent the growth during antiquity. A check on this can be obtained by also measuring the additive growth characteristic (Figure 8.3) and seeing whether the sensitivity (in the linear region) is the same. Also, since changes in sensitivity are likely to differ in amount in other regions of the glow-curve, the occurrence of such changes will cause the sample to fail the plateau test.

One reason for sensitivity change is that the laboratory spectrum used for bleaching is different than that experienced by the sample prior to deposition; for instance, if one or the other contains too much ultraviolet there is the possibility of transferred thermoluminescence being present after 'bleaching'. Use of a light source which is a good match to the solar

8.2 Methods for Evaluation of Paleodose

spectrum (see Figure 8.9) is therefore desirable, though it must be remembered that grains which go high into the atmosphere experience a different spectrum to those which are bleached at ground level or under water. Another possible reason is that the temperature of the sample during bleaching has an effect.

The effect of such sensitivity change is circumvented in 'method b' of Wintle and Huntley (1980), though many more measurements are required. Having found the dose R required to regenerate a level equal to the natural thermoluminescence the procedure is repeated for portions which have been given an additive dose β. Then as in the additive dose method, a plot is made of R_N and R_{N+A} versus β and the equivalent dose is given by the intercept on the horizontal axis.

8.2.4 THE PARTIAL BLEACH METHOD

In this method, introduced as 'method c,' the $R - \Gamma$ method, by Wintle and Huntley (1980), the glow-curves of some portions are measured after artificial irradiation as in the additive dose procedure while for another set of portions a short bleach is administered between irradiation and measurement, giving a partially-bleached growth line as illustrated in Figure 8.5. On the basis that the amount of thermoluminescence removed by the

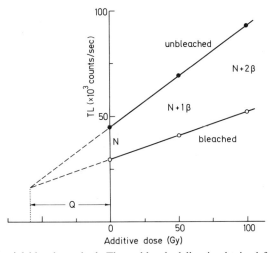

Figure 8.5 Partial bleach method. The unbleached line is obtained following the same procedure as in the additive dose method. For the bleached line, portions are exposed to sunlamp illumination before measurement; the bleaching caused by this illumination must be less than the bleaching prior to deposition. Q is the equivalent dose.

bleach is a given fraction of the bleachable thermoluminescence, the intercept of the two lines indicates the equivalent dose. Alternatively, this can be seen by regarding the difference between the pair of points at each irradiation level as the response to dose of the easily bleached component of the thermoluminescence; this includes the difference for zero laboratory dose, that is, the difference between (N) and $(N + \text{Bleach})$. If these differences are plotted against dose then the equivalent dose is given by the intercept on the horizontal axis in the usual way.

It does not matter if the bleaching prior to deposition was incomplete as long as it was more complete than that effected by the short laboratory bleach. If it was less complete, then the short laboratory bleach will be able to reach components of the natural thermoluminescence that were acquired prior to deposition thence causing the evaluated equivalent dose to be overestimated. Indication as to whether the laboratory bleach was too long can be obtained by recreating the bleached line using different bleaching times; as long as the completeness of the bleaching does not exceed that of the bleaching prior to deposition, the same equivalent dose should be obtained irrespective of bleaching time. If the laboratory bleaching is too strong, an erroneously large equivalent dose will be obtained. Tests on zero-age samples have been reported by Huntley (1985).

Of the three methods this is the only one that accommodates partial resetting. On the other hand, because it is a subtle method dependent on the concept that some parts of the thermoluminescence are more easily bleachable than others, it is unlikely that it would be applicable to a partially reset sample for which glacial grinding or weathering had been dominant.

Another drawback is that because it relies on subtraction (sometimes of two quantities of comparable size) it is not applicable to samples showing non-linear response and in particular to the onset of saturation. Unfortunately this is most likely to be the case for quartz, and it is for quartz, being difficult to bleach, that this method is most needed.

As with the regeneration method there is the possibility that the bleaching will alter the thermoluminescence efficiency, in this case only in respect of the probability that a released electron finds a luminescence centre, the actual trapping having occurred prior to the bleaching.

8.2.5 MEASUREMENT BY PHOTOTRANSFER

The potential advantage of this technique, in which light is used to transfer electrons from deep traps to shallow ones before measurement (see Section 6.2), is that unbleachable traps would be ignored. It can be used in conjunction with any of the methods just discussed. Unfortunately in trials so far, although satisfactory dating was achieved, the residual level after a long bleach remained appreciable (Mobbs, 1978); this casts doubt on the simple model of bleachable and unbleachable traps. It is possible that the residual level achieved during bleaching represents an equilibrium between trap filling and trap emptying; an alternative explanation is in terms of a feeder mechanism from difficult-to-bleach traps to the traps reached in transfer.

8.2.6 OPTICAL DATING

This technique does not involve thermoluminescence but *photostimulated luminescence* (PSL); it is an alternative way of measuring paleodose. As in phototransfer the electrons are evicted from traps by photons, but in this case the prompt luminescence is measured, that is, the light emitted by electrons that go immediately to a luminescence centre. Using green light from a laser (rather than the ultraviolet that was used by Mobbs in phototransfer) it has been found (Huntley, Godfrey-Smith and Thewalt, 1985) that very low residual levels can be achieved, at any rate in quartz; consequently the technique is particularly relevant to young sediments.

8.3 Laboratory Procedures

8.3.1 SAMPLE PREPARATION

Sample preparation follows the same general lines as for the techniques of earlier chapters, except of course that the crushing step is not necessary. In the fine-grain technique the 4–11 μm fraction is selected by settling in de-ionized water to which a peptizer, such as sodium oxalate, has been added to prevent flocculation. The water is removed with methanol and deposition onto discs is from acetone. Prior washing in hydrochloric acid is essential for sediments which are rich in carbonates.

The fine-grain approach is well-established as being satisfactory for loess; a strong argument for using fine grains rather than the larger grains of around 20–40 μm, usually more plentiful, is that the former are airborne longer. On the other hand, for samples such as from sand dunes there is the risk that fine grains may have percolated downwards in the deposit and consequently use of coarser grains is to be preferred for such samples. Consideration then needs to be given to the question of attenuation, within the grain, of the alpha particle dose. If the alpha activity of the grains being used for measurement is the same as for other grains, then there is no problem, though of course it is necessary to evaluate the alpha effectiveness by means of fine grains obtained by crushing; the alpha contribution in feldspar grains is likely to be relatively more important than in quartz grains because of the much higher alpha effectiveness in feldspar.

If the alpha activity in the sediment is predominantly external to the grains used for measurement, then the situation is similar to that pertaining for quartz grains in pottery. For samples in which grains of around 100 μm are present the procedure is to etch away the alpha-irradiated outer layer of grains so that the thermoluminescence observed from the inner cores remaining is due only to beta, gamma, and cosmic radiation. The situation of the alpha activity being predominantly external could arise from a coating deposited on the grains or because the alpha activity is located preferentially in fine grains, for instance. If it is necessary to use coarse grains smaller than 100 μm, then etching away of the alpha-irradiated layer is not practical; a possible approach then is to assess the alpha contribution to the annual dose by alpha irradiation of the coarse grains without crushing (see Appendix K).

Although separated quartz grains can be obtained by dissolving[5] away the feldspar it is not possible to obtain separated feldspar grains (except for the alkali feldspar fraction, which can be obtained using the heavy liquid technique, see Figure 6.14). However the thermoluminescence from a polymineral sample is predominantly from feldspar because that mineral is so bright; also, as discussed shortly, discrimination against quartz thermoluminescence can be further enhanced using an ultraviolet colour filter.

8.3.2 THERMOLUMINESCENCE MEASUREMENT; CHOICE OF COLOUR FILTER

Measurement techniques for sediment follow standard lines except that a slower heating rate is used, for example, 2–5°C/sec, particularly in measuring feldspar. This is because in the plateau region the glow-curve (see Figure 8.1) is sharply falling and a slight difference in thermal lag between glow-curves can cause serious mismatch, with consequent upset to the plateau. Feldspar samples are so bright that the reduction in thermoluminescence intensity due to the slower rate is acceptable, though a greater degree of patience is required on the part of the operator. Thermal lag can also be reduced, or at any regularized from disc to disc, by use of a heat transmitting paste such as Wakefield compound.

For normalization between portions, as is required in all of the methods of Section 8.2, it is advantageous to use the zero-glow monitoring technique mentioned in Section 5.2. There is a suitable peak at around 150°C

8.3 Laboratory Procedures

from feldspar, and for samples predominantly composed of quartz there is the well-known peak at around 100°C (but see note 13 of Chapter 5).

Choice of Colour Filter (for Thermoluminescence Measurement). Until 1983 it was common practice to interpose a blue-pass filter (such as Corning type 5-58) between sample and photomultiplier. However there are strong advantages to be gained by judicious choice of filter; it has been shown (Debenham and Walton, 1983) that by use of a filter that does not transmit blue light but only ultraviolet, the small quartz contribution to the thermoluminescence from a polymineral sample can be reduced by an order of magnitude. Although the quartz contribution is small, it is advantageous to make it negligible since, being more resistant to bleaching, it is a relatively large component in the thermoluminescence remaining after a long illumination. Also, the comparatively early onset of saturation in quartz means that its presence may upset the simplicity of the linear growth that is obtainable from feldspar up to a fairly high level of dose (around 2000 Gy). On the other hand, if the feldspar is afflicted by anomalous fading then a quartz-rich sample needs to be used.

Figure 8.6 shows spectra for a polymineral sample and for the quartz extracted from it; note the absence of thermoluminescence emission from quartz in the wavelength region below 400 nm. Figure 8.7 shows the effect on glow-curve shape of using an ultraviolet filter (Schott type UG 11) which cuts off wavelengths longer than 380 nm; the commonly observed shoulder on the glow-curve in the 350°–400° region is removed, this shoulder corresponding in temperature to the peak observed from the separated quartz (the shoulder also contains a blue-green emission from feldspar). The more rapid bleaching of the thermoluminescence in the ultraviolet region compared to that in the blue region is illustrated in Figure 8.8.

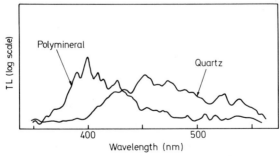

Figure 8.6 Thermoluminescence spectra for fine-grains of wind-blown sand (from Debenham and Walton, 1983); uncorrected for wavelength dependence of spectrometer response.

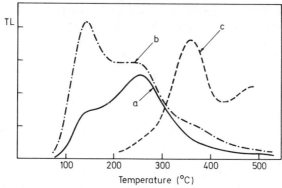

Figure 8.7 Thermoluminescence from fine-grains of loess (from Debenham and Walton, 1983). a, polymineral sample observed using a UV (Schott UG 11) filter; b, polymineral sample using a blue (Corning 5-58) filter; c, separated quartz fraction in the blue region. Note that the shoulder in the 350°–400° region on curve b corresponds to the peak in the quartz glow-curve. All curves are first-glow, with a dose of 100 Gy added to the natural dose.

In discussion throughout this chapter we assume that there are no other thermoluminescence minerals present in the samples except feldspar and quartz. When other minerals contribute significantly and have undesirable characteristics (e.g., resistance to bleaching, non-linearity) their interference may be removable by means of a suitable colour filter. Debenham and Walton (1983) cite the case of a recent sediment which had an abnormal glow-curve shape and a residual thermoluminescence equivalent to 700 Gy when a blue filter was used for measurement; the residual

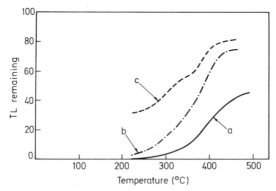

Figure 8.8 Percentage of thermoluminescence remaining after bleaching for 16 hr with an artificial sunlight source (from Debenham and Walton, 1983). a, polymineral sample observed using a UV (Schott UG 11) filter; b, polymineral sample using a blue (Corning 5-58) filter; c, separated quartz fraction in the blue region.

8.3 Laboratory Procedures

equivalent dose was reduced to 20 Gy when the ultraviolet filter was used. They also note that, for polymineral samples from a variety of sites, the glow-curve shapes were fairly uniform when the ultraviolet filter was used, in contrast to the wide variation in shape obtained with the blue filter; they suggest that the thermoluminescence emissions in the ultraviolet are dominated by a single type of trap.

8.3.3 THE LIGHT SOURCE FOR BLEACHING

The spectrum of sunlight extends from ultraviolet to infrared, with maximum intensity in the visible region (see Figure 8.9). It is the shorter wavelengths that are most effective in evicting electrons from traps, but as can be seen from Figure 8.10, for this particular alkali feldspar, some bleaching occurs even for orange light. With respect to quartz Figure 8.11 shows spectra for excitation of phototransfer in quartz; it is to be expected that the spectrum for bleaching will be similar.[6] The very sharp rise with decreasing wavelengths in the case of one of the quartz samples is in marked contrast to the flat dependence of the other and the substantial bleaching throughout the visible spectrum for the feldspar. While it is evidently dangerous to generalize, the sharp rise seen for one of the

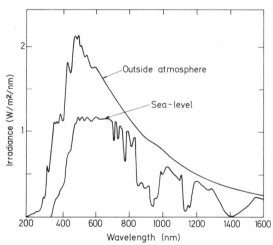

Figure 8.9 Spectra of sunlight: outside the atmosphere and at sea-level with the sun 60° below the zenith (from Solar Simulator handbook of Oriel Corporation, based on work by M. P. Thekaekara). The vertical scale gives the power received per unit area per unit wavelength interval. Absorption in the ultraviolet is due primarily to ozone and in the infrared to water vapour and carbon dioxide; the sea-level spectrum assumes that average amounts of these gases are present in the atmosphere.

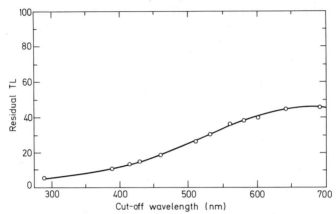

Figure 8.10 Residual thermoluminescence after sunlight exposure for 4 hr, for sample of feldspar (330°C peak). Samples were covered by Wratten filters that transmitted only light of longer wavelength than indicated; the steady rise across the spectrum indicates that a wide range of wavelengths are effective in bleaching (from Kronborg, 1983).

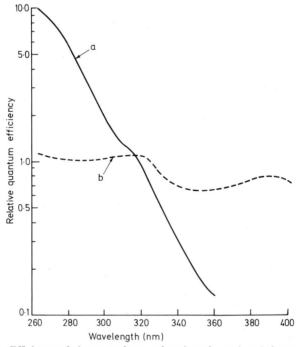

Figure 8.11 Efficiency of phototransfer as a function of wavelength for two quartz samples: a, for Norwegian alpha-quartz; b, for quartz extracted from Romano-British pottery (from Bailiff et al., 1977). The vertical scale is a measure of the phototransfer per incident photon.

8.3 Laboratory Procedures

quartz samples does suggest the possibility that quartz's reputation of being hard to bleach could be due to the absence from sunlight of short-enough wavelengths, at any rate at sea-level.

For laboratory bleaching it is common to use the type of sunlamp sold for indoor sunbathing. These are usually medium-pressure mercury lamps, and although there is a weak continuous spectrum most of the energy is emitted as a discontinuous spectrum, see Figure 8.12. Although three of the emission wavelengths are within the wavelength range of the main maximum of the solar spectrum, the strongest wavelength is marginal. A more representative spectrum is obtained by interposing a colour filter which transmits only the three wavelengths mentioned.[7]

A better but more expensive light source is the Solar Simulator.[8] In this a xenon discharge lamp is used and most of the emitted energy is carried in a continuous spectrum; after modification by means of filters a good match to the solar spectrum is obtained, see Figure 8.13. Different filters can be inserted according to the amount of atmospheric absorption it is desired to simulate. From the 300-W model the intensity also is approximately the same as for sunlight, that is, around 1 W/m^2/nm interval of wavelength; the total intensity at sea-level with the sun at an altitude of 30° is about 850 W/m^2. It is usual to place the samples on a water-cooled plate.

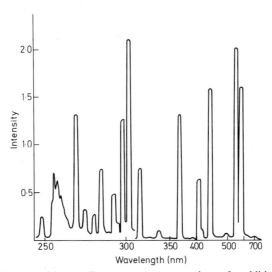

Figure 8.12 Spectrum from medium-pressure mercury lamp. In addition to the discrete emissions, there is a weak continuous spectrum; the relative intensities of the different emissions vary with pressure and other details of the particular type of lamp. There is a scale change at 310 nm, the right-hand section being expanded by a factor of 16; no correction has been made for detector response (from Parker, 1968).

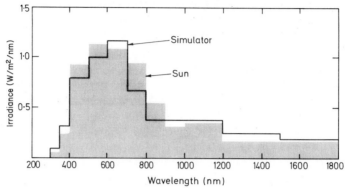

Figure 8.13 Spectrum from solar simulator using 'Air Mass Two' filtering which corresponds to sunlight at sea-level with the sun 60° below the zenith, the spectrum for which is also shown (from Solar Simulator handbook of Oriel Corporation). The spectra have been averaged over broad wavelength intervals for convenience of comparison.

Although a reasonable match between the laboratory light source spectrum and the solar spectrum seems desirable intuitively, it is not at present established how good this match needs to be. Using either the additive dose method or the regeneration method the presence in the laboratory source of a strong ultraviolet component of shorter wavelength than the limit of the solar spectrum gives risk of removal of thermoluminescence from what is residual as far as sunlight is concerned; this would lead to an over-estimate of the paleodose. With the partial bleach method the effect of the stronger ultraviolet component a rough compensation can be obtained by using a shorter illumination time; however it might be expected that the plateau will be poor.

Transmission through Water. For sediments deposited from water, whether in lakes or in the sea, there will have been some modification of the spectrum because the green to yellow region is better transmitted than longer and shorter wavelengths, see Figure 8.14. However the depth at which most of the bleaching occurred is unknown and so it is not feasible to think of matching spectra. Direct experimental confirmation that bleaching of feldspar thermoluminescence occurs at a depth of 7 m has been reported by Kronborg (1983).

Effect of Temperature. It is to be expected that elevation of temperature will allow bleaching to proceed more rapidly because the trapped electrons then have a higher initial energy and less is needed for eviction. Data for 'phonon-assisted' phototransfer in calcium fluoride (Sunta, 1971; Bailiff *et al.*, 1977) imply a 12% increase in transfer efficiency for illumina-

8.4 Stability; Age Range

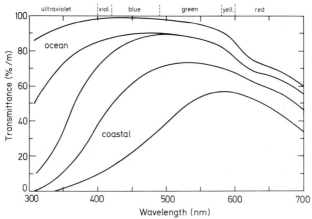

Figure 8.14 Transmittance of downward irradiance for high solar latitudes in the upper layers of several types of oceanic and coastal water (from Jerlov, 1970). The downward irradiance is the energy received per unit time on a horizontal surface facing upwards.

tion at 30°C instead of 20°C, and an increase by a factor of 2 at 100°C. Temperature dependence dictates that during laboratory bleaching the sample should be kept at constant temperature. There is also possible relevance to natural bleaching, for example, for sediment that has been high up in the atmosphere the bleaching would have occurred at relatively low temperature.

8.4 Stability; Age Range

The predicted lifetimes, for storage at 15°C, of the high temperature thermoluminescence traps in quartz are of the order of 100 million years (see Appendix E). Consequently there are no doubts about quartz having adequate stability; however beyond about 50,000 yr there are complications due to onset of saturation except in circumstances where the annual dose is abnormally low. The data available for the high temperature peak at *circa* 320°C exhibited by coarse grains of feldspar indicate a more than adequate lifetime for that mineral too. So here also the age range is limited by the onset of saturation, but in this case there is linearity up to around 2000 Gy, implying that a million years should be within reach. This assumes that the feldspars concerned are not afflicted with anomalous fading; although the results currently being obtained in Scandinavia are encouraging in this respect it can not be taken as the general rule and stability tests should be rigourously applied.

Since the thermoluminescence from polymineral fine-grain samples is presumed to be predominantly from feldspar the same stability and age range may be available, as long as anomalous fading is absent. However, the 320°C feldspar peak is not dominant in the fine-grain glow-curve and it is not clear to what extent the thermoluminescence in that temperature region represents the tail of a lower temperature peak with correspondingly shorter lifetime. The predicted lifetime for the peak at 220°C exhibited by the ocean sediment samples of Wintle and Huntley (1980) is 10,000–100,000 yr for storage at the ocean floor temperature of 4°C but lower by a factor of 5 for storage at 10°C. Although the practical evidence that good plateaux are obtained for fine-grain loess samples suggests that this estimate is pessimistic, there are indications that such samples will not reach beyond 100,000 yr; this is a vital aspect needing further investigation.[9]

8.5 Annual Dose

The unusual feature of sediments is that sample and surroundings are one and the same as long as the layer is sufficiently thick. Standard techniques for radioactivity measurements can be used but an alternative is the use of on-site measurements with a gamma spectrometer (see Appendix L) to derive the alpha and beta components in addition to the gamma; however laboratory measurements are still necessary in testing for disequilibrium effects. As with the standard techniques a much more important source of error is the usual one of uncertainty about the average water content since deposition; appropriate procedures have been discussed by Rendell (1985).

As concerns ocean sediments there are two unusual features. One is the very high water content (around 50%), and the other is the effect of precipitation of thorium-230 and protactinium-231 from sea water. On the ocean bed the activity of these two are in excess of that of the respective parents but deeper in the sediments there has been time for decay, the respective half-lines being 75,000 yr and 33,000 yr. Thus the annual dose will have decreased with time; appropriate equations have been given by Wintle and Huntley (1980).

8.6 Application

This technique is currently in such a rapid stage of development that any detailed discussion of results achieved and difficulties encountered would quickly become obsolete. For instance it would be misleading to set aside certain types of sediment on account of a substantial paleodose having been reported for zero-age samples; on present knowledge this could be because there was too much ultraviolet in the laboratory lamp used for bleaching or because the spectral response of the thermoluminescence detector was not appropriately restricted so as to discriminate in favour of the easy-to-bleach component. Conversely the apparently poor stability of fine grains from loess may be alleviated if difficult-to-bleach components are utilized. Methods for dating sediments that are only partially bleached have been discussed by Mejdahl (1985b).

The proceedings of the third and fourth *Specialist Seminars on Thermoluminescence and ESR Dating* held respectively at Helsingor in 1982 and Wörms, near Heidelberg, in 1984 contain accounts of recent application; these are published in *PACT volume 9* and in *Nuclear Tracks and Radiation Measurements volume 10*. A review of earlier work has been given by Wintle and Huntley (1982); two recent application reports of particular interest are those by Prescott *et al.* (1983) and Wintle *et al.* (1984).

Technical Notes

[1] Some geologists favour a non-aeolian origin. A compendium of papers on loess formation and transportation has been edited by Smalley (1975) and a recent review given by Pye (1984).

[2] In early research only the alpha contribution to the thermoluminescence was taken into account but with no allowance for reduced effectiveness or for attenuation in the 5–50 μm grains used. Discussions of this work have been given by Dreimanis *et al.* (1978), Hutt and Smirnov (1982), and Wintle and Huntley (1982).

[3] Following on from earlier thermoluminescence dating at Dartmouth College, U.S.A., of ocean sediment using the calcitic remains of foraminifera (Bothner and Johnson, 1969), a study was being made of the thermoluminescence of siliceous remains of radiolaria (Huntley and Johnson, 1976); in both cases there was an increase of paleodose with depth. In extension of this latter study it was found that the thermoluminescence was not coming from the siliceous material itself but from adhering grains of sediment.

[4] This 'hardening' of the residual thermoluminescence parallels the time dependence of anomalous fading. As there, reported data often shows the thermoluminescence remaining to be initially dependent on log t, implying that the probability of eviction is proportional to t^{-1}. The simplest explanation is that within the sample there are different types of traps having different degrees of bleachability. An alternative model, paralleling that proposed for anomalous fading, is that there is a single type of trap but that eviction becomes progressively more difficult as hypothetical acceptor centres for the evicted electron become more and more difficult to reach, the easily accessible ones having been used up.

[5] For coarse grains this is done with hydrofluoric acid, as for pottery. For fine grains a quartz-rich sample can be obtained by means of immersion for several days in hydrofluosilicic acid (H_2SiF_6) as described by Berger et al., (1980); the acid needs to be pretreated by adding 4–88 μm commercial silica and storing at 4°C for at least three days with occasional stirring. After removing the commercial silica by centrifuging, the acid is added to the sample to give a weight ratio of liquid : solid of 40 : 1. This mixture is stored for 3–6 days for each 50 mg of feldspar present. Prior to the treatment with H_2SiF_6 the sample should be washed in HCl to remove iron oxides.

[6] It is sometimes noted that use of a laboratory light source with a strong ultraviolet component gives rise to phototransferred thermoluminescence in the glow-curve region being used for dating, that is, from deep donor traps not seen in a glow-curve terminating at 500°C; this is not seen for the same degree of bleaching when using a source comparatively weak in ultraviolet.

[7] The effect of interposing different filters has been investigated by Berger et al. (1984). Use of a Corning CS3-73 cuts off wavelengths shorter than 420 nm, thereby transmitting only the three lines that are within the main solar maximum. Other filters used by these authors were the Corning types CS0-52 (360 nm) and CS3-67 (550 nm).

[8] The version for which the spectrum is shown in Figure 8.13 is manufactured by the Oriel Corporation, 15 Market St., Stamford, Connecticut 06902.

[9] Recent reports on the stability of the thermoluminescence signal from fine-grain loess have been made by Wintle (1985) and by Debenham (1985); the latter suggests that the upper limit to the age range may be in the region of 100,000 yr. On the other hand, geologically acceptable older ages have been obtained for coarse-grain Scandinavian feldspar sediments (e.g., Kronborg, 1983). There is relevant continuing discussion in Vol. 3 of *Ancient TL*, 1985.

APPENDIX A

The Age Equation

As is explained in Chapter 5, besides having different thermoluminescence effectiveness, alpha particles also differ from less heavily ionizing radiation with respect to the linearity of the thermoluminescence growth characteristic; the supralinear region does not occur with alpha particles because of the absence of track overlap. If the latent thermoluminescence acquired during antiquity is written as $G = G_\alpha + G_1$, where G_α is the component due to alpha particles and G_1 the component due to the lightly ionizing radiations (beta, gamma, and cosmic), then

$$G_\alpha = \chi_\alpha D_\alpha A, \qquad (A.1)$$
$$G_1 = \chi_1(D_1 A - I), \qquad (A.2)$$

where A is the age and χ and D are the thermoluminescence effectiveness and the annual dose of the radiations indicated. Supralinear growth has been illustrated in Figure 2.2, and so long as the natural thermoluminescence is above the initial non-linear portion (and below the onset of saturation), the second equation is valid, I being the intercept of the linear portion on the dose axis. Hence we have

$$A = \frac{G + \chi_1 I}{\chi_\alpha D_\alpha + \chi_1 D_1}. \qquad (A.3)$$

Denoting G/χ_1 by Q, the equivalent dose, and χ_α/χ_1 by k, the alpha particle effectiveness, we have

$$A = \frac{Q + I}{kD_\alpha + D_1} = \frac{P}{D'_\alpha + D_1} \qquad (A.4)$$

since the paleodose P has been defined as being equal to $Q + I$ and the effective alpha dose-rate D'_α is equal to kD_α. If the a-value system is being used then, as discussed in Appendix K, $D'_\alpha = 1.28a\alpha$ Gy/kyear, where α is the number of alpha counts per kilosecond for a thick sample of diameter 42 mm.

APPENDIX B

Age Evaluation and Assessment of Error Limits

This appendix follows the lines developed in two papers (Aitken and Alldred, 1972; Aitken, 1976) which had as objectives the collation of the various equations on which a thermoluminescence date is based and the formalization of a systematic procedure for calculating the overall error in the date corresponding to quantifiable uncertainties in the various measurements and parameters on which the date is based. The sections on error evaluation are not recommended at a first reading, but the equations given for age calculation may be found useful at an earlier stage.

An alternative approach in error-limit assessment is the more empirical one of carrying out a test programme on known-age samples and observing the degree of accuracy achieved. The drawback to this approach is that the accuracy attainable varies from sample to sample and site to site, depending on the thermoluminescence and radioactive characteristics of the sample. For example, if sample and soil are relatively rich in potassium, then the error in age arising from uncertainty in the degree of radon escape is likely to be unimportant; similarly, the date for a sample having a high moisture-holding capacity will have a higher error due to uncertainty about the wetness of the site than that for a sample which is relatively impervious. Test programmes are certainly an important step in the technique's development, but they should be seen in the role of determining whether the observed degree of accuracy corresponds to the predicted degree; if it is greater, then evidently there are unsuspected sources of error or the known sources have been underestimated. One source of error that is particularly difficult to assess is the effect of leaching and deposition of radionuclides by groundwater (see Section 4.2.5).

RANDOM AND SYSTEMATIC ERRORS

When it comes to averaging the dates of a group of samples known to be coeval, it is necessary to know which sources of error are correlated from sample to sample and which are uncorrelated. For instance, the error due to inaccuracy in radioisotope source calibration is the same for all samples and averaging the dates for a context cannot reduce it. On the other hand, measurement errors are likely to be in one direction for some samples and in the opposite direction for others; consequently, this type of error is reduced by averaging. Whatever other arguments may be adduced about the meaning of the words 'random' and 'systematic', in what follows the criterion for division is whether or not it is to be expected that averaging will produce reduction.

AGE CALCULATION

The thermoluminescence age is given by

$$A = \frac{P}{D'_\alpha + D_\beta + D_\gamma + D_c}, \tag{B.1}$$

where P is the paleodose and D represents the annual dose for the type of radiation indicated by the subscript. D'_α is the effective dose to alpha particles after making allowance for the reduced effectiveness of that type of radiation in inducing thermoluminescence. In the two basic methods $P = Q + I$ where Q is the equivalent dose and I the supralinearity correction (see Appendix A).

In the discussion that follows it will be convenient to subdivide D_β into $D_{\beta,K}$ and $D_{\beta,Th,U}$, and likewise for D_γ. It is also useful to use D for the total effective annual dose, that is, for the denominator of the above equation, and to denote the fractional components of this by

$$f_\alpha = D'_\alpha/D, \qquad f_\beta = D_\beta/D, \qquad \text{etc.}$$

Values for f corresponding to the 'typical' sherd and soil of Tables 1.1 and 4.5 are given in Table B.1.

The annual dose values to be inserted into the age equation need to be corrected for the effect of absorbed water in sample and soil; as discussed in Section 4.2.3, a given water content affects the different types of radiations to differing degrees. The correction is made in terms of

$$W = \text{(saturation wet weight-dry weight)/(dry weight)}$$

Appendix B: Age Evaluation

TABLE B.1
Typical Values for Fractional Components of Annual Dose

Fractional component	Fine-grain dating	Inclusion dating
f_α	0.45	0
f_β	0.30	0.56
$f_{\beta,K}$	0.16	0.30
$f_{\beta,Th,U}$	0.14	0.26
f_γ	0.22	0.39
$f_{\gamma,K}$	0.05	0.08
$f_{\gamma,Th,U}$	0.17	0.31
f_c	0.03	0.05

and

F = (average water content during burial)/(saturation water content).

Whereas laboratory measurement can be made of W, for F it is a matter of best estimate having regard to such indications about past climate as the paleoclimatologist and the soil scientist may be able to provide. It is usually necessary to place rather wide error limits on the value of F in most circumstances, for example, ±0.2, and unless W is small, for example, less than 0.05, this gives rise to a serious basic limitation on the dating accuracy that can be achieved.

There are a variety of techniques for evaluation of annual dose, as discussed in Chapter 4. This appendix will be presented in terms of the 'traditional' methods: alpha dose by means of alpha counting, beta dose either by thermoluminescence dosimetry or alpha counting (plus chemical analysis of potassium), and gamma dose by on-site thermoluminescence dosimetry. The equations developed should be adaptable to other methods by reference to the tables given in Appendix G; see also Mangini *et al.* (1983). It is assumed that all laboratory measurements are made in the dry state; with respect to the gamma dose, correction needs to be made for the difference in wetness between the average assumed for antiquity and the value pertaining while the capsule was buried.

Alpha Dose. From Table G.4 we see that for a sample having equal thorium and uranium activities, the dose-rate corresponding to an alpha count-rate of 1 count/ksec (for a 42-mm diameter scintillator and an electronic threshold of 0.835) is 1.56 Gy/per kyear (ka); hence on the k-value system

$$D'_\alpha = \frac{1.56 k \alpha_B}{1 + 1.5WF}, \qquad (B.2)$$

where α_B ks^{-1} is the count-rate estimated as appropriate to the degree of radon retention during burial. For the purpose of error analysis, we need to have regard to the way in which k depends on measured quantities: from Section 2.2 we see that $k = Q/Q_\alpha$, where Q is the equivalent dose evaluated with a beta source and Q_α that evaluated with an alpha source. Thus in the age equation Q appears in the denominator as well as the numerator.

If the preferred a-value system (see Appendix K) is in use, then

$$D'_\alpha = \frac{1.28 a \alpha_B}{1 + 1.5 WF}. \tag{B.3}$$

In this system a is determined from the time of alpha irradiation, y, that induces a level of thermoluminescence equal to the natural thermoluminescence, being measured by the additive dose method of Figure 2.2 except that the horizontal axis now represents time of irradiation. If S μm^{-2} per unit time is the strength of the alpha source, then

$$a = Q/13Sy \tag{B.4}$$

It is assumed in Equations B.3 and B.4 that the absorbed water has full effect on the alpha dose-rate. As discussed in Section 4.2.3, alpha particles do not necessarily traverse water-filled pores; if not, the effect of wetness will have been overestimated.

Beta Dose. If thermoluminescence dosimetry is used, then we have

$$D_\beta = \frac{bD_B}{1 + 1.25 WF}, \tag{B.5}$$

where D_B is the dose-rate measured for the sample in the dry state but with a correction made for any difference in the degree of radon escape in this condition and the degree estimated to pertain during burial and b is the beta attenuation appropriate to the grain size used (for fine-grains $b = 1.00$; for the etched cores of 100-μm quartz grains, it is taken as 0.90; see Appendix C).

If the beta dose is derived from alpha counting and chemical analysis for potassium, then

$$D_{\beta,Th,U} = \frac{0.072 b \alpha_B}{1 + 1.25 WF}, \tag{B.6}$$

$$D_{\beta,K} = \frac{0.853 bm}{1 + 1.25 WF}, \tag{B.7}$$

where m is the percentage by weight of potassium present in the sample; if potassium has been assessed as K_2O, then the percentage given needs to

Appendix B: Age Evaluation

be multiplied by 1.205 to obtain m. A contribution from rubidium has been included in Equation B.7 on the basis that the K : Rb ratio is 200 : 1.

Gamma Dose. Although burial of a thermoluminescence capsule in the soil might be thought to give a direct measure of the gamma dose-rate, in fact allowance needs to be made for differences between the attenuation due to the capsule wall and the average attenuation in the portion of the sample used for thermoluminescence; there is also a small effect due to the intrinsic difference in low-energy response between the thermoluminescent phosphor and the sample. As discussed in Appendix L, for a copper capsule having 0.07-mm thick walls, the dose recorded by fluorite within it is 10 ± 5% lower than the true infinite matrix dose to quartz-like grains. As regards attenuation in the sample, the principle of superposition tells us that if the external gamma dose is attenuated by a percentage p, then the self-dose from the internal radioactivity of the sample must make a contribution equal to $10^{-2} p$ times the infinite matrix gamma dose corresponding to that radioactivity. Hence the internally generated gamma dose-rate, in grays per kiloyear, is given by

$$D_{\gamma,\mathrm{i}} = \frac{10^{-2}p(0.085\alpha_\mathrm{B} + 0.241m)}{1 + 1.14WF}. \tag{B.8}$$

The numerical factors in the numerator, obtained from Table G.4, assume equal thorium and uranium activities. Estimation of p is discussed in Appendix H.

The value of 10% for attenuation in the capsule wall does not apply to the cosmic-ray component. Hence denoting this attenuation factor by g (i.e., $g = 1.10$ for the case considered) the external gamma dose-rate is given by

$$D_{\gamma,\mathrm{e}} = \left(\frac{100 - p}{100}\right) g(D_\mathrm{p} - D_\mathrm{c}) \frac{(1 + 1.14W'F_\mathrm{p})}{(1 + 1.14W'F)}, \tag{B.9}$$

where D_p is the capsule dose-rate, W' is the fractional saturation water content for the soil, and F_p is defined similarly to F but relates to the period of capsule insertion.

If sample and soil have the same (wet) radioactivity, then

$$\frac{(0.085\alpha_\mathrm{B} + 0.241m)}{(1 + 1.14WF)} = g(D_\mathrm{p} - D_\mathrm{c}) \frac{(1 + 1.14W'F_\mathrm{p})}{(1 + 1.14W'F)}, \tag{B.10}$$

and in this case

$$D_\gamma = g(D_\mathrm{p} - D_\mathrm{c}) \frac{(1 + 1.14W'F_\mathrm{p})}{(1 + 1.14W'F)}. \tag{B.11}$$

The same holds to a good approximation for typically sized samples having a radioactivity level that is not too different to that of the soil; for example, if it is 30% higher and $p = 10$, then use of Equation B.10 underestimates the gamma dose-rate by only 3% and this has a negligible effect on the age.

Cosmic Radiation. Except for sites of exceptionally high altitude (see Appendix I) or very low radioactivity, D_c is less than 20% of D_p, and it is adequate to omit D_c from the denominator of Equation B.1 and to compensate for this by omitting its subtraction from the capsule dose-rate in Equations B.9 and B.11. This is equivalent to assuming that cosmic rays are attenuated by capsule wall and moisture to the same extent as gamma rays, whereas in fact the attenuation is negligible.

RANDOM ERRORS

The procedure for error-limit assessment, whether the sources are random or systematic, is to calculate the percentage error in the date corresponding to a given error in each of the quantities or parameters on which the date is based and then to obtain the overall error as the square root of the sum of the squares of the individual errors. This assumes that the sources are uncorrelated and that the scatter of values for a quantity has a Gaussian distribution, the standard deviation σ being used as a measure of the error limit. Hence there will be a 68% probability that the true date lies within the error limits so derived, and a 95.5% probability that it lies within limits twice as wide.

Thermoluminescence Measurements. If δQ, δI, and δa are the operator's estimate of measurement uncertainties in Q, I, and a, at the 68% level of confidence, as elsewhere, then the corresponding percentage error in age, σ_1, is given by

$$\sigma_1^2 = \left\{100\left(1 - \frac{Q+I}{Q}f_\alpha\right)\frac{\delta Q}{Q+I}\right\}^2 + \left\{\frac{100\,\delta I}{Q+I}\right\}^2 + \left\{100 f_\alpha \frac{\delta a}{a}\right\}^2. \quad \text{(B.12)}$$

This forbidding expression is obtained by differentiation of Equation B.1 after substitution from the succeeding equations. Note that in the composite equation so obtained Q appears also in the denominator through its involvement in Equation B.4; the uncertainty in a resulting from uncertainty in Q is thus included in the first term of Equation B.12 and $\delta a/a$ should be put equal to $\delta y/y$. In practice, σ_1 is usually around 5%.

Annual Dose. Taking the random uncertainty in alpha count-rate to be ±5% (composed of 3% statistical counting error plus allowance for scintil-

Appendix B: Age Evaluation

lation efficiency variation from screen to screen) and the uncertainties in beta and gamma thermoluminescence dosimetry measurements to be δD_L and δD_p, respectively, the consequent percentage uncertainty σ_2 in the age is given by

$$\sigma_2^2 = 25f_\alpha^2 + \left\{100f_\beta \frac{\delta D_L}{D_L}\right\}^2 + \left\{100f_\gamma \frac{\delta D_p}{D_p}\right\}^2. \tag{B.13}$$

In practice the errors in D_L and D_p are usually about 5% also, in which case

$$\sigma_2^2 = 25\{f_\alpha^2 + f_\beta^2 + f_\gamma^2\}, \tag{B.14}$$

so that after substituting 'typical' values from Table B.1 we find σ_2 is equal to 3.0% for fine-grain dating and 3.6% for inclusion dating. That it is less than 5% is because the contributory uncertainties are uncorrelated. No allowance has been included for measurement error in moisture content determination because the resultant uncertainty in age is usually negligible; the same holds with respect to the cosmic-ray dose. If the beta dose-rate is derived from the alpha count-rate and potassium analysis, then the age error σ_2' is given by

$$(\sigma_2')^2 = 25\{(f_\alpha + f_{\beta,\text{Th},U})^2 + f_{\beta,K}^2 + f_\gamma^2\}. \tag{B.15}$$

If there are stones present in the burial soil, there is additional random uncertainty; a rough estimate of this uncertainty is given by

$$\sigma_3 = 100rf_\gamma \left(\frac{D_1 - D_2}{D_\gamma}\right), \tag{B.16}$$

where D_1 and D_2 are the individual gamma dose-rates for soil and stone and r is the proportion by weight of stones present.

SYSTEMATIC ERRORS

Calibrations. Because thermoluminescence dating gives absolute dates in the sense of not being dependent on any pre-existing chronology, the accuracy of the calibrations on which dates are based is a very important consideration. The paleodose is dependent on calibration of the beta (or gamma) source used; with respect to annual dose the measurement devices are usually calibrated by means of standard sands for which the radioactive content has been established by some form of analysis. In principle, source calibration can be related to a standard sand by evaluation of the dose received by phosphor grains which have been immersed in the sand for a known length of time; the dates are then independent of any calibration except that of the clock used. There has been some progress in this approach (see, for instance, Aitken, 1968; Aitken, Huxtable, Wintle and Bowman, 1975; Fleming, 1979, p. 53; Wang, 1983), but it is impeded by unsuspected problems once high accuracy is sought. In practice most dating systems are still 'calibration dependent', at any rate to some extent, and the following error analysis is appropriate.

As regards the alpha contribution to the thermoluminescence, it should be noted that because Q enters into the numerator of Equation B.4 as well as the numerator of Equation B.1 there is independence of beta source calibration so long as I is small, as will be assumed. For this contribution there is, of course, dependence on the alpha source calibration and the alpha counter calibration.

As regards the beta contribution, even though the dose-rate is evaluated by thermoluminescence dosimetry, calibration independence does not usually obtain; this is because the dosimetry facility is calibrated by means of a radioactive sand, though direct immersion of phosphor grains in the sand (Bailiff and Aitken, 1980; Fleming, 1979) can avoid this need. If thermoluminescence dosimetry is not used, then there is, of course, dependence on the alpha counter calibration and the potassium analyser. It is only with respect to the gamma contribution that calibration independence pertains in most systems. Essentially, the dose received by the phosphor grains during burial of the capsule is evaluated in terms of the same beta (or gamma) source calibration that is used for evaluation of paleodose. Even if the attenuation factor g or the self-dose percentage p are derived using a sand of known composition rather than by calculation, the intrusion is only minor.

Given that the thermoluminescence age does depend on source calibrations, etc., in most systems, we are then faced with the question of making some assessment of the accuracy of these calibrations. In carrying out calibration of a reference beta source in the Oxford Laboratory, Murray (1981; see also Murray and Wintle, 1979) claimed an error limit of 3.5%; the main contributions to this were the 2% uncertainty in the gamma exposure against which comparison was made and a 2% uncertainty in the relative mass absorption coefficients which enter into the derivation of the absorbed dose in the thermoluminescent phosphor used. A number of laboratories base their source calibration on intercomparison with the Oxford calibration and, allowing for some additional uncertainty being introduced, it would seem that ±5% is a reasonable value to take. As yet there has been no comprehensive study of the uncertainty involved in alpha source calibration, but from the author's experience in this the same level of error limit seems reasonable, as is also the case for the other

Appendix B: Age Evaluation

calibrations involved. On this basis it can be shown that the resulting percentage error in the age σ_4 is given by

$$\sigma_4^2 = 25\{2f_\alpha^2 + f_\beta^2 + (f_\beta + f_c)^2\} \qquad (B.17)$$

in the case where thermoluminescence dosimetry is used for the beta dose (and for the gamma dose). The alpha source and the alpha counter calibrations contribute equally to the first term, the beta thermoluminescence dosimetry calibration gives rise to the second term, and the beta source calibration the third term. If thermoluminescence dosimetry is not used for the beta dose, then we have

$$(\sigma_4')^2 = 25\{f_\alpha^2 + (f_\alpha + f_{\beta,\text{Th},\text{U}})^2 + f_{\beta,\text{K}}^2 + f_c^2\}. \qquad (B.18)$$

In both of these equations it has been assumed that I can be neglected. Inserting 'typical' values, we obtain $\sigma_4 = 3.9\%$ and $\sigma_4' = 4.1\%$ for fine-grain dating, 3.1% and 3.8%, respectively, for quartz inclusion dating.

Parameter Uncertainties. Contained in the equations on which the dose-rates are based are several numerical factors and parameters, and some allowance needs to be made for uncertainties in these. For instance, the numerical factor of 1.28 in the expression for the alpha dose-rate is proportional to the value of η, as discussed in Appendix K, and there is a ±5% uncertainty in this. There is the same level of uncertainty in the parameter g by which allowance is made for gamma attenuation in the capsule wall, as discussed earlier in this appendix. There is similar uncertainty in the parameter by which allowance is made for gamma attenuation in the sample; for small samples having a level of activity not too dissimilar from the soil this is unimportant, but it should not be forgotten as a source of error in other circumstances. Finally, there is uncertainty in the coefficient b by which allowance is made for beta attenuation in the etched cores used in quartz inclusion dating; a limit of ±5% is appropriate here as well. The age error σ_5 corresponding to these three sources is given by

$$\sigma_5^2 = 25\{f_\alpha^2 + f_\beta^2 + f_\gamma^2\}, \qquad (B.19)$$

where the second term should be omitted in the case of fine-grain dating.

Thorium/Uranium Ratio. In deriving the beta dose-rate from the alpha count-rate, it is assumed that there is equal thorium and uranium activity in the sample. If one or other of these series dominates, the beta dose-rate so derived may be in error by 20%. Complete domination is unlikely, and if an activity ratio of no greater than 2:1 is taken as having a 68% probability, the corresponding error in beta dose-rate is ±10% and the corresponding error in age is

$$\sigma_6' = 10 f_\beta. \tag{B.20}$$

Of course, this applies only when the beta dose-rate is not determined by thermoluminescence dosimetry and no measurement is made of the ratio.

Radon Escape; Leaching; Radioactive Disequilibrium. Assessment of the uncertainty introduced by this source of error is highly complex. The degree of gas escape is likely to be different during laboratory measurement from what it was during burial, and during burial it may have varied since it is affected by wetness. Additionally, as discussed in Section 4.3.2, radon escape is one likely cause of 'overcounting' in measurement of the alpha count-rate. Finally, as discussed in the same section, radon escape may be symptomatic of radionuclide transportation by groundwater, in which circumstance the assumption that the annual dose has been constant has to be questioned. Whereas formerly attempts have been made to make a routine assessment of the age error associated with radon escape, it is now considered that this may give a false sense of security and that any sample or soil for which there is evidence of appreciable radon escape, that is, greater than 10%, should be set aside for special investigation.

Wetness. Finally, there is the error term corresponding to the uncertainty δF in the fractional water uptake F. The value to be used for δF depends on what is known about site rainfall and drainage in the past, and adequate allowance should be made for possible variations. With some approximation the corresponding percentage error in the age is given by

$$\sigma_7 = 100 \, \delta F \{(1.5 f_\alpha + 1.25 f_\beta) W + 1.14 f_\gamma W'\}. \tag{B.21}$$

For the 'typical' case and for $W = 0.2$, $W' = 0.3$, and $\delta F = 0.2$, the value of σ_7 is 5.9% for fine-grain dating and 5.8% for inclusion dating. On the other hand, if the sample and soil have somewhat lower saturation values, say, $W = 0.05$ and $W' = 0.15$, the error limits are only 1.9% and 2.2% for the two methods, respectively.

AVERAGING AND CITATION

The overall error for sample i from a context may be written as

$$\sigma_i = \{(\sigma_i)_r^2 + (\sigma_i)_s^2\}^{1/2}, \tag{B.22}$$

where $(\sigma_i)_r$ is square root of the sum of the random error variances σ_1^2, σ_2^2, and σ_3^2 for sample i, and similarly for the systematic errors. The best value for the age of the context is obtained by weighting the individual ages

Appendix B: Age Evaluation 251

according to

$$A = \frac{\sum A_i/\sigma_i^2}{\sum 1/\sigma_i^2}, \tag{B.23}$$

and the standard error on this average is given by

$$\sigma^2 = \sigma_s^2 + \sigma_r^2, \tag{B.24}$$

where

$$\sigma_s = \frac{\sum (\sigma_i)_s/\sigma_i^2}{\sum 1/\sigma_i^2} \tag{B.25}$$

and

$$\sigma_r^2 = \frac{1}{\sum 1/(\sigma_i)_r^2}. \tag{B.26}$$

The quantity

$$e = \sigma A/100 \tag{B.27}$$

gives the standard error in years. This is the predicted error on the basis of the foregoing discussion; it is also of interest to calculate the experimental root-mean-square deviation of the individual ages as given by

$$s = \frac{100}{A} \sqrt{\frac{\sum (A_i - \bar{A})_2}{(N-1)}}, \tag{B.28}$$

where N is the number of samples and $\bar{A}$ the linear mean of the individual ages. If it were valid to assume that all causes of error were random between samples, then the quantity

$$q = \bar{A}s/100 \; N^{1/2} \tag{B.29}$$

could be considered as the standard error on the average age. Although the causes of error are not, in fact, random, q is nevertheless a useful precision parameter for a context. Since the causes of error are not all random between samples, one expects q to be smaller than e, and if this is not the case then some unaccounted for source of error is intruding. Both e and q should be quoted in the citation if the date is based on three or more coeval samples, otherwise only e. The value of q can be considered as a minimum for the error limit and is relevant only when comparing different contexts of the same site or locality, whereas e is more realistic in comparisons of thermoluminescence dates with radiocarbon dates or other chronologies.

APPENDIX C

Attenuation of Annual Dose within Grains

As introduced in Chapters 1 and 2, the basic techniques of pottery dating are based on the simplifying assumptions that the fine-grains have received a full alpha dose, but that the etched cores of quartz grains (~ 100 μm) have received a negligible alpha dose and a beta dose attenuated by 10%. We are concerned here with the annual dose rather than the dose administered in artificial irradiation; the two differ substantially in energy spectrum, and therefore in attenuation, because the former is from radioactivity distributed in the pottery while the latter is from a source thin enough for self-absorption to be unimportant. Because of practical problems involved in the replication of the annual dose spectra by means of artificial sources, the treatment of annual dose attenuation is mainly theoretical; however, as we shall see, with respect to the etching process in particular, actuality is often formidably irregular, and the extent to which precise reliance can be placed on theoretical prediction is still open to question.

This appendix includes related topics in addition to attenuation itself; the sections are

1. attenuation of alpha dose;
2. etching of quartz grains;
3. attenuation of beta dose;
4. internal activity in quartz grains;
5. internal activity in feldspar grains.

Appendix C: Attenuation of Annual Dose

ATTENUATION OF ALPHA DOSE

The highest-energy alpha particle emitted in the thorium and uranium series is from Po-212, and it has a range, in quartz, of 47 μm; the lowest-energy alpha particle is from Th-232, with a range of 14 μm. The average ranges for the two series are 27 and 23 μm, respectively. However, in considering the attenuation of alpha particles within a radioactive-free grain embedded in a radioactive matrix, for example, a quartz grain in pottery, we must take into account that the particles incident upon the grain have already used up some of their energy in reaching the grain. For each energy of alpha emission the spectrum of particle ranges incident upon the grain spreads from a maximum corresponding to particles emitted very close to the grain down to zero for particles that can only just reach the grain; calculation shows that the range spectrum is approximately rectangular; that is, there are the same number of particles in each interval of range.

The basis of the various calculations for alpha attenuation relevant to thermoluminescence dating (e.g., Fleming, 1969, 1979; Bell, 1978, 1979a, 1980) are the tabulated values given by Howarth (1965) for the dose at various depths within spherical grains of different diameters; these values take into account dependence on particle energy of the energy loss per unit length of track (LET, linear energy transfer), assuming that the residual range of a particle of energy E is proportional to $E^{1.5}$. Figure C.1 shows, for a spherical quartz grain of diameter 100 μm, the decrease of dose with depth into the grain, and Figure C.2 shows the way in which the average dose in a spherical quartz grain depends on the diameter of the grain. These values have been calculated for the average energy of each series, and hence they only approximate reality; obviously, alpha particles that have higher-than-average energy will penetrate more deeply than implied by Figure C.1 and particles that have less-than-average energy will penetrate less. The variation shown applies strictly only to quartz; for feldspars the attenuation is slightly less so that for a given particle energy the range is slightly greater. However, it seems likely that any errors due to these effects are negligible compared to those arising from the actual grains being irregular in shape, particularly after etching.

Fine-Grain Dating. Included in Figure C.2 is an expanded plot for grain diameters up to 40 μm. From this we see that the average alpha dose for a 10-μm grain is 10% less than for a negligibly small grain, and that for a uniformly distributed spread of grains between 1 and 8 μm, as used for fine-grain dating, the average attenuation is close to 5%. It must be remembered, however, that the fine-grains are not necessarily free of alpha radioactivity, and therefore this value is an upper limit to the actual dimi-

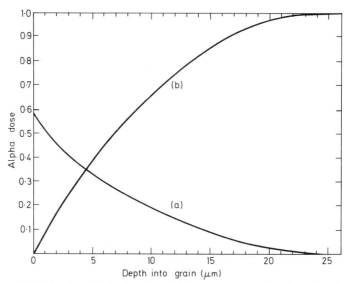

Figure C.1 (a) Alpha dose in the interior of 100-μm quartz grains, free of radioactive impurities, embedded in a matrix having equal uranium and thorium activities; the dose is expressed as a fraction of the dose that would be experienced by grains small enough for attenuation to be negligible (i.e., of diameter less than a few microns). Note that even at the surface the fraction is less than unity; this can readily be understood by considering a plane interface between matrix and quartz when alpha particles arrive from one side only so that the fraction in that case is 0.5. The values given are the average of the values given by Bell (1978, 1979) for the two series, and these latter were calculated assuming a mean range of 25.6 μm for the thorium series and 21.7 μm for the uranium series. (b) Fraction of the grains' overall alpha dose removed by etching away the surface (derived from Bell, 1978, 1979a).

nution of the alpha dose; in any case the alpha particle contribution to the thermoluminescence is less than half of the total, and so any such diminution is diluted sufficiently to become negligible in normal circumstances. Note, however, that if for some reason, for example, to reduce spurious thermoluminescence, somewhat larger grain-size limits are employed (by reducing the settling times), or if the initial grain-size distribution is biassed toward the larger limit, the attenuation may be non-negligible.

Coarse-Grain Dating. We see from Figure C.2 that the average alpha dose received by grains of diameter 100 μm is 23% of that received by fine-grains. Using the data of Table 1.1 and taking the alpha-particle effectiveness of quartz to be within the range 0.03 to 0.1, we find that the alpha-particle contribution to the thermoluminescence in 100-μm quartz grains will be within the 4–12% range of the total thermoluminescence for the 'typical' pottery and soil specified. As generally understood with respect

Appendix C: Attenuation of Annual Dose

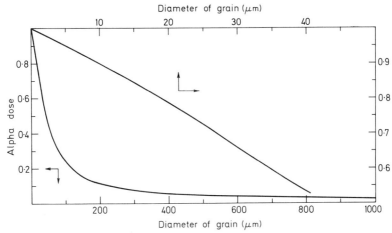

Figure C.2 Average alpha dose for radioactivity-free quartz grains embedded in a matrix having equal thorium and uranium activities; the dose is expressed as a fraction of the dose that would be experienced by a grain small enough for attenuation to be negligible (derived from Bell, 1980).

to the quartz inclusion technique, this contribution is made negligible by etching away the outer surface of the grains with hydrofluoric acid: for instance, Fleming (1979) mentions the removal of a 16-μm layer by 150 min of acid attack, and according to Figure C.1 this will remove nearly 90% of the alpha contribution, that is, reducing it to less than $1\frac{1}{4}$% of the total thermoluminescence in the case of the 'typical' pottery and soil just mentioned.

Further discussion of etching is given shortly. Because of problems and uncertainties in this process, the alternative procedure of making allowance for the alpha contribution has been proposed (Valladas and Valladas, 1982). To assess the alpha contribution (Guérin and Valladas, 1980; Guérin, 1982a; Valladas and Valladas, 1982) a monolayer of the grains is irradiated by a calibrated alpha source, the grains being carried on a vibrating pan so that the jumping up and down of the grains ensures that they receive an omnidirectional flux of particles as occurred during antiquity. The energy of the incident alpha particles is adjusted, by interposition of absorbers, to be 3.4 MeV, the effective average energy of the natural dosage.

ETCHING

Besides reduction of the alpha-particle contribution, acid treatment also eliminates most other minerals from the sample, giving a glow-curve that is 'cleaner' and more quartz-like. In sample preparation dilute hydrochloric acid should be used as a preliminary in order to remove calcite before the use of concentrated hydrofluoric acid for removal of feldspars. The HF does not remove zircons; zircons are undesirable because their high alpha activity would upset the validity of the assumption made in inclusion dating that there is negligible alpha contribution to the observed thermoluminescence; if they are present in sufficient quantity to make a significant contribution to the thermoluminescence, the glow-curve will not be quartz-like, and this indicates that heavy liquid separation should have been used for their removal.

Etching also improves the transparency of the quartz grains from pottery by removing the thin outer layer, estimated as 2 μm in thickness by Fleming (1979), into which impurities are likely to have diffused from the clay matrix during kiln firing; experimental evidence for this diffusion has been reported by Valladas (1977), who observed an enhanced magnetic susceptibility resulting from the iron oxide component. This impurity layer is undesirable also because of the additional glow-curve peaks resulting from the thermoluminescence traps and luminescence centres introduced. Besides making the glow-curve more complex, the associated alpha effectiveness is likely to be appreciably higher than for the more pure interior of the grain. Finally, the diffusion may bring in thorium and uranium from the clay (see the section on internal activity in quartz grains), weakening still further the validity of the assumption that the observed natural thermoluminescence does not contain an alpha contribution.

However, although etching is highly desirable, it remains questionable as to whether it can be achieved in a satisfactory manner. This is because of the tendency of quartz to etch preferentially along dislocation lines in the crystal lattice structure (Lang and Miuscov, 1967). This means that the assumption of isotropic etching may not be valid, and if so it will be erroneous to use the observed loss of weight during etching to calculate how much of the alpha-irradiated surface layer has been removed; indeed, the work of Bell and Zimmerman (1978) using optical and electron microscopy makes it doubtful whether there is any straightforward solution. They found that when viewed through a binocular microscope quartz grains extracted from the Lake Mungo hearths, and also from pottery from a variety of sources, could be divided into 'shiny' and 'frosty' grains; observation with the scanning electron microscope confirmed this division and revealed very deep etch pits in frosty grains after 40 min of treatment with 40% hydrofluoric acid, some grains being much more severely attacked than others, with some even losing a substantial part of their inner volume. In terms of weight loss they found that for the treatment just specified a sample of frosty grains lost 50% of its initial weight (which would correspond to removal of the outer 10 μm if the etching was isotropic) but a sample of shiny grains lost only 18% (corresponding to a layer of 3 μm). There was no difference in glow-curve shape between the two types, nor in thermoluminescence sensitivity.

The authors concluded that even for very long etching times it could not be guaranteed that the remaining material would not contain some grains which still carried an alpha contribution; quite apart from the practical difficulty of not having much sample left, this approach runs into the difficulty that there is appreciable diminution in beta dose in the core of the grain (see the next section), and uncertainty about the residual grain sizes therefore introduces uncertainty into the beta attenuation factor that should be used in the age equation. The alternative approach suggested by the authors, of etching for a very short time, just long enough to remove feldspars and surface discolouration from the quartz grains, would produce grains with minimum deviation from isotropic removal, thereby allowing a reliable evelution of the alpha dose itself. The difficulty with this, as the authors note, is that diffusion of impurities during firing may be deeper and may give rise to a gradient in alpha effectiveness; to make a reliable evaluation of the alpha dose contribution to the thermoluminescence would then require measurement of alpha response along the lines employed by the French authors mentioned above. It may be noted incidentally that Valladas and Valladas (1982) and Goedicke (1984) reported experimental studies of the effect of etching on alpha-irradiated quartz grains (100–160 μm).

Recapitulating from the foregoing paragraphs, the problems in etching are (i) that for a given time of immersion in acid some grains etch more rapidly than others and (ii) that the etching is not isotropic, so that even if the loss of weight were exactly known calculations of alpha dose removal and beta dose attenuation are only approximate. The difficulty in measuring the weight loss for the actual sample being measured is that there may be feldspar grains present, in which case the observed loss will be greater than the actual loss of quartz. In default of a better course of action, it is usual to estimate the quartz loss for a given time of acid immersion by reliance on data obtained for geological quartz; according to the data of Fleming (1969), an hour's immersion in 40% HF produces a weight loss corresponding, for isotropic etching, to the removal of a 9-μm layer, and Bell (1979a) suggests this as a reasonable compromise between the risks of inadequate elimination of alpha dose on the one hand and too much deviation from isotropic removal on the other. Even assuming isotropic etching, removal of a 9-μm layer only makes the alpha contribution comfortably negligible if the potassium content of the clay is above average; however, in view of the other uncertainties involved, it would be misleading to recommend making allowance for the residual alpha dose in the case of clay which has low potassium content.

At the time of this writing it is evident that there is the need for more research into the various unsatisfactory aspects, and in particular there is the need to develop routine techniques for checking the response to etching of a particular sample being dated. On the other hand, the quartz inclusion technique has had wide utilization, and the general consonance of the results with other dating techniques suggests that the various adverse effects, if present, are mutually compensating.

BETA ATTENUATION

As with alpha dose, there are two factors to be evaluated with regard to beta attenuation: first, the factor by which the average dose to a grain of given size is smaller than the dose to a fine-grain (for which there is negligible attenuation); and second, the factor by which that dose is reduced if a given thickness is removed by etching (because the outer layer receives a higher-than-average dose). Of course, these factors are very much closer to unity than for alpha dose because of the much greater penetrating power of beta particles. Bell (1979a) evaluated both factors with respect to 100-μm grains; Mejdahl (1979) evaluated the first factor (i.e., for unetched grains) for grains having diameters in the range 5 μm to 10 mm and considered the effect of etching with respect to 100-μm grains. There is excellent agreement between the results obtained by the two authors with respect to the beta dose itself, though there are differences with respect to the combined contribution from internal conversion electrons and Auger electrons (this contribution, usually denoted IC, is included with the beta dose-rate in the age equation, being 13% of the whole in the case of the thorium series and 7% in the case of the uranium series).

The calculations of Mejdahl are based on those of Berger (1971, 1973) for water-equivalent media with conversion to quartz grains in pottery by means of the scaling procedure of Cross (1968). The approach used is to begin by calculating the self-dose of a grain containing beta radioactivity that is embedded in a non-radioactive matrix, that is, the reverse of actuality. This self-dose is, of course, dependent on grain diameter because the larger the grain the greater is the proportion of the beta emission energy that is deposited within the grain. The ratio of that self-dose to the dose within the infinitely large grain having the same level of radioactivity is termed the absorbed beta-dose fraction ϕ. Values for ϕ as a function of grain diameter are calculated for the beta spectrum of each relevant nuclide and also for the IC contributions; then, by appropriate summing the values for each series are obtained, and these have been tabulated by Mejdahl. The attenuation factors that we actually need, that is, the ratio between the average beta dose within a non-radioactive grain and the point dose within the radioactive matrix in which it is embedded, are given by $1 - \phi$. This can be seen to be true by considering a radioactive grain embedded in a matrix having the same level of radioactivity.

The values obtained by Mejdahl are shown in Figures C.3a and b. Because of the greater attenuation of the IC contributions, and also because of the lower average beta energies, the overall attenuation factors for the Th-232 series are greater than for the U-238 series, and both are greater than for K-40. Hence there is some weak dependence on the relative abundance of the nuclides in the samples; for 12 ppm of Th and 3 ppm of U, Mejdahl calculates that in the case of a 100-μm diameter grain the average attentuation factor is 0.921 for 1% K_2O and 0.935 for 2% K_2O; for a 1-mm diameter grain the corresponding factors are 0.381 and 0.367. Similarly, the attenuation factors are slightly affected by escape of radon and consequent loss of daughter nuclides; however, Mejdahl calculates that for a 100-μm grain even if 50% of the radon escapes the attenuation factors for pottery of the compositions quoted above do not change by more than 10%.

The removal of the surface layer by etching reduces strongly the IC and low-energy beta contributions from the thorium and uranium series but has little effect on the beta dose from potassium. The results of calculations published by Bell (1979a) are shown in Figures C.4 and C.5; specifically, he calculated that removal of a 9-μm layer reduces the average beta dose in a 100-μm grain by a factor of 0.915 in the case of the Th-232 series and by a factor of 0.940 in the case of the U-238 series. Combining these factors with the attenuation factors quoted above for the two clay compositions and weighting the etching factors appropriately, the overall factors for 100-μm grains from which a 9-μm layer has been removed are (0.921 × 0.974) = 0.89 for the 1% K_2O composition and (0.935 × 0.974) = 0.91 for the 2% K_2O composition. In view of other uncertainties in the etching process, it is pointless to go to the extent of making allowance for composition variations, and accordingly a factor of 0.90 is recommended for general use, as in Equation 2.1.

INTERNAL ACTIVITY IN QUARTZ GRAINS

Values for the alpha activity of quartz grains extracted from pottery have been reported by Fleming (1970) and by Sutton and Zimmerman (1978a). Fleming obtained alpha count-rates for a thick sample on a scintillation screen of diameter 42 mm in the range of 0.15 to 4.5 counts/ksec

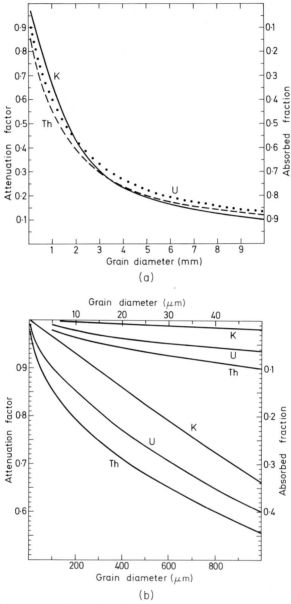

Figure C.3 Absorption of beta dose in quartz grains (drawn from tabulated values of Mejdahl, 1979). The left-hand scale gives the average beta dose within radioactivity-free quartz grains embedded in a clay matrix containing the radioactivity indicated. The right-hand scale gives the average beta dose within grains containing the radioactivity indicated embedded in a radioactivity-free matrix. In all cases the dose is expressed as a fraction of the

Appendix C: Attenuation of Annual Dose

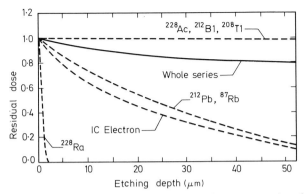

Figure C.4 Average beta (and IC) dose in radioactivity-free quartz grains of initial diameter 100 μm after etching away the surface to the depth indicated, the grains having acquired their dose from a clay matrix containing the thorium series in equilibrium. The dose is expressed as a fraction of that before etching began (from Bell, 1979a).

for the unetched cystalline extract and in the range of 0.05 to 1.4 for the etched extract, some 13 sherds being investigated; he also reported potassium contents for the etched crystalline extract from 20 sherds, these lying in the range of 0.03 to 0.14%. In all cases the values for the etched extract were less than one-tenth (usually much less) of the corresponding

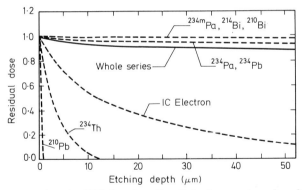

Figure C.5 Average beta (and IC) dose in radioactivity-free quartz grains of initial diameter 100 μm after etching away the surface to the depth indicated, the grains having acquired their dose from a clay matrix containing the U-238 series in equilibrium (from Bell, 1979a).

---

relevant infinite-matrix beta dose. The upper three curves in part (b) refer to the upper scale and the lower three curves to the lower one. Internal conversion (IC) electrons are included in the Th-232 and U-238 series, and these series are assumed to be in equilibrium; the U-235 series is not included.

values for the clay matrix in which the grains had been embedded, and in view of the low alpha effectiveness for quartz, typically 0.02 to 0.05, Fleming concluded that little effective dose-rate would be provided by the self-activity; it is only the alpha component that needs to be considered because the grains are small compared to the range of beta particles.

Sutton and Zimmerman obtained count-rates in the range of 0.8 to 3 counts/ksec (for the same area of screen) for the 4 etched extracts investigated and concluded that the self-radioactivity could provide a small additional dose-rate of the order of 5% of the total received by the grains from the matrix and the soil; obviously, in cases when the latter is high the percentage is decreased and likely to be negligible. These authors used heavy liquid separation with bromoform to remove dense grains so as to avoid the possibility of zircons being present, zircons, of course, being rich in uranium. They also investigated the uniformity of the uranium distribution by fission track mapping and of the luminescence efficiency by cathodoluminescence. Some grains, out of several hundred examined, had uranium-rich areas (~several hundred ppm) of less than 10 μm in size, but elsewhere in the grain the content was usually very low (less than 0.1 ppm), except in the case of the small fraction of grains that were microcrystalline—these contained about 0.5 ppm of uniformly distributed uranium. (It may be useful to note here that the alpha count-rate corresponding to 1 ppm of uranium is 1.7 counts/ksec). The cathodoluminescence was uniform for the majority of the grains, including those with high uranium areas. In calculating the percentage contribution noted above, the authors assumed a homogeneous distribution of radioactivity on the grounds that the dimensions of the areas of high uranium are less than the average alpha range, and the cathodoluminescence is predominantly uniform; however, there is the possibility that these areas will be in thermoluminescence saturation, in which case their estimate of the percentage contribution will be erroneously high.

It appears, therefore, that there are circumstances in which the internal radioactivity may make a non-negligible contribution to the thermoluminescence, and hence it is desirable to make an evalution of the alpha activity of the etched quartz grains routinely. However, this is often not practicable by alpha counting because of the small amount available, and induced fission track evaluation is then necessary. If a check (and allowance, where appropriate) cannot be made, then we have to note this as an additional source of systematic error, systematic in the sense that like inadequate etching of the outer layer of a grain it will always give rise to a date that is erroneously too ancient.

INTERNAL ACTIVITY OF FELDSPAR GRAINS

In dating plagioclase grains extracted from lava of the Chaîne des Puys, France, Guérin and Valladas (1980) reported that, although the thorium and uranium contents as determined by induced fission track counting were low (<0.2 ppm), the contribution to the total thermoluminescence was not negligible. For grains in the size range 80–125 μm, the radioactivity within the grains is estimated to contribute around 25% of the alpha thermoluminescence, and the overall alpha thermoluminescence sometimes contributes as much as half of the total thermoluminescence (etching was not employed in this work). This strong alpha contribution is surprising, even allowing for the grains being unetched; it is due to the rather high alpha thermoluminescence effectiveness, relative to beta radiation, of feldspar, being about five times that in quartz.

For the feldspar grains studied by Guérin and Valladas, the potassium content did not differ substantially from the average for the lava, being in the range of 1.5 to 2%. On the other hand, for grains of potassium feldspar, as used by Mejdahl (1983) in the dating of Scandinavian pottery and burnt stones (see Section 6.5), the internal potassium content can be as high as 14%, much higher than the matrix in which they were embedded. For such grains the 'intrinsic' beta dose-rate becomes the dominant source of thermoluminescence, and it is evaluated by means of the values for absorbed dose fraction given in Figure C.3a and b; the bigger the grains the higher is the proportion of internally emitted beta energy that is absorbed within it.

APPENDIX D

Sample Collection Instructions

POTTERY AND BURNT FLINTS

Number. The ideal number for each level is 6 to 12. Flints need to be well-burnt, and this is often a difficult requirement. In the case of pottery we prefer a variety of fabric types, if available. Surface decoration does not matter. Please, above all, avoid samples whose inclusion in a set is in any way doubtful. The method gives the date of heating. Consequently, we do not want residuals from earlier periods nor samples which have been burnt at some later period.

Size. The outer 2-mm layer of sample must be removed in the laboratory, and we must then be left with a disc of sample measuring at least $\frac{1}{2}$ cm by 3 cm. This is the very minimum size we need, and with irregular sample shapes it is no good submitting shapes which will not yield such a disc. This is the very minimum; bigger pieces are better.

Context

1. Only samples that have been buried to a depth of 0.2 m (8 in.) or more for at least two-thirds of their burial time are acceptable. This means that pits and ditches that have been filled up fairly quickly (either by silting or by ancient man) are ideal sources.

2. The samples should be at least 0.3 m from any boundary (e.g., edge of pit, change of soil-type, wall, floor, rock surface).

Appendix D: Sample Collection Instructions

3. The best situation is a uniform soil which is relatively free of other materials (e.g., rock, building debris, shell, or bone). A small scatter of stone does not matter as long as none of the samples selected were close up against a large one; the bigger the stone the more serious will be the effect.

Treatment

1. In the case of flint avoid prolonged exposure to light; in fact, try to avoid any exposure at all. Particularly avoid sunlight or fluorescence light. Put the samples in an *opaque* bag (black plastic bags are available on request). In the case of pottery exposure to light is relatively unimportant but avoid unnecessary direct sunlight.

2. For both types of samples avoid excessive heating (this means not beyond the boiling point of water). Avoid exposure to ultraviolet, infrared, X-rays, beta rays, or gamma rays; however, the level of dosage used in postal or airport examinations is unimportant.

3. Pottery: for sites where there is little doubt about the moisture content (e.g., typical lowland sites in northwest Europe, where it can be assumed that both sherds and soil are saturated), no other precautions are necessary except that detergents or other additives should not be in the water used for washing; shade the sherds from brilliant sunlight when drying them. For sites where there is doubt about the degree of saturation (e.g., high, well-drained sites and sites in regions of intermediate rainfall), the sherds should not be washed but put directly in a plastic bag (plus any lumps of earth attached) within a few minutes of removal from the soil and tied up tightly. This bag should be put inside a second outer bag, which should also be tied tightly. This will allow us to measure the water content of the sherds as found in the ground. The bags should not be too thick as this makes it difficult to get a water-tight seal. If washing is necessary to confirm the identity of the sherds, then this consideration should take priority.

Soil. About $\frac{1}{2}$ kg of soil that is typical of that in which the samples were buried should be double-bagged as described above. Exposure of the soil to sunlight, ultraviolet, etc. *does not matter*. A sample of each type of material occurring in large proportions within 30 cm of the sample is also required. In the case of a scatter of small stones in the soil, these should be included in the soil sample in correct proportion. Soil is particularly important in the case of flints. A preliminary check of feasibility can be made with a small amount of soil (e.g., an egg-cup full) but more should be available on request.

On-Site Measurements. The soil is required for assessment of the environmental gamma dose which the samples have received. If possible, laboratory specialists should visit the site and make corroborative on-site measurements. One procedure is to bury small copper capsules (about 1 cm in diameter by 3 cm in length) for some months, preferably a year, in situations which are similar to the likely location of the flints. Another procedure is to use a portable gamma spectrometer; this involves making 30-cm long auger holes (7 cm diameter) into the relevant levels. With the spectrometer each measurement is completed in an hour.

If it is not feasible for a laboratory member to visit the site, the capsules can be transmitted by post. On the day of burial they need to be heated to 400°C, upon transmission after retrieval they need to be accompanied by a 'travel monitor' capsule, which is heated to 400°C on the day of retrieval.

It is necessary to know the water content of the soil at the time of spectrometer measurement and the average content during the period of capsule burial; appropriate samples should be collected.

General. Information about burial conditions is essential; this should include a sketch section of the context (and, if possible, photographs) showing very roughly the points from which the samples were taken and the deposits for at least 30 cm around.

Please try to give a rough estimate of how the average water content of the soil relates to that of the soil sample supplied. It is also useful to know how the water content of the soil varies between contexts and with respect to surface conditions (e.g., 'though bone dry at the surface in these hot climatic conditions, by a depth of 2 m the soil was pretty well saturated'). Obviously it is also important to know if the water table is (or has been) anywhere near to the contexts concerned. If there is any seasonal or long-term information about variations in rainfall, we would like to know it.

The samples are destroyed in the course of measurement.

Acceptance Criteria. Acceptance criteria include

    1. whether the present accuracy (between $\pm 5\%$ and $\pm 10\%$ of the age) is good enough for the problem concerned;
    2. suitability of site and material for the technique;
    3. archaeological importance; and
    4. lack of, or ambiguity in, other dating evidence.

As a general rule it is not easy to date flints which are more recent than 10,000 years old, but it is worth submitting a trial sample (plus soil) for a feasibility test. The upper limit also is dependent on the type of flint and the radioactivity of soil, but certainly the method can reach back to

Appendix D: Sample Collection Instructions

around 300,000 years. For pottery and baked clay the range of applicability is from the present back to around 20,000 years, the actual limit being highly dependent on the individual sample.

SEDIMENT DATING

The same general considerations apply as with pottery and burnt flints except that precautions against exposure to light must be much more stringent (see below).

Quantity. For the thermoluminescence measurements themselves two samples from each level are desirable, preferably not close together; each should be about a quarter kilogram. Unless on-site radioactivity measurements with a gamma spectrometer are made, it is desirable in addition to collect between $\frac{1}{2}$ and 1 kg for radioactive analysis (e.g., a lump about 10 cm across from each level). The radioactive analysis sample does not need to be kept dark, nor does it need to be sealed for water content measurement.

Context. The ideal sample is one which is taken from the *middle* of a thick (i.e., $\frac{1}{2}$ m or more) layer. However, in the case of thin layers it can be checked by radioactive measurements whether neighbouring layers have differing radioactivities, and to some extent allowance can be made by calculation. Of course, if the sample is within 0.2 m of the ground surface, the situation is not good and the accuracy attainable is severely limited.

Treatment. It is absolutely vital that there be no exposure to direct sunlight; exposure to daylight or fluorescent light should be minimal (i.e., not more than about 10 min). If there is a need to examine the samples, this should be done with a bulb light or with a torch. Exposure to sunlight, etc., is likely to make the thermoluminescence data erroneously too recent.
Therefore:

1. Scrape off at least 1 cm of the already exposed surface; this is done in order to get deeper than where the light has penetrated. Shade from direct sunlight. If the surface is too hard for scraping, make a mark on the exposed face.

2. Still shading, collect the sample in an opaque or semi-opaque container; straightaway put the container in a black plastic bag (preferably of *photographic quality*) or other light-tight container.

3. Because it is necessary to get information about water content, the container used should be moisture-tight (e.g., tins tightly sealed with tape, or two tightly sealed thin plastic bags—as in the case of pottery above). Sometimes measurement of the water content can be made by the site

excavator on a separate comparable sample; this is highly preferable and, of course, the need for watertight sealing is then eliminated.

Soil. In the case of sediment dating sample and soil are the same material, but the same considerations apply as for the material collected for radioactivity analysis (see above). The considerations for on-site measurement, burial conditions, and acceptance criteria are the same as for burnt flint, though the range of applicability is not yet well established.

APPENDIX E

Kinetic Studies and Evaluation of Trapped-Electron Lifetimes

Although the evidence of the plateau test takes precedence for an individual sample, prediction of lifetime for the thermoluminescence peak of a mineral is helpful in establishing the likely time range over which it will be useful; in either case it is also necessary to test for anomalous fading (see Section 3.4) by storage of irradiated portions and comparison of the level of thermoluminescence with that from freshly irradiated portions. If the peak is not affected with anomalous fading, then the lifetime in the case of first-order kinetics is given by

$$\tau = s^{-1} \exp(E/kT), \tag{E.1}$$

where s (sec^{-1}) is the frequency (or pre-exponential) factor, E (eV) the trap depth, T the absolute temperature, and k Boltzmann's constant (at 17°C, $kT = 0.025$ eV). Besides checking that a dating peak has adequate stability, determination of lifetime is also relevant with respect to the kinetic methods of annual dose evaluation discussed in Section 6.4.

There are various methods for laboratory determination of E and s, of which three are commonly used in studies relevant to dating:

1. initial rise,
2. isothermal decay,
3. peak shift with heating rate (Hoogenstraaten's method).

In addition, there are methods based on the parameters defining the shape of the peak and on curve-fitting techniques. References for further reading are Halperin and Braner (1960), Braunlich (1968), Chen (1969, 1976), Shenker and Chen (1971), Shalgaonkar and Narlikar (1972), and Ganguly and Kaul (1984).

In all methods an accurate measurement of sample temperature is important, and because of thermal lag this may be appreciably different to what is indicated by the thermocouple. Useful here are some careful measurements of peak temperatures for a number of phosphors (Gorbics *et al.,* 1968). The amount of lag is determined by observing the thermocouple temperature corresponding to one of these peaks when the phosphor sample replicates the form of the mineral sample under study. For natural fluorite (MBLE) the temperature of the principal peak is at 260°C for a heating rate of 1°C/sec, 283°C for 5°C/sec, and 294°C for 10°C/sec.

THE INITIAL RISE METHOD
(THE 'ARRHENIUS PLOT')

At the start of a glow-peak the thermoluminescence is proportional to $\exp(-E/kT)$, irrespective of whether first-order kinetics are obeyed or not. This temperature dependence continues until the number of electrons remaining trapped decreases significantly—see Figure 3.3. Hence by plotting log(thermoluminescence intensity) versus T^{-1} the value of E can be obtained from the slope of the straight line obtained. To avoid error due to a decrease in the number of electrons remaining trapped, it is necessary to restrict the temperature range such that the thermoluminescence intensity reached does not exceed one-tenth of the peak intensity. Of course, overlapping peaks on the lower edge of the peak under study need to be removed by prior thermal treatment.

A drawback with this method is that if the relevant luminescence centres are subject to thermal quenching, too low a value is obtained for E. As discussed in Section 3.3, this can give rise to a value for the lifetime that is totally erroneous. In the case of the 325°C peak of quartz, Wintle (1975b) obtained a value for E that was 0.64 eV lower than that obtained by other methods. Having determined E, the value for s is obtained from Equation E.3.

ISOTHERMAL DECAY METHODS

If an irradiated sample is held at elevated temperature T_1, the phosphoresence will decay with time t according to

$$I = I_0 \exp(-t/\tau_1), \quad (E.2)$$

where I_0 is the intensity of the phosphorescence at $t = 0$ and τ_1 is the lifetime of the trapped electrons at T_1. This is repeated for several temperatures, and log τ_1 is plotted against T^{-1}. From Equation E.1 this plot should yield a straight line of slope E/k and have an intercept log s^{-1} on the T^{-1} axis. If several types of trap are present having different values for τ, then the plot will be curved; however, if the lifetimes are sufficiently different, it is possible to analyse the plot by successive subtractions, starting with the longer-lived components. An alternative reason for curvature is that the peak does not follow first-order kinetics and the method is not applicable in simple form. In this case it is useful to plot the product It against log t; this product reaches its maximum value at $t = \tau_1$ (Visocekas et al., 1976; Chen and Kirch, 1981, p. 174), irrespective of the order. A difficulty with using phosphorescence is its low intensity except when close to the peak temperature. An alternative is to measure the residual thermoluminescence after different storage times at each temperature since, of course, this too shows exponential decay with time. Though much more laborious, this has the advantage of avoiding interference by a peak occurring at different temperature which has a similar lifetime.

Since each set of measurements at a given temperature is self-contained, yielding a value of τ, there is no upset by thermal quenching.

PEAK SHIFT WITH HEATING RATE

As illustrated in Figure 3.4, the temperature T^* of the maximum of a glow-peak depends on the heating rate β. Equation 3.6 may be rearranged as

$$\ln(T^{*2}/\beta) = E/kT^* - \ln(sk/E), \quad (E.3)$$

and so if T^* is measured for a number of different heating rates, E can be found from the slope of the straight line obtained by plotting the logarithmic term on the left-hand side against $1/T^*$; the value of s is found from the intercept.

This method has the advantage that it can deal with weaker peaks than the initial rise method or the phosphorescence method; also, it is less prone than the latter to interference from other peaks. Although it is strictly applicable only to peaks obeying first-order kinetics, it appears that it yields values of E to a good approximation in general (Chen and Kirch, pp. 168, 276). In the particular case of the quartz studied by Wintle (1975b), it is not significantly affected by thermal quenching.

TABLE E.1
Some Values for Trap Parameters and Electron Lifetimes at 15°C[a]

Mineral and peak Temp.[b] (°C)	Trap depth (eV)	Frequency factor (sec^{-1})	Method[c]	Lifetime[d] (years)	Reference
Quartz					
85°	0.84	1.9×10^{11}	IR	0.13×10^{-3}	Strickertsson (1985)
110°	0.80	2.7×10^9	IR	1.2×10^{-3}	Strickertsson (loc. cit.)
110°	0.98	8×10^{12}	Several	0.8×10^{-3}	Fleming (1969, 1979)
110°	0.99	—	IR,ID,PS	—	Wintle (1975)
190°	1.42	3.4×10^{14}	IR	0.7×10^3	Strickertsson (loc. cit.)
230°	1.79	5×10^{18}	PS	130×10^3	Wintle (1974, 1975a)
240°	1.60	9.2×10^{14}	IR	340×10^3	Strickertsson (loc. cit.)
310°	1.68	1.8×10^{13}	IR	450×10^6	Strickertsson (loc. cit.)
325°	1.69	1×10^{14}	ID,PS	100×10^6	Wintle (1975a, 1977)
375°	1.66	1.5×10^{13}	IR	$\gtrsim 10^8$	Fleming (loc. cit.)
Flint					
330°	1.66	6×10^{12}	PS	600×10^6	Bowman (1982a)
370°	1.74	$\sim 10^{13}$	IR	10×10^9	Wintle & Aitken (1977)
Calcite					
275°	1.75	4×10^{15}	IR,ID,PS	30×10^6	Wintle (1977)
280°	1.56	5×10^{13}	ID	1×10^6	Debenham (1983)
Limestone					
285°	1.52	2×10^{13}	IR,PS	650×10^3	Wintle (1974)
350°	1.94	1.4×10^{15}	IR,PS	200×10^9	Wintle (1974)
K-feldspar					
90°	0.76	6×10^9	IR	0.16×10^{-3}	Strickertsson (loc. cit.)
110°	1.10	1.3×10^{13}	IR	43×10^{-3}	Strickertsson (loc. cit.)
210°	1.40	2.8×10^{13}	IR	3.6×10^3	Strickertsson (loc. cit.)
280°	1.62	4.1×10^{13}	IR	3.9×10^6	Strickertsson (loc. cit.)
320°	1.6	1×10^{13}	IR	1×10^9	Hutt and Smitnov (1983 and pers. comm.)
350°	1.68	2.8×10^{12}	IR	9.2×10^9	Strickertsson (loc. cit.)
Labradorite					
200°–300°	1.6	—	IR	—	Wintle (1977a)
Fluorite					
200°	1.40	7×10^{14}	—	100	Fleming (1969)
200°	1.42	3×10^{14}	CF	700	Ganguly and Kaul (1984)
280°	1.75	2×10^{15}	CF	70×10^6	Ganguly and Kaul (loc. cit.)
300°	1.80	6×10^{15}	—	150×10^6	Fleming (loc. cit.)
320°	1.90	3×10^{15}	CF	20×10^9	Ganguly and Kaul (loc. cit.)

[a] Data for synthetic phosphors will be found in Oberhofer and Scharman (1981), as well as elsewhere.

[b] The peak temperatures are for the heating rates used in the work concerned, usually 2–10°C/sec.

[c] IR, initial rise; ID, isothermal decay; PS, peak shift with heating rate; CF, curve fitting.

[d] The lifetimes quoted have been calculated from the E and s values assuming first-order kinetics. If, in fact, the kinetic order is higher, then the lifetime will be longer.

TESTING FOR FIRST-ORDER KINETICS

The simplest test for first-order kinetics is to check that the temperature of the peak maximum, T^*, is not dependent on dose; for kinetics of higher order, T^* shifts downward in temperature as the dose is increased. Another test is that the time dependence of isothermal decay should be exponential, as given by Equation E.2 above. However, whereas both of these tests are straightforward to apply in the case of an isolated narrow peak, such as the 100°C peak of quartz, a more complex approach using curve-fitting techniques is necessary when the glow-curve consists of overlapping peaks.

SOME RESULTS

Listed in Table E.1 are some reported values, by no means exhaustive, for trap parameters relevant to dating application. When such very long lifetimes are predicted from short-term laboratory measurements, it is not unreasonable to keep in mind the possibility that some unforeseen effects may invalidate the extrapolation. Thermal quenching, as already discussed, is an effect that can invalidate the determination of E and anomalous fading an effect that can make the effective lifetime substantially less than predicted by values of E and s; hence, once again, it is relevant to emphasize the importance of the plateau test.

It should be noted that the lifetimes quoted in the table are for storage at 15°C; temperatures in caves are likely to be in the range of 10 to 15°C and those on the ocean floor about 4°C, giving a substantially longer lifetime. Equation E.1 above predicts that for $E = 1.8$ eV the lifetime at 10°C will be greater than that at 15°C by a factor of 3.

APPENDIX F

Anomalous Fading

EXPERIMENTAL STUDIES

Although thermoluminescence ages can be affected substantially, for most minerals the amount of fading that occurs during practical laboratory storage times is not sufficient to make study of it an easy matter. Consequently, our experimental knowledge of the phenomenon is largely confined to minerals that are acutely affected; it does not necessarily follow that less seriously affected minerals follow the same behavior, but in default of better guidance, that is what we have to assume.

Feldspar. In Wintle's (1977a) classic study of labradorite the thermoluminescence in the 300°C region of the glow-curve, for which the initial decay at room temperature was by 20% in 2 hr, was interpreted as being composed of three roughly equal components; two of these components decayed exponentially with time, with lifetimes at 10°C of 1 hr and 90 hr respectively, and the third showed no measurable decay. The dependence of lifetime on storage temperatures between −20°C and 160°C implied values for E of 0.2 and 0.3 eV for the two fading components, with frequency factors of ~ 1 sec^{-1}, whereas kinetic study of the thermoluminescence using the initial rise method indicated a trap depth of 1.6 eV. However, the weak dependence on storage temperature did not extend down to −200°C and −250°C, where a small fading was still observed, the same for both temperatures; $E = 0.2$ eV predicts a lifetime of about a million years at −200°C. Other conclusions drawn from this study were (Wintle, 1978) (i) that the fading did not show strong dependence on the type of ionizing radiation used, though there was a tendency for alpha-irradiated samples to fade faster than beta-irradiated ones (see Wintle *et al.*, 1971, with regard to $CaF_2:Mn$), (ii) that the wavelength spectrum of the thermoluminescence was the same shape for both an immediate and a delayed measurement, and (iii) that negligible phosphorescence was observed during storage and hence the fading process was assumed to be non-radiative, at any rate, in the wavelength range of observation, 300–550 nm. (In a subsequent study of the same sample Visocekas, 1985, detected phosphorescence at a wavelength beyond the upper limit). Three possible mechanisms were considered, namely, wave-mechanical tunnelling, diffusion of defects, and decay of effective luminescence centres; emphasis was put on the first.

Studies by Mejdahl (1983) on feldspar grains from Scandinavian pottery, burnt stones, and glacial sediments indicate that fading is not a serious problem for these materials, at least for samples up to 100,000 years old. Work by Wintle (1985) and Debenham (1985) on feldspar thermoluminescence from fine-grain samples of European loess suggests that fading may be a limitation for older samples; attempts to eliminate the fading component by thermal pre-treatment have not so far been successful. Note 9 of Chapter 8 is relevant here.

Volcanic feldspars can show serious fading in the 300–400°C region of the glow-curve, but not in the 600° region (Guérin and Valladas, 1980).

Zircon. Studies by Bailiff (1976) using phototransfer showed that whereas there was substantial fading in traps associated with the thermoluminescence emitted at around 350°C, the fading in deep traps corresponding to a glow-curve temperature of greater than 500° was less than a few percent in 6 months; this was confirmed by Zimmerman (1978).

In their single-grain dating studies, Sutton and Zimmerman (1976) noted that whereas some grains faded appreciably, others did not fade at all. Zimmerman (1979a) noted a rough correlation between low ages (by up to a factor of 6) and the amount of fading observed in one day.

From 100-day storage experiments at 20, 125, and 175°C using fine-grain discs, Templer (1985b) concluded that it is possible to eliminate the fading component without thermally draining the stable component; he also reported that exposure to long-wavelength light (greater than 550 nm) accelerated the fading at room temperature. From a study of the spectrum of the emitted thermoluminescence he found that the fading occurred equally at all wavelengths (280–650 nm).

Other Minerals. The fading in *apatite* can be as strong as in zircon and labradorite; studies have been reported by Wintle (1973, 1977a), Bailiff (1976), and Sutton and Zimmerman (1978a). The fading over a year was followed by Zimmerman and this has been reported by Visocekas et al. (1983), who point out the logarithmic dependence on time.

With the *pumice* samples studied by Sutton and Zimmerman (1978a) there was no further decay after 20 days at 160°C, measurement being continued for 100 days; on the other hand, a slow decrease was still observed at that time for storage temperatures of 20, 80, and 120°C. Because there was a 20% decrease in the natural thermoluminescence after 1 week at 160°C, presumably due to thermal detrapping, the authors concluded that such treatment is unable to separate stable and unstable components. However, they overlooked the possibility that the natural thermoluminescence itself contained an unstable component (this would be expected for that part of the natural thermoluminescence acquired recently, over a time period corresponding to the lifetime of the unstable component). Also, so far as dating is concerned, it is not essential that the pretreatment does not erode the natural thermoluminescence; so long as the same pretreatment is applied before measurement of natural thermoluminescence as before natural plus artificial thermoluminescence, the stable component is eroded by the same percentage in both.

Zimmerman (1979a) also investigated the possible use of phosphorescence as an indicator of anomalous fading; he concluded that although it was useful with some minerals this was not the case in general. However, it is possible that the phosphorescence was outside the wavelength range of his detection system.

Fading observations incidental to dating reports have been included by various authors, some of which are mentioned in the main text; these are usually on polymineral fine-grains.

TUNNELLING

Escape of electrons from traps through overlap of the wave function with that of charged centres nearby is a mechanism that has been discussed and investigated experimentally with respect to synthetic crystals for some decades (e.g., Dexter, 1954; Hoogenstraaten, 1958). Such escape was later found to be accompanied by 'tunnel afterglow' (see, for instance, Riehl, 1970), that is, the phosphorescence observed after cessation of irradiation with the sample held at low temperature. This afterglow was interpreted as being emitted when the escaped electron undergoes radiative recombination at the centre.

Time Dependence. The intensity of the afterglow, and therefore the rate of detrapping also, is proportional to t^{-1} within limits, where t is the time elapsed since irradiation; by integration it follows that the number of electrons detrapped is proportional to $\log t$, and this is in agreement with

Appendix F: Anomalous Fading

the theoretical expectation (Delbecq, Toyozawa and Yuster, 1974), which will be discussed shortly. Visocekas (1979; Visocekas, Ouchene and Gallois, 1983) pointed out that from reported data on anomalous fading, for most natural minerals the *thermoluminescence lost* is in fact proportional to log t and that this gives strong support to explanation in terms of tunnelling; thermal detrapping would result in a drastically different time dependence, that is, exponential. Whereas in the latter the probability of detrapping does not change with time, for the former it becomes progressively more difficult for detrapping to occur. This is because the probability of a transition is very strongly dependent on the distance between trap and centre: in a theoretical example given by Visocekas (1979), the life expectancy of a pair separated by 2 nm is only 10^{-4} sec, whereas if the separation is 4 nm it is about a year, and if 5 nm about a million years. Thus traps which have a centre very close to them lose their electrons almost instantaneously, and then as time proceeds detrapping occurs between pairs of increasing separation; in the example quoted we would expect that after about a year there will be no trapped electrons left in traps that have a centre within 4 nm.

The expression for the lifetime of pairs separated by a distance R may be written as

$$\tau = s^{-1} \exp(R/R_0), \tag{F.1}$$

where s and R_0 are constants. This follows from the exponential dependence of tunnelling probability on barrier thickness. Hence after a time t all pairs separated by less than R_c will have recombined, where

$$R_c = R_0 \ln(st). \tag{F.2}$$

If there is random distribution of traps and centres, each with concentration N, then so long as $NR_c^3 \ll 1$ the concentration of pairs having a separation less than R_c is given by

$$(4\pi/3)R_c^3 N^2. \tag{F.3}$$

If L_1 is the number of electrons detrapped by time t_1, and similarly for L_2, then

$$L_2^{1/3} - L_1^{1/3} = (4\pi/3)^{1/3} N^{2/3} R_0 \ln(t_2/t_1). \tag{F.4}$$

For the condition that $(L_2 - L_1) \ll L_1$, it follows that

$$\frac{L_2 - L_1}{L_1} \approx \ln^2(st_1) \ln(t_2/t_1). \tag{F.5}$$

Thus the number of electrons detrapped, and therefore the thermoluminescence lost, in the time between t_1 and t_2 is proportional to $\ln(t_2/t_1)$

during times for which the number detrapped is small compared to the number already detrapped by time t_1. Alternatively, the domain of applicability may be defined so that for which the change in R_c is relatively small; in the example quoted above from Visocekas (1979), this appears to be the case for times extending to millions of years, so long as t_1 is not impracticably short.

The above is a simplified version of the theoretical treatment given by Delbecq *et al.*

Practical Implications of Logarithmic Decay. Although the limitations in applicability must not be forgotten, it is of interest to see what implications this time dependence has with respect to the efficacy of the storage tests discussed in the main text. In numerical terms if the lost thermoluminescence is proportional to log t, then the loss between 1 hr and 10 hr would be the same as that lost between 10 hr and 100 hr and between 100 hr and 1000 hr (40 days) and so on.

Suppose that for the 'immediate' measurement the delay t_i between middle of the irradiation and measurement is 2 min, and for the delayed measurement the storage duration t_s is $4\frac{1}{2}$ months; then we have $t_s/t_i = 100{,}000$, and so $\log(t_s/t_i) = 5$. If we denote by $g\%$ the fading that occurs between two times differing by a factor of 10, then the observed fading between t_i and t_s will be $5g\%$; hence g can be evaluated.

Suppose that for evaluation of the paleodose there is overnight storage, that is, for 16 hr, between irradiation and measurement; substituting $t_i = 2$ min and $t_s = 16$ hr, we find the fading loss is $(g \log 480)\%$, that is, $2.7g\%$. For a hypothetical short irradiation in antiquity at 200 years ago, the fading relative to $t_i = 2$ min will be $[g \log(480 \times 10^5)]\%$, that is, $(2.7 + 5)g\%$; thus the loss suffered by the 'ancient' thermoluminescence is $5g\%$ more than that occurring during overnight storage. Of course, the natural dose has been received continuously throughout the sample's antiquity so that the storage time for some of the dose is much less than the age; it will be shown shortly that it is appropriate to multiply the 200-yr elapsed time by the mathematical constant e, equal to 2.7, giving 500 years as the age for which $5g\%$ extra loss will have occurred. For each factor of 10 by which the age is higher, the loss will have been greater by $g\%$, that is, for 5000 yers, $6g\%$; for 50,000 years, $7g\%$; and for 500,000 years, $8g\%$.

The upper limit that can be set to the possible loss is determined by the precision of the $4\frac{1}{2}$ month *vs* 2 min comparison; if this is taken to be 5%, then 1% is the minimum value of g that is detectable, that is, this corresponds to an 8% loss in a 500,000-year-old sample. Note that the precision is not much reduced if the $4\frac{1}{2}$ months is shortened by a factor of 10, to 14 days; the loss during storage is then $4g\%$ and the minimum value of g

Appendix F: Anomalous Fading

detectable is 1.25%. Of course, it has been assumed in the above that laboratory storage has been at the same temperature as during burial. If elevated temperatures are used; then the upper limit is reduced, but by an unknown factor unless *ad hoc* measurements of temperature dependence are made for each sample.

In conclusion it must be emphasized that even for a mineral that has been shown to fade with logarithmic dependence during laboratory storage tests, the validity of extrapolating into the past is unproven; there may be other mechanisms operative too. However, the above is the best that can be done so far as estimation is concerned. Undoubtedly the preferred course is to develop an adequate pretreatment to eliminate the effect of the fading along the lines proposed for zircon by Templer (*loc. cit*). However, this is not necessarily possible for all minerals.

One situation which might lead to non-logarithmic fading would be a deviation in the distribution of inter-pair distances from that corresponding to random distribution. If the pair is part of a complex defect with a fixed separation R_f, the lifetime will be given by

$$\tau_f = s^{-1} \exp(R_f/R_0) \qquad (F.6)$$

and the fraction of pairs that have recombined after time t by

$$1 - \exp(-t/\tau_f). \qquad (F.7)$$

So long as τ_f is short then, like logarithmic fading, this will show up in laboratory fading tests; but if τ_f is of the order of 10 years or more, this will not be so. However, it is unlikely that the thermoluminescence from different parts of the glow-curve will be due to pairs of the same separation; consequently, there will be no plateau and the sample will be rejected.

Although logarithmic predictions may be useful as a guide in predicting upper limits to fading, there can be no question of making a quantitative correction. Undoubtedly more research in this area would be timely.

The Factor Relating a Short Irradiation to an Extended One. Suppose that for a pulse of irradiation the percentage fading loss after time t is $g \log(t/t_0)$, where t_0 is an arbitrary constant. Then for a continuous irradiation starting at t_2 and finishing at t_1, both t_1 and t_2 being time before measurement, the percentage fading loss is given by

$$g \left\{ \int_{t_1}^{t_2} \log(t/t_0) \, dt \right\} \Big/ (t_2 - t_1), \qquad (F.8)$$

that is,

$$g[t_2 \log(t_2/t_0) - t_1 \log(t_1/t_0) - 0.43(t_2 - t_1)]/(t_2 - t_1). \qquad (F.9)$$

On putting

$$t^* = \frac{1}{2.7} \frac{t_2^{t_2/(t_2-t_1)}}{t_1^{t_1/(t_2-t_1)}}, \tag{F.10}$$

the expression reduces to

$$g \log(t^*/t_0). \tag{F.11}$$

For $t_2 < 10\, t_1$, as is usually the case in a laboratory irradiation, $t^* \simeq t_1 + \frac{1}{2}(t_2 - t_1)$, whereas for $t_2 > 100 t_1$, as for the natural dose, $t^* \simeq t_2/2.7$. Hence the amount of fading in a natural sample of age t_2 is the same as for a hypothetical short irradiation occurring $t_2/2.7$ years ago.

SOME OTHER MECHANISMS

Decay of Effective Luminescence Centres. The attraction of this mechanism is that it gives a good explanation of why, with certain minerals, anomalous fading does not destroy the plateau, that is, why the fading is to the same degree for a substantial portion of the glow-curve; it is presumed that the same type of centre is used by the whole range of traps involved. However, it is not easy to envisage how the mechanism operates in other respects.

The first difficulty is that the kinetics describing the decay must be such that there is no substantial loss of effective centres in the course of the glow-curve before the thermoluminescence peak is reached. Both the frequency factor and the activation energy need to be abnormally low, for example, $s \simeq 1$ sec^{-1} and $E \simeq 0.2 - 0.3$ eV, as inferred by Wintle for labradorite.

Another requirement is that for the number of luminescence centres to be a controlling factor in the amount of thermoluminescence these centres need to be in competition for the electrons evicted from traps; if the evicted electron had no other fate than annihilation at luminescence centres, then there would be no control by the number of the latter. The requirement of control implies that the unfaded number of effective centres is proportional to the dose received since last heating, and that after fading it is restored to the unfaded value by re-dosing; if this were not the case, then the thermoluminescence sensitivity measured subsequent to read-out of the faded thermoluminescence would show a diminution, which it does not. Also, there should be an increase in sensitivity with dose and this is not observed, at any rate, for fading minerals in general.

Thus, although decay of centres may play a part in fading, the construction of a model for their role requires considerable ingenuity.

Pseudo-fading. It is also relevant to mention two effects associated with sample preparation, both of which give rise to erroneously low ages. Wintle, Aitken and Huxtable (1971) hypothesized that in preparing fine-grain samples from the pottery of one site extra traps ('vice traps') were produced when the sample was crushed in the vice. These gave rise to an enhanced sensitivity in the first subsequent irradiation. However, the effect could be removed by holding at 150°C for 10 min before measurement. The other effect was with respect to calcite samples obtained with a power drill; the low ages obtained were explained by Wintle (1975a) in terms of competition for detrapped electrons during the thermoluminescence process by non-radiative centres on the grain surface which had been activated during the drilling process. Fortunately, the now standard procedure of washing calcite grains in dilute acetic acid eliminates the effect, as also does washing in hydrogen peroxide.

Defect Modification. In LiF (TLD 100) used for biological dosimetry, there is another type of time-dependent effect. Following thermal annealing there are sensitivity increases in some peaks and decreases in others, depending on treatment; interpretation is in terms of changes in the complex defect structure responsible for the thermoluminescence (Johnson, 1974). The effect reinforces the emphasis in thermoluminescence dating on the desirability of measurements being 'first-glow' as far as possible.

Defect diffusion would be expected to lead to similar experimental observations as found for LiF.

APPENDIX G

Annual Dose Evaluation: Summary of Radioactivity Data

Various tables are given here which enable the annual dose to be derived from the radionuclide content or from the alpha activity. These are based on measurements of the energy of emitted particles and radiations as published in nuclear data tables. To the extent that the infinite matrix assumption, discussed in Section 4.2, is applicable in the situation concerned, the annual dose is equal to the rate of energy emission. Tables G.1–G.3, together with Figures G.1–G.3, give details for the radioactive series and Tables G.4 and G.5 summarize the energy release in forms convenient for routine use. Further radioactivity details specifically relevant to alpha counting and gamma spectrometry are given in Appendices J and M, respectively.

Appendix G: Annual Dose Evaluation

TABLE G.1
Energy Emission Details: Thorium Series

Radioisotope	α	β_{max}	β_{av}	IC	γ
Th-232	4.006	—	—	0.0099	—
Ra-228	—	0.055	0.0141	0.0015	—
Ac-228	—	1.2051	0.3929	0.0589	0.8750
Th-228	5.382	—	—	0.0190	0.0024
Ra-224	5.674	—	—	0.0021	0.0099
Rn-220	6.288	—	—	—	0.0002
Po-216	6.7785	—	—	—	—
Pb-212	—	0.3713	0.1067	0.0652	0.1055
Bi-212 (64.1%)	—	1.3360	0.4901	0.0015	0.1013
(35.9%)	2.1799	—	—	0.0074	0.0039
Po-212 (64.1%)	5.6295	—	—	—	—
Tl-208 (35.9%)	—	0.5804	0.2055	0.0098	1.2070
Positions not shown in decay schemes				0.0006	0.0021
X-rays					0.1197
Auger electrons				0.0094	
	35.9379	3.5478	1.2093	0.1858	2.5060

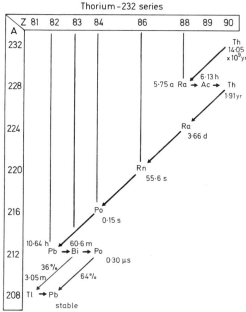

Figure G.1 Decay schemes for the thorium-232 radioactive series; data are from Lorenz (1983). A long arrow indicates alpha decay and a short one beta decay. Branching is shown except when the branching involved is less than 1%; thus At-216 (0.04%) has been omitted.

TABLE G.2
Energy Emission Details: Uranium-238 Series

Radioisotope	α	β_{max}	β_{av}	IC	γ
U-238	4.185	—	—	—	0.0111
Th-243	—	0.1625	0.0456	—	0.0149
Pa-234m	—	2.2647	0.8372	—	0.0112
Pa-234 (0.13%)	—	0.0009	0.0003	—	0.0016
U-234	4.759	—	—	—	0.0001
Th-230	4.658	—	—	0.0119	0.0004
Ra-226	4.775	—	—	0.0033	0.0090
Rn-222	5.490	—	—	—	0.0004
Po-218	6.004	—	—	—	0.0001
Pb-214	—	0.7237	0.2280	0.0698	0.2285
Bi-214	—	1.7829	0.6606	0.0112	1.5088
Po-214	7.689	—	—	—	0.0001
Pb-210	—	0.0243	0.0062	0.0253	0.0019
Bi-210	—	1.1610	0.3945	—	—
Po-210	5.306	—	—	—	0.00001
X-rays					0.0332
Auger electrons				0.0074	
	42.883	6.1201	2.1446	0.1535	1.8213

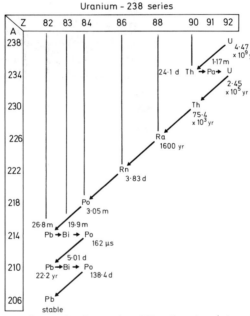

Figure G.2 Decay schemes for the uranium-238 radioactive series; data are from Lorenz (1983). A long arrow indicates alpha decay and a short one beta decay. Branching is shown except when the branching involved is less than 1%; thus At-218 (~0.02%), Rn-218 (very weak), Tl-210 (0.04%), and Tl-206 (0.0001%) have been omitted.

Appendix G: Annual Dose Evaluation

TABLE G.3
Energy Emission Details: Uranium-235 Series

Radioisotope	α	β_{max}	β_{av}	IC	γ
U-235	4.3897	—	—	0.0131	0.1485
Th-231	—	0.2193	0.0627	0.0503	0.0151
Pa-231	4.9225	—	—	0.0888	0.0287
Ac-227	0.0689	0.0398	0.0102	—	0.0099
Th-227	5.8379	—	—	0.0002	0.0988
Fr-223 (1.4%)	—	0.0159	0.0040	—	0.0007
Ra-223	5.8224	—	—	—	0.0865
Rn-219	6.7590	—	—	—	0.0576
Po-215	7.3864	—	—	—	0.0003
Pb-211	—	1.3234	0.4564	0.0025	0.0764
Bi-211	6.5680	0.0017	0.0004	0.0071	0.0491
Po-211 (0.28%)	0.0209	—	—	—	0.00002
Tl-207	—	1.4340	0.5054	—	0.0024
	41.7757	3.0341	1.0391	0.1620	0.5740

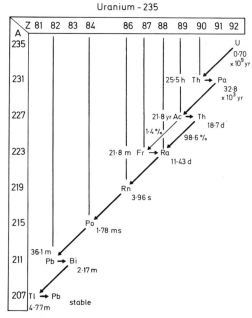

Figure G.3 Decay schemes for the uranium-235 radioactive series; data are from Lorenz (1983). A long arrow indicates alpha decay and a short one beta decay. Branching is shown except when the branching involved is less than 1%; thus At-215 (very weak) and Po-211 (~0.3%) have been omitted.

NOTES TO TABLES G.1–G.3

1. All energies are given in MeV.

2. The values given represent the rate of energy release per unit activity of parent. Thus the gamma contribution may be the summation of quite a large number of separate gamma emissions released during a single transition. Also, as at bismuth-212, there may be two branches in the decay chain and the value given is then that obtained after multiplying by the fractional probability concerned; for example, although the energy of the alpha particle emitted by polonium-212 is 8.78 MeV the energy release per unit activity of parent is $8.78 \times 0.64 = 5.62$ MeV.

3. Beta particles have a continuous spectrum of energies from zero to β_{max}. The average energy, β_{av}, depends on the shape of the spectrum for the emitter concerned but it is approximately $\frac{1}{3}\beta_{max}$. When a particle having energy less than β_{max} is emitted, the missing energy is carried away by a neutrino but this is so weakly absorbed in matter that it makes no contribution to the annual dose.

4. The abbreviation IC signifies internal conversion electrons, an alternative mode of nuclear de-excitation to gamma emission; IC emission is usually accompanied by Auger electron or X-ray emission. In evaluating annual dose the two former are added to the beta particle contribution and the latter to the gamma contribution.

5. Bell (1979a) has commented on the difficulty of making a proper assessment of the accuracy of the data but considers that any associated error is likely to be outweighed by errors from other sources in thermoluminescence dating; he regards 5% as the maximum error likely in the overall factors for conversion from parts per million of parent to annual dose.

6. The data are from Bell (1976, 1977, 1979a); see also Lederer and Shirley (1978) and ICRP no. 38 (1983, *Annals of the ICRP* vols. 11–13, Pergamon Press).

TABLE G.4
Annual Dose Data for Thorium and Uranium Series[a]

	Th-232	U-238	U-235	nat. U
1. Atomic abundance	100%	99.28%	0.72%	—
2. Half-life ($\times 10^9$ yr)	14.05	4.47	0.70	—
3. Parent activity (Bq/mg)	4.08	12.5	80	13.0 (238: 95.6%; 235: 4.4%)
Energy release (MeV) per parent disintegration:				
4. α, full chain	35.9	42.9	41.8	42.8
5. α, pre-Rn[b]	15.0	18.4	(41.8)	19.4
6. β + IC, full chain	1.39	2.30	1.20	2.25
7. β + IC, pre-Rn	0.50	0.93	(1.20)	0.94
8. γ + X, full chain	2.50	1.83	0.57	1.77
9. γ + X, pre-Rn	1.01	0.064	(0.57)	0.086
To obtain dose-rate (Gy/ka) per unit specific activity of parent (Bq/kg)				
10. Divide lines 4–9 by	198	198	198	198
To obtain dose-rate (Gy/ka) per 1 ppm				
11. Divide lines 4–9 by	48.6	16.0	2.47	15.4
Average alpha ranges (μg/mm^2)				
12. Full chain	69.5	58.2	68.6	58.5
13. pre-Rn	52.8	46.1	(68.6)	46.2
Effective number of full alpha emissions per parent disintegration:				
14. Full chain	6	8	7	7.96
15. pre-Rn	3	4	(7)	4.13
Alpha count-rate (ksec^{-1}) from 42-mm dia. scintillator for unit activity of parent (Bq/kg)[c]				
16. Full chain	0.123	0.132	0.136	0.132
17. pre-Rn	0.0439	0.0498	(0.136)	0.0515
Alpha count-rate (ksec^{-1}) from 42-mm dia. scintillator for 1 ppm of parent				
18. Full chain	0.502	1.65	10.9	1.72
19. pre-Rn	0.179	0.622	(10.9)	0.669
Dose-rate (Gy/ka) for alpha count-rate = 10/ksec, from 42-mm dia. scintillator				
20. α, full chain	14.8	—	—	16.5
21. α, pre-Rn	17.3	—	—	19.1
22. Effective α for a = 0.1, full chain	1.28	—	—	1.28
23. Effective α for a = 0.1, pre-Rn	1.32	—	—	1.32
24. β + IC, full chain	0.574	—	—	0.865
25. β + IC, pre-Rn	0.578	—	—	0.927
26. γ + X, full chain	1.03	—	—	0.681
27. γ + X, pre-Rn	1.17	—	—	0.085

[a] Bq = 1 disintegration per sec = 27 pCi; 1 gray (Gy) = 1 J/kg; 1 ka = 1000 years = 31.5 $\times$ 10^6 ksec; 1 MeV = 1.602 $\times$ 10^{-13} J.

[b] The lines labelled 'pre-Rn' give the values for 100% escape of radon in the case of the thorium-232 and uranium-238 series, but because of the short half-life of radon-219 the values given in the uranium-235 and nat. U columns include the contribution of that gas and its daughters.

[c] The alpha count-rate is calculated for an electronic threshold setting such that for a pottery sample containing only the thorium-232 chain, 85% of the counts are recorded; with this setting the corresponding figure for natural U is 82%, and for the pre-Rn parts of the chain the figures are 80% and 78% for thorium-232 and natural U, respectively.

TABLE G.5
Annual Dose Data for Potassium and Rubidium

	K-40	Rb-87
1. Atomic abundance	117 ppm	27.8%
2. Half life ($\times 10^9$ yr)	1.25	49
3. Specific activity (Bq/kg) for concentration of 1% natural K and 50 ppm natural Rb[a]	β: 284 γ: 33.3	β: 43.4 —
4. Energy release (MeV)[a]	β: 0.583 γ: 1.461	β: 0.104 —
5. Annual dose (Gy/ka) for concentrations as in 3.[a,b]	β: 0.830 γ: 0.241	β: 0.023 —
6. As for 5 but for 1% K_2O and 50 ppm Rb_2O	β: 0.689 γ: 0.200	β: 0.019 —

[a] The relative concentrations used in lines 3, 5, and 6 correspond to the recommendation by Warren (1978) that when the rubidium content of a sample has not been measured, a potassium to rubidium ratio of 200:1 should be assumed. However because of the short range of the beta particles the rubidium contribution should be excluded in the case of quartz inclusion dating.

[b] The values quoted for the potassium-40 annual gamma dose are 3.5% lower than those given by Bell (1979a). This is because the energy carried by the neutrino emission associated with electron capture has now been omitted.

APPENDIX H

Gamma and Beta Gradients

LAYER TO LAYER VARIATION IN
SOIL RADIOACTIVITY

In the energy range of 0.1 to 3 MeV, which covers nearly all natural gamma emissions, the dominant mode of interaction of gamma photons with matter is through the Compton process, in which part of the photon's energy is transferred to an ejected electron and part carried away by a gamma photon of lower energy. In an extended medium about half the dose is delivered by this secondary flux and about half by the primary, or 'uncollided', flux. Whereas the latter can be dealt with analytically this is not the case for the former because of the difficulty of evaluating its spectrum. One approach is to assume that the build-up factor, that is, the ratio of total dose to primary dose, increases linearly with distance from source; this is the basis on which Figure 4.1 showing the dose at the centre of a sphere of radioactive material was calculated (see Evans, 1955, p. 739). It can also be used for the case of layer to layer variations, as by Fleming (1979). The alternative approach is to model the situation with a computer, feeding in details of atomic composition, cross-sections, primary emission spectra, etc. This is the basis for Figures 4.3–4.6 which have been derived from computations kindly carried out by L. Løvborg of the Risø National Laboratory.

Table H.1 gives the dose in an active medium (I) at a distance z from its boundary with an inert medium (II), both media having the composition given. According to Løvborg, the dose in the active medium has

TABLE H.1
Gamma Dose in Radioactive Soil Adjacent to Inert Soil[a]

z (cm)[b]	Fractional Dose			Weighted average	z (cm)[b]	Fractional Dose			Weighted average
	K	Th	U			K	Th	U	
0	0.5000	0.5000	0.5000	0.5000	17	0.9404	0.9438	0.9510	0.9453
0.5	0.5555	0.5577	0.5607	0.5582	18	0.9465	0.9495	0.9564	0.9510
1.0	0.5938	0.5974	0.6022	0.5981	20	0.9570	0.9592	0.9655	0.9607
1.5	0.6258	0.6306	0.6366	0.6314	22	0.9655	0.9669	0.9727	0.9684
2.0	0.6536	0.6594	0.6663	0.6603	24	0.9722	0.9732	0.9784	0.9746
2.5	0.6782	0.6849	0.6924	0.6858	25	0.9752	0.9759	0.9808	0.9772
3.0	0.7003	0.7076	0.7156	0.7085	26	0.9778	0.9783	0.9829	0.9796
3.5	0.7202	0.7281	0.7364	0.7290	28	0.9822	0.9824	0.9865	0.9836
4.0	0.7382	0.7466	0.7552	0.7475	30	0.9858	0.9857	0.9893	0.9868
4.5	0.7547	0.7634	0.7721	0.7643	31	0.9873	0.9871	0.9905	0.9882
5.0	0.7697	0.7787	0.7876	0.7796	32	0.9886	0.9884	0.9916	0.9894
5.5	0.7836	0.7927	0.8016	0.7936	33	0.9898	0.9895	0.9925	0.9905
6.0	0.7964	0.8055	0.8145	0.8064	34	0.9909	0.9905	0.9934	0.9915
6.5	0.8083	0.8174	0.8263	0.8183	35	0.9919	0.9915	0.9941	0.9924
7.0	0.8193	0.8283	0.8373	0.8292	35.5	0.9923	0.9919	0.9944	0.9927
7.5	0.8295	0.8384	0.8473	0.8393	36	0.9928	0.9923	0.9948	0.9932
8.0	0.8391	0.8478	0.8567	0.8487	36.5	0.9932	0.9927	0.9950	0.9935
8.5	0.8480	0.8565	0.8653	0.8574	37	0.9935	0.9930	0.9953	0.9938
9.0	0.8564	0.8646	0.8734	0.8656	37.5	0.9939	0.9934	0.9956	0.9942
9.5	0.8643	0.8722	0.8809	0.8732	38	0.9942	0.9937	0.9958	0.9944
10	0.8716	0.8793	0.8879	0.8803	38.5	0.9946	0.9940	0.9961	0.9948
11	0.8851	0.8921	0.9006	0.8933	39	0.9949	0.9943	0.9963	0.9950
12	0.8971	0.9035	0.9118	0.9047	39.5	0.9951	0.9946	0.9965	0.9953
13	0.9078	0.9135	0.9216	0.9148	40	0.9954	0.9949	0.9967	0.9955
14	0.9174	0.9224	0.9303	0.9238	40.5	0.9957	0.9951	0.9969	0.9958
15	0.9259	0.9304	0.9380	0.9318	41	0.9959	0.9953	0.9971	0.9960
16	0.9335	0.9375	0.9449	0.9389					

[a] The dose is expressed relative to that for an infinite extent of the radioactive soil. The weighted average is (0.2K + 0.5Th + 0.3U). This corresponds approximately to the radioactive composition of Table 4.5, that is 1% K, 10 ppm Th and 3 ppm U. Both media are taken to consist of 1.6 g/cm^3 of soil plus 0.4 g/cm^3 of water, that is, wet density 2 g/cm^3. Data for wet density ρ can be obtained to a reasonable approximation by scaling z; the values given for z should be multiplied by $(2/\rho)$. The composition used for the soil is taken from Leeper (1964), namely: O 50%, Si 36%, Al 6.9%, Ca 0.5%, Fe 3.5%, Mg 0.4%, K 1.5%, Na 0.6%, and Ti 0.6%. However the data are not much affected by soil composition; also it is to be expected that substitution of air for the inert soil will not appreciably affect the dose in the radioactive soil. The spectra used, inclusive of X-rays, are those given in ENSDF 1981 (see Ewbank, 1979); this latter is the Evaluated Nuclear Structure Date File maintained by the Nuclear Data Project at Oak Ridge National Laboratory, Oak Ridge, Tennessee 37830. Computation was by L. Løvborg of the Risø National Laboratory using the transport code developed by Kirkegaard and Løvborg (1980).

[b] z is the distance into the radioactive soil, from the boundary.

Appendix H: Gamma and Beta Gradients

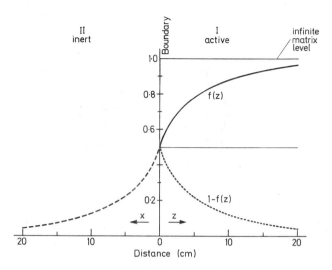

Figure H.1 Principle of superposition. If II had the same level of activity as I then the total dose in I would be unity (i.e., the infinite matrix dose). Hence the radioactivity in II must supply a dose in I equal to $1 - f(z)$, shown dotted. This must also be the dose that I supplies in II. Hence if II is inert the dose in it is as shown by the dashed curve. $f(z)$ is taken from Table H.1.

only a weak dependence on the composition of the inert medium; hence these data give the dose in soil as the ground surface is approached, see Figure 4.3.

When medium II is an inert soil the dose at a distance x into it can be obtained as

$$\{1 - f(x)\}D_1, \tag{H.1}$$

where D_1 represents the infinite matrix dose for I and $f(x)$ is obtained from Table H.1 by putting $z = x$. This follows from the principle of superposition (see Figure H.1): if II was not inert but had equal activity to I then the dose in it would have a constant level of D_1; hence the dose in an inert II must be a mirror image of the dose in I about a line through $f = 0.5$.

The dose in and adjacent to a layer, as in Figures 4.4 and H.3, can be obtained similarly. Consider first the situation with A and B equally active and L inert (see Figure H.2). The dose in L and B due to A is given by

$$\{1 - f(x)\}D_A, \tag{H.2}$$

and the dose in A and L due to B is given by

$$\{1 - f(t - x)\}D_B, \tag{H.3}$$

where $f(x)$ and $f(t - x)$ are obtained from the table by putting $z = x$ and $z = t - x$, respectively. Hence the dose in L due to A and B is given by

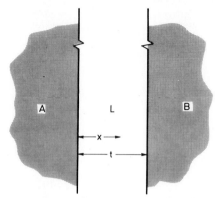

Figure H.2 Dose diagram for A and B equally active and L inert.

$$\{2 - f(x) - f(t - x)\}D_{AB}, \tag{H.4}$$

where $D_{AB} = D_A = D_B$. Consider next the situation of having L active with A and B inert; by the principle of superposition the dose in L must be

$$\{f(x) + f(t - x) - 1\}D_L. \tag{H.5}$$

Similarly it can be shown that in medium B the dose due to A and B is

$$\{1 - f(x) + f(x - t)\}D_{AB} \tag{H.6}$$

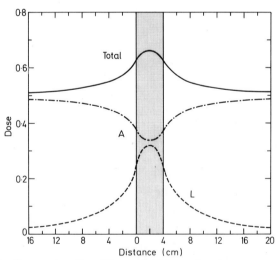

Figure H.3 Gamma dose for soil layer of thickness 4 cm between two regions having half as much radioactivity, expressed as a fraction of the infinite matrix dose in the layer. Curve L shows the contribution from the layer itself and curve A shows the contribution from the adjacent soil. Densities and compositions are as for Table H.1.

Appendix H: Gamma and Beta Gradients

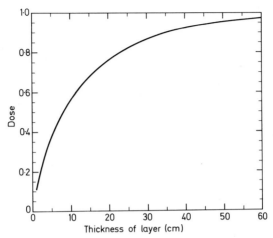

Figure H.4 Gamma dose in median plane of radioactive layer between two inert regions, as functions of layer thickness; expressed as a fraction of the infinite matrix dose for the layer. Based on Table H.1. For an inert layer between two active regions the median plane dose is $(1 - f')$, where f' is the fraction shown.

and that due to L is

$$\{f(x) - f(x - t)\}D_L. \tag{H.7}$$

Figures 4.4 and H.3 have been obtained by putting $D_{AB} = \frac{1}{2}D_L$, with $t = 40$ and 4 cm respectively; other values can be dealt with appropriately. Figure H.4 shows the dose in the median plane of the layer as a function of t. The data given are applicable strictly only for soil of composition, density, and water content as specified in the Table. For other soils an approximation can be obtained by scaling z up or down proportionally to the amount by which the wet density differs from 2 g/cm³.

GAMMA DOSE WITHIN A SAMPLE

According to Løvborg the computations used for the data of Table H.1 are not appropriate for the smaller distances involved in this context. Instead we use the results of experimental measurements of gamma attenuation made by Clark (Aitken *et al.*, 1985); in these, thermoluminescence phosphors were exposed to the gamma flux from a large concrete block doped with uranium, various thicknesses of glass absorber being interposed. These measurements indicated that for an absorber of density 2 g/cm³ having comparable composition (i.e., mainly SiO_2), the dose at a distance x cm into the absorber (both absorber and radioactive medium being semi-infinite) could be approximated, for $x > 0.2$ cm, by:

$$0.47\, D^e \exp(-\mu x), \tag{H.8}$$

where D^e is the infinite matrix gamma dose of the concrete and $\mu = 0.15$ per cm. Strictly this is for the uranium spectrum but it will be seen from Table H.1 that the values appropriate to a composite spectrum will not be significantly different.

For a parallel-sided sample of thickness t cm and infinite extent, with the same radioactive medium on both sides, the dose at a distance x from either face will be, for $x > 0.2$ cm,

$$0.47\, D^e \{\exp(-\mu x) + \exp(-\mu(t - x))\}. \tag{H.9}$$

By integration it follows that the average dose in the central region (i.e., excluding the discarded surface layers) is given by

$$D^e_{av} = \frac{6.5\, D^e}{(t - 0.4)} \{0.94 - \exp(-0.15 t)\} \tag{H.10}$$

for $\mu = 0.15$ per cm.

By substitution of numerical values it will be found that this is well fitted by the expression

$$D^e_{av} = D^e \exp(-0.06 - 0.07 t). \tag{H.11}$$

By the principle of superposition it follows that the average self-dose is given by:

$$D^i_{av} = D^i \{1 - \exp(-0.06 - 0.07 t)\}, \tag{H.12}$$

where D^i is the infinite matrix gamma dose of the sample. This is effectively the same as the self-dose percentage given in Section 4.2.2, namely,

$$p = 100\{1 - \exp(-0.06 - 0.07 t)\}.$$

This rule of thumb is an approximate one, but given the irregular shape of most samples there is no point in attempting to be more precise. If the lateral extent of the sample is not large compared to the thickness, then the self-dose percentage will be less than given by the rule. If substitution of the value given into Equation 4.4 indicates that there is an appreciable effect on the date, and the date is important, then a replica experiment may be appropriate.

Spherical Samples

Mejdahl (1983) has utilized the absorbed dose percentages calculated for spherical samples of water-equivalent material, converting to pottery

Appendix H: Gamma and Beta Gradients

and rock on the basis of electron and mass densities. Backscatter effects are included in the calculations but the X-rays were not included in the primary photon spectrum used, which was taken from Table 2-14 of NCRP report No. 50 (NCRP, 1976). The relative proportions of radioactive components assumed were 12 ppm Th, 3 ppm U, and 2% K. The values given by Mejdahl for various sample diameters d cm of pottery (density taken as 2 g/cm^3) and stone (2.6 g/cm^3) up to 12 cm are well-fitted by the rule of thumb $p = 2\ d\%$ for pottery and $p = 2.6\ d\%$ for stone.

BETA PENETRATION FROM THE SOIL

Measurements by Gaffney (Aitken *et al.*, 1985) using various thicknesses of aluminium absorber interposed between a beta-thick source and a thermoluminescence phosphor dosemeter indicated exponential decrease of dose as the beta flux penetrates the sample. The mass attenuation coefficients were 6.6 per g/cm^2 for the thorium spectrum (using monazite) and 5.7 per g/cm^2 for the uranium spectrum (using pitchblende). For the yttrium-90 spectrum from the standard thin strontium-90 beta source the decrease was not well fitted by an exponential, presumably due to the beta build up effect discussed in Section 5.2; the effective coefficient was ~6 per g/cm^2 for the first mm, falling to ~3 at 2 mm. In all cases appropriate subtraction of the bremsstrahlung dose was made.

Taking 2 g/cm^3 as the density of pottery, the linear coefficient for the thorium series is 1.32 per mm and that for the uranium series 1.14. Hence the soil beta dose in the sherd at a depth of 2 mm is reduced to $\exp(-2.64) = 0.07$ of its surface value for thorium and $\exp(-2.28) = 0.10$ for uranium. Thus even for the more penetrating uranium beta flux the soil beta dose at a depth of 2 mm is only 5% of the infinite matrix dose D_β^e (the surface value is half of this); hence as long as the outer 2-mm layers of sample are discarded the average beta contribution from the soil is negligible. The potassium spectrum will be less penetrating because its maximum energy of beta emission is 1.3 MeV whereas in the uranium series 40% of the beta dose is carried by the 2.26 MeV emission from Pa-234m.

It is of interest to estimate the maximum error introduced if there is no removal of surface layers (and if the beta dose is taken as the infinite matrix dose for the sample). The equations are similar to those of the last section, though simpler because the integration is now over the whole thickness. It can be shown that the average beta dose in the sample is

$$D_\beta^i - \frac{(D_\beta^i - D_\beta^e)}{\mu t}\{1 - \exp(-\mu t)\}, \qquad (H.13)$$

where D_β^i and D_β^e are the sample and soil infinite matrix beta doses respectively, μ per mm is the linear attenuation coefficient, and t mm is the thickness. If we insert $\mu = 1.14$ as for the uranium spectrum, for the case in which $D_\beta^i = 2 D_\beta^e$ the average beta dose is less than D_β^i by 9% for a 5-mm sample, 5% for a 10-mm sample and 2.5% for a 20-mm sample. In the extreme case such as might occur with flint or calcite, where the sample has negligible radioactivity so that its dominant dose is the gamma dose from the soil, the upper limit to the average beta dose in the sample is 18% of D_β^e for a 5-mm sample, 9% for a 10-mm sample, and 4.5% for a 20-mm sample.

Of course it must not be forgotten that removal of surface layers may be necessary on account of possible bleaching effects as well as soil contamination in the case of pottery.

APPENDIX I

Cosmic Ray Dose

Cosmic radiation usually contributes only a few percent of the annual dose and in the majority of circumstances it can be taken as 150 μGy/a. However, for low radioactivity samples and soil at high altitude or beneath thick overburden, a better estimate is desirable. This Appendix summarizes the data given for this purpose by Prescott and Stephan (1982).

The 'soft' component of cosmic radiation, largely electrons, is absorbed by about $\frac{1}{2}$ metre of soil, and so for thermoluminescent samples we are concerned with the more penetrating 'hard' component, composed mostly of muons. There is dependence on latitude because the earth's magnetic field gives some shielding in equatorial and low latitudes. However for the hard component at sea level the latitude effect is only slight; the intensity rises by 7% on going from the equator to latitude 40° and then remains effectively constant for higher latitudes. This is because the dominant influence is absorption by the atmosphere (which has a thickness in terms of weight of 1 kg/cm^2). Figure I.1 shows the dependence on altitude; at high altitudes the latitude effect becomes much stronger because atmospheric absorption is weaker.

The vertical intensity of the muon flux at sea-level, for latitudes above 40°, is close to 0.56 per cm^2/min/sr and the omnidirectional flux is 1.1 per cm^2/min. The corresponding annual dose for material of pottery-like composition ($Z \sim 11$, $A \sim 22$), under 100 g/cm^2, $\sim\frac{1}{2}$ metre of soil, is estimated at 185 ± 11 μGy/a. At shallower depths there is a contribution from the soft component, and for zero thickness the total annual dose is approximately 280 μGy/a.

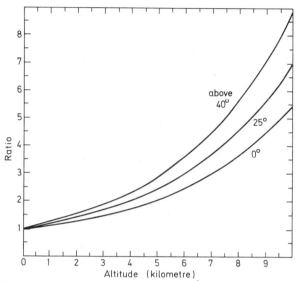

Figure I.1 Variation with altitude of the intensity of the cosmic ray hard component for the equator, latitude 25°, and latitude 40° and above. Expressed as a ratio to the sea-level value at latitude 40° and above, for which the corresponding annual dose is 185 μGy/a. (Derived from Figure 3 of Prescott and Stephan, 1982).

Measurement with thermoluminescence capsules (see Figure 4 of Prescott and Stephan) indicate a value of 150 μGy/a at depths of around 1 metre. For greater depths it is estimated that the attenuation should be at the rate of 8.5% per 100 g/cm³ of rock (10% per 100 g/cm² of water). Taking a density of 2 g/cm³ as typical of sedimentary rock this gives the following attenuation factors relative to the 1 metre value as unity: 0.84 at a depth of 2 metres, 0.71 at 3 m, 0.60 at 4 m, 0.43 at 6 m, 0.22 at 10 m, 0.09 at 15 m. In igneous rock with a typical density of 2.6 g/cm³ the fall-off will of course be more rapid.

APPENDIX J

Alpha Counting: Derivation of Formulae; Ranges; Standards

The first two sections of this Appendix are based mainly on the classic paper by Turner, Radley and Mayneord (1958).

BASIC FORMULAE

Consider a thick sample on top of a scintillation screen of area A. The particles from a small element of volume $dA \times dh$ of sample that can reach the screen are those emitted within a cone such that $\cos\theta = h/R$, where R is the range (see Figure J.1). Since the solid angle of a cone is given by $2\pi(1 - \cos\theta)$, that is, $2\pi(1 - h/R)$ in this case; the probability that an alpha particle emitted without preferential direction within the 4π solid angle of a sphere will reach the screen is given by $\frac{1}{2}(1 - h/R)$. Hence if c is the number emitted per unit time per unit mass, and ρ is the density of the sample, the number reaching the screen from a column of area dA is given by

$$\tfrac{1}{2}c\rho \, (dA) \int_0^R (1 - h/R) \, dh = \tfrac{1}{4}(dA)R\rho c.$$

Since the diameter of the screen is very much larger than R, edge effects will be negligible and the total number reaching the screen from the whole

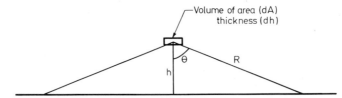

Figure J.1 Alpha counting with a thick sample on top of a scintillation screen.

sample will be $\frac{1}{4}AR\rho c$ as in Equation 4.8. As shown there, if the sample contains a radioactive series of n alpha emitters in equilibrium and c is the specific activity of the parent then the number reaching the screen is $\frac{1}{4}A\bar{R}\rho nc$, where $\bar{R}$ is the average range.

If the thickness t of the sample is less than R, then t replaces R as the upper limit of the integral given above and on evaluation the formula given in Note 6 of Chapter 2 is obtained.

THRESHOLD FRACTION

Because of the finite electronic threshold only those particles which reach the screen with enough energy to produce a pulse of sufficient height will be counted. If this minimum energy corresponds to a residual range δR then the number actually counted is obtained by replacing R in the above derivation by $(R - \delta R)$. The threshold setting usually employed is such that for the thorium series the count-rate is 0.85 times the hypothetical count-rate corresponding to zero threshold. Since in terms of residual range the spectrum is rectangular and for the thorium series $\bar{R} = 69.5\ \mu g/mm^2$, it follows that $\delta R = 0.15 \times 69.5 = 10.4\ \mu g/mm^2$. With this threshold setting the count-rate for the uranium series, for which $\bar{R} = 58.5\ \mu g/mm^2$, is lower than the zero threshold count-rate by the factor

$$\frac{58.5 - 10.4}{58.5}, \quad (\text{i.e., } 0.82).$$

As in the main text if we substitute $A = 1385\ mm^2$, corresponding to a 42-mm diameter screen, for the thorium series we obtain

$$\dot{\alpha}_h = 0.123 c_h, \tag{J.1}$$

and for the uranium series

$$\dot{\alpha}_u = 0.132 c_u, \tag{J.2}$$

where $\dot{\alpha}$ per ksec is the count-rate resulting from a specific parent activity of c Bq/kg; as elsewhere c_u includes the contribution from the uranium-235 series.

Appendix J: Alpha Counting

PAIRS COUNTING

If the new nucleus formed by alpha emission is also an alpha emitter, and is of short half-life, there will be a higher-than-random probability of observing two alpha particles in quick succession, that is, a 'pair'. We are concerned principally with the decay of radon-220 into polonium-216 in the thorium series; both are alpha emitters and the latter has a half-life of 0.145 seconds (Diamond and Gindler, 1963). Suppose R_1 and R_2 are the ranges of the successive alpha particles, then the probability that both will reach the screen is

$$\frac{1}{2}(1 - h/R_1) \times \frac{1}{2}(1 - h/R_2).$$

Hence if c is the specific parent activity of the series concerned, the number of pairs reaching the screen per unit time is

$$\frac{1}{4} A\rho c \int_0^{R_1} \left(1 - \frac{h}{R_1}\right)\left(1 - \frac{h}{R_2}\right) dh,$$

where R_1 is taken to be less than R_2. After integration and simplification this becomes

$$\frac{1}{12} A R_1 \rho c \left(1 + \frac{R_2 - R_1}{2 R_2}\right),$$

and the number producing an above-threshold pulse will be

$$\frac{1}{12} A (R_1 - \delta R) \rho c \left(1 + \frac{1}{2}\frac{R_2 - R_1}{(R_2 - \delta R)}\right).$$

The number actually recorded as pairs depends on the window duration τ of the coincidence circuitry. If λ is the decay constant of the nuclide from which the second alpha particle is emitted, then the probability that the emission will take place within time τ of formation (i.e., emission of the first alpha particle) is given by $(1 - \exp(-\lambda\tau))$. For the pairs due to the nuclei in the thorium chain mentioned above it is usual to set $\tau = 0.21$ sec so that $\lambda\tau = 1$, and the probability that the two emissions take place within time τ of each other is $(1 - e^{-1})$, that is, 0.632. The ranges of the successive emissions are 72.5 and 82.0 $\mu g/mm^2$, and taking $\delta R = 10.4$ $\mu g/mm^2$ as before, the pairs count-rate for the thorium series is obtained as

$$\dot{p}_h = \frac{0.632}{12} Ac_h\, 62.1 \left(1 + \frac{1}{2} \times \frac{9.5}{71.6}\right)$$
$$= 0.0048\, c_h. \tag{J.3}$$

Hence from Equation (J.1)

$$\dot{\alpha}_h = 25.5 \, \dot{p}_h. \tag{J.4}$$

Thus by measuring $\dot{p}_h$ the contribution of the thorium series to the total alpha count can be determined.

There are also 'fast' pairs arising from the decay of radon-219 into polonium-215 in the uranium-235 series. The half-life of polonium-215 is only 0.002 sec so that there is effectively 100% probability that the second emission will occur within the coincidence time of 0.21 sec. The respective ranges are 82 and 94 μg/mm^2 so that the pairs count-rate $\dot{p}_u$ associated with specific activity c_u of the uranium series is given by

$$\dot{p}_u = \frac{1}{12} A \, (0.044 \, c_u) \, 71.6 \left(1 + \frac{1}{2} \times \frac{12}{83.6}\right)$$
$$= 0.39 \times 10^{-3} \, c_u, \tag{J.5}$$

where the factor 0.044 allows for the contribution of the uranium-235 series to the overall uranium activity. Hence from Equation J.2 we have

$$\dot{p}_u = \dot{\alpha}_u / 340 \tag{J.6}$$

The experimentally obtained quantities are $\dot{\alpha}$ and $\dot{p}$, the total count-rate and the total pairs count-rate, and from Equations J.4 and J.6 we can evaluate the contribution to $\dot{\alpha}$ from the thorium series in terms of these as

$$\dot{\alpha}_h = 28\dot{p} - \dot{\alpha}/12. \tag{J.7}$$

In some types of coincidence circuitry being used the 'fast' pairs are counted separately and not included in the slow pairs; in this case the value of $\dot{\alpha}_h$ can be obtained directly from the slow pairs. Such circuitry is used in *Daybreak* equipment (for instance), the coincidence window for the slow pairs being open from 0.020 to 0.400 sec after each count; the correct numerical factor in Equation J.4 is then 21.1. Measurement of fast pairs has been used by Huntley and Wintle (1981) for determination of protoactinium-231 higher up in the series; the build-in of this nuclide, with a half-life of 34,000 yr, is used for radioactive dating in the same way as thorium-230. These authors note that in some commercially available coincidence units there is a random delay between each pulse and the opening of the slow pairs coincidence window of up to 0.02 sec, depending on the time relationship of the pulse to the mains supply cycle. This has the effect of reducing the number of pairs counted by 4%.

RANDOM PAIRS

In addition to the 'true' pairs there are also random coincidences. Each alpha pulse generates a coincidence window of duration τ and the probability that another alpha pulse will arrive within this period is $\dot{\alpha}\tau$; hence the number of random pairs occurring per unit time is $\dot{\alpha}^2\tau$. If we use $\dot{d}$ for the total 'doubles' rate observed then the true pairs rate is given by

$$\dot{p} = \dot{d} - \dot{\alpha}^2\tau. \qquad (J.8)$$

If we take as example $\dot{\alpha} = 10$ counts/ksec and $\tau = 0.21$ sec, then the random pairs rate is 0.021 counts/ksec. Assuming the thorium and uranium series contribute equally to the alpha count-rate, Equation J.7 gives the expected true pairs rate as

$$\frac{5 + 0.83}{28} = 0.21 \text{ counts/ksec},$$

that is, the random pairs contribute 10% of the total.

The full formula for obtaining $\dot{\alpha}_h$ is given in the main text Equation 4.17.

ALPHA RANGES

If the a-value system is being used then there is direct conversion of alpha count-rate to annual alpha dose, with no dependence on range to a good approximation. On the other hand if the alpha count-rate is being used to evaluate the thorium and uranium activities as an intermediate step to obtaining the beta and gamma doses then knowledge of the range is necessary. When there is radioactive equilibrium then the average ranges as given in Appendix G are appropriate, but for other circumstances (e.g. in dating calcite when initially there are only the two alpha emissions preceding thorium-230) it is necessary to know the individual ranges. These have been evaluated by Bowman (1982a) for pottery taking the effective atomic number Z to be 10.6 and the average atomic weight A to be 21.4 and interpolating on the basis of proportionality to $A^{\frac{1}{2}}$ between the ranges given for neon ($Z = 10$, $A = 20$) and for sodium ($Z = 11$, $A = 23$) in nuclear data tables (Northcliffe and Schilling, 1970; Williamson, Boujot and Picard, 1966). Bowman's values are given in Tables J.1–J.3; see also Figure 5.8.

For substances other than pottery, further reference should be made to nuclear data tables; alternatively to within $\pm 15\%$, the ranges may be calculated from the Bragg–Kleeman rule which states that $R\rho$ is approxi-

TABLE J.1
Alpha Ranges: Thorium Series[a]

Nuclide	Energy[b]	Intensity[c]	$R\rho$ ($\mu g/mm^2$)[d]
Th-232	4.01	77	37.5
	3.95	23	
Th-228	5.42	72.7	58.5
	5.34	26.3	
Ra-224	5.69	95	62.5
	5.45	5	
Rn-220	6.29	100	72.5
Po-216	6.78	100	82.0
Bi-212	6.05	25.2	68.5
	6.09	9.8	
Po-212	8.78	64	124
Average			69.5

[a] The data given were evaluated by Bowman (1982a) for pottery taking the effective Z to be 10.6 and the effective A to be 21.4, and interpolating between the ranges given for neon and for sodium in nuclear data tables (Northcliffe and Schilling, 1970; Williamson et al. 1966).
[b] Alpha particles energies are given in MeV.
[c] 'Intensity' gives the emission rate expressed as a percentage of the decay rate of the parent of the series.
[d] When there is more than one energy of alpha emission for a given nuclide the value given for $R\rho$ is the weighted average.

mately proportional to $A^{\frac{1}{2}}$. The effective value for A is given by

$$A^{\frac{1}{2}} = \frac{(n_1 A_1 + n_2 A_2 + \cdots)}{(n_1 A_1^{\frac{1}{2}} + n_2 A_2^{\frac{1}{2}} + \cdots)},$$

where $n_1, n_2, \ldots$ are the appropriate atomic fractions. As is seen in Table J.4 based on data given by Bowman (1976, 1982a), most mineral inclusions in pottery have effective atomic weights close to the value of 21.4 used for the range tables. For these the ±15% uncertainty in the correction does not introduce significant uncertainty in the final value; on the other hand, in respect to pitchblende, for which the effective A is 133, the correcting factor by which the pottery ranges need to be multiplied is 2.5; ranges derived in this way should be regarded only as a rough guide.

CALIBRATION STANDARDS

For testing agreement between predicted and observed count-rates the standards chosen should have alpha activity of the same order of magni-

TABLE J.2
Alpha Ranges: Uranium-238 Series[a]

Nuclide	Energy[b]	Intensity[c]	$R\rho$ ($\mu g/mm^2$)[d]
U-238	4.20	77	39.5
	4.15	23	
U-234	4.77	72	48.5
	4.72	28	
Th-230	4.69	76.3	48.0
	4.62	23.4	
Ra-226	4.78	94.5	48.5
	4.60	5.5	
Rn-222	5.49	100	59.5
Po-218	6.00	100	66.0
Po-214	7.69	100	100.0
Po-210	5.30	100	56.0
Average			58.2

[a] The data given were evaluated by Bowman (1982a) for pottery taking the effective Z to be 10.6 and the effective A to be 21.4, and interpolating between the ranges given for neon and for sodium in nuclear data tables (Northcliffe and Schilling, 1970; Williamson et al. 1966).

[b] Alpha particles energies are given in MeV.

[c] 'Intensity' gives the emission rate expressed as a percentage of the decay rate of the parent of the series.

[d] When there is more than one energy of alpha emission for a given nuclide the value given for $R\rho$ is the weighted average.

tude as the samples being dated and be in 'thick-source' form so that there is similarity of alpha spectra; there should either be one standard having equal thorium and uranium activities, or a separate standard for each series. Obviously each standard should have been accurately analysed by several other techniques (e.g., neutron activation, fission track analysis, alpha spectrometry, gamma spectrometry) and it must be in radioactive equilibrium (which implies too that gas escape is negligible). The average Z and A should be the same as for the samples being dated, or at any rate not too far different so that any correction made is a small one; the standard should be in powder form with about the same grain size as the samples, and hence there is the need for homogeneity so that size-differentiation effects (such as discussed, for instance, by Mangini, Pernicka and Wagner, 1983) are absent. Finally the colour should not be far different to the samples so that reflectivities are similar.

It is not easy to fulfill all these requirements, and it is doubtful if the ideal standard exists; glassy standards such as the tektite used by Mangini

TABLE J.3
Alpha Ranges: Uranium-235 Series[a]

Nuclide	Energy[b,c]	R_ρ (μg/mm²)[d]
U-235	4.40	42.5
Pa-231	5.01	51.1
Th-227	6.33	67.5
Ra-223	5.72	62.5
Ra-219	6.82	82.0
Po-215	7.39	94.0
Bi-211	6.62	80.0
Average		68.6

[a] The data given were evaluated by Bowman (1982a) for pottery taking the effective Z to be 10.6 and the effective A to be 21.4, and interpolating between the ranges given for neon and for sodium in nuclear data tables (Northcliffe and Schilling, 1970; Williamson et al. 1966).
[b] Alpha particles energies are given in MeV.
[c] Only the energy of the most intense decay is shown.
[d] When there is more than one energy of alpha emission for a given nuclide the value given for $R\rho$ is the weighted average.

et al. are one approach. It is more usual however to use a diluted radioactive ore such as the reference standards issued by the Canadian Mines Branch, Ottawa or the New Brunswick Laboratory of the U.S. Dept. of Energy, Argonne. The latter are pitchblende or monazite diluted in silica and ground to a fine powder. Sand 105A has a specified uranium content of 10.2 ppm (with an error limit of ±2% at the 95% level of confidence); the radium content is estimated to be within 2% of the equilibrium value. Using line 18 of Table G.4, and making allowance for the range in silica being 3% less than in standard pottery due to a slightly lower value of $\bar{A}$, the predicted count-rate is 17.0 counts/ksec (42-mm diameter screen, standard threshold). The observed value for this sand, unsealed, in the author's laboratory, is 18.3 ± 0.1 counts/ksec; measurements with a gas cell indicate the 'lost count-rate' due to radon escape is 1.3 ± 0.4 counts/ksec. Evidently some overcounting occurs, although some of the excess may arise because the alpha particles expend part of their range in the grain of pitchblende within which they are emitted, this will make the effective range slightly higher due to the higher value of $\bar{A}$.

Sand 109 has a specified thorium content of 104 ppm (with 95% confidence limit of 3%) plus 3.7 ppm of uranium, giving a predicted count-rate of (50.7 + 6.2) = 56.9 counts/ksec. The observed value is 58.6 ± 0.3

TABLE J.4
Effective Z and A for Various Substances[a]

Sample	Formula	Z	A
Quartz	SiO_2	10.2	20.2
Feldspar	$KAlSi_3O_8$	10.5	21.3
Hectorite	$NaMg_6Si_6O_{20}$	10.0	20.8
Illite	$KAl_4Si_7AlO_{20}$	10.6	21.4
Brick dust	Turner et al (1958)	10.6	—
GSP sand	Flanagan (1969)	10.6	21.5
Igneous rock	Turner et al (1958)	11.0	—
Dunite	$\begin{Bmatrix} 76\% \ Mg_2SiO_4 \\ 19\% \ Fe_2SiO_4 \\ 5\% \ FeCr_2O_4 \end{Bmatrix}$	11.4	22.3
Zircon	$ZrSiO_4$	16.2	35.3
Monazite	$(Ce, La, Th) PO_4$	22	52
Pitchblende	UO_2	50	133
Lithium fluoride	LiF	6.4	13.8
Aluminium oxide	Al_2O_3	10.1	20.7
Calcium carbonate	$CaCO_3$	10.5	21.1
Calcium sulphate	$CaSO_4$	11.8	23.6
Calcium fluoride	CaF_2	13.2	26.9

[a] These values are 'effective' in the sense of being appropriate in respect of alpha range, that is, where there is dependence on $A^{\frac{1}{2}}$. Both Z and A have been calculated on the basis of the Bragg–Kleeman rule given in the text. They are not appropriate with respect to gamma absorption. Data are from Bowman (1982a).

counts/ksec with a lost count-rate of 1.1 ± 0.1 counts/ksec, so that in this case any overcounting is barely significant. In the same series of silica-based sands there are a variety of concentration levels, ranging up to 1% thorium or uranium. Pitchblende itself is available (sand 6A: U content 58%) and also monazite sand (sand 7A: Th content 8.5%, U content 0.3%). These two are useful for quick daily checks of electronic performance and for testing scintillation screen efficiency.

Another low activity standard in use is GSP-1 (Flanagan 1969, 1972) with a thorium content of 104 ppm and a uranium content of 2 ppm; this standard has the advantage that its average atomic weight is within $\frac{1}{2}$% of that of standard pottery, but on the other hand there is a spread of about ± 10% in the concentration values obtained by other techniques. The observed count-rate is 65 counts/ksec.

APPENDIX K

The *a*-Value System of Assessing the Alpha Particle Contribution

As illustrated in Figures 5.10 and 5.12 the energy loss per unit length of path increases sharply as an alpha particle slows down, whereas the thermoluminescence per unit dose decreases. Thus, if the strength of the alpha source is specified in grays it is necessary to know the energy spectrum of the particles as they penetrate the sample together with the energy dependence of the k value. On the other hand the thermoluminescence per unit length of alpha particle track is independent of particle energy to a first approximation and allows development of the so-called *a-value* system; this is simpler to use—once it has been assimilated—and less affected by uncertainty about various parameters which change the effective alpha spectrum seen by the sample, for example, thickness of the protective coating or foil on the source (this thickness may change slowly due to corrosion), obliquity of the particles, and thickness of sample (usually unknown). Also, as is seen in the next paragraph, the alpha count-rate for a sample is a direct measure of its effective alpha dose-rate that is independent of density and atomic weight, and of particle energy (in so far as the approximation of constant thermoluminescence per unit length of track is valid). A variant of this system is the *b-value* system which is also discussed below. On the basis of this approximation the latent thermoluminescence generated in a sample by alpha particles during antiquity is simply given by the total track length multiplied by the thermoluminescence per unit track length. Thus the rate at which latent thermoluminescence was being produced during antiquity may be written as

$$\dot{G}_\alpha = \xi v (N_1 R_1 + N_2 R_2 + \cdots), \tag{K.1}$$

where v is the volume of the sample, N_1, N_2, ... are the disintegration rates per unit volume of the alpha emitting isotopes in the sample, R_1, R_2, ... are the alpha particle ranges, and ξ is the thermoluminescence per unit length of track. The above equation is of course rather similar to Equation 4.9 for the alpha count-rate (extrapolated to zero electronic threshold); this may be written as,

$$\dot{\alpha}_t = \tfrac{1}{4} A (N_1 R_1 + N_2 R_2 + \cdots), \tag{K.2}$$

where A is the area of the scintillator. The quantity $(N_1 R_1 + N_2 R_2 + \cdots)$ occurs now because particle range determines the depth in the sample from which particles can reach the scintillator. It is because this quantity occurs in both equations that the relationship between $\dot{G}_\alpha$ and $\dot{\alpha}$, which may be written as

$$\left(\frac{\dot{G}_\alpha}{v}\right) = 4\xi \left(\frac{\dot{\alpha}_t}{A}\right) \tag{K.3}$$

is independent of density, atomic weight, and alpha particle energies; however this is only to the extent that the assumption of constant thermoluminescence per unit length of track is valid, as is discussed further below.

EVALUATION OF THERMOLUMINESCENCE PER UNIT LENGTH OF TRACK

The calibration of an alpha source in terms of length of track delivered to unit volume of sample is discussed in Section 5.3. Hence by measuring the thermoluminescence induced in the sample (of volume v) by exposure to a source we can obtain $v\xi_s$, the subscript s being used to distinguish between the value relevant to artificial irradiation and the value relevant to acquisition of latent thermoluminescence during antiquity as required for Equations K.1 and K.3. That there is a difference is due to the assumption of constant thermoluminescence per unit length of track being only approximately true. Using the alpha particle beam from a Van de Graaff accelerator, Bowman (1976) has obtained extensive data (up to 7 MeV) on this question with respect to ten types of sample. As defined in

TABLE K.1
Values[a] for η

		No Radon loss		100% Radon loss	
		U	Th	U	Th
Feldspars	Microcline	0.91	0.92	0.88	0.90
	Orthoclase	0.92	0.93	0.89	0.91
Pottery	Knossos (150r16)	0.92	0.93	0.89	0.91
	Stephania (5a1)	0.93	0.93	0.90	0.92
	Bronze core (SC 587)	0.87	0.88	0.84	0.86
Pre-dosed quartz		0.85	0.88	0.83	0.85
Phosphors	CaF$_2$:Tb	0.95	0.95	0.94	0.95
	CaF$_2$:Mn	0.90	0.91	0.87	0.89
	Fluorite (MBLE/S)	0.84	0.86	0.79	0.82
	CaSO$_4$:Dy	0.86	0.87	0.82	0.84
Average		0.895	0.906	0.865	0.885
Standard Deviation		0.037	0.031	0.045	0.041
Average		0.901		0.875	
Standard Deviation		0.034		0.043	

[a] The values are based on experimental determination of thermoluminescence per unit length of track ξ_E as a function of particle energy, using the alpha beam from a Van de Graaff generator (Bowman, 1976). This energy dependence has been folded in with the range spectra of alpha particles seen by a fine grain in pottery, that is,

$$\eta = \int_0^{R_{max}} \left(\xi_E \frac{dn}{dR} \right) dR \bigg/ \xi_s \int_0^{R_{max}} \frac{dn}{dR} dR$$

Table K.1, $\eta = \xi/\xi_s$, and for dating purposes the value of η should be taken as 0.90 ± 0.05 when using an artificial source for which the energy of the emitted alpha particles is in the range 4–6 MeV; η is less than unity because of the decrease of thermoluminescence per unit track length at low energy (see Figure 5.11) and the fact that the alpha spectrum relevant during antiquity is a thick source spectrum; the rather wide error limits reflect the variation from material to material (see Table K.1) and also uncertainty as to whether the fall-off at low energy is really the result of conglomeration of fine grains causing particle penetration to be incomplete. Bowman has also evaluated the dependence of ξ_s on the energy of the alpha particles used for measurement; she finds that the average value for the ten phosphors studied is within a few percent of the 5 MeV value from 4 to 6 MeV, and that for lower energies there is, as might be expected, a just significant fall-off: $(4 \pm 3)\%$ at 3.5 MeV, $(6 \pm 5)\%$ at 3 MeV, $(8 \pm 6)\%$ at 2.5 MeV.

Appendix K: The *a*-Value System

INSERTION OF *a*-VALUE

Both $\dot{G}_\alpha$ and ξ_s are in terms of 'thermoluminescence', and in order to remove intrusion of instrumental and sample sensitivity it is convenient to make substitution of the beta dose that would induce the same amount of thermoluminescence. Thus, if χ_β is the thermoluminescence per gray of beta radiation the rate at which equivalent dose is generated in the sample per minute by an alpha source is given by

$$\dot{x} = \xi_s v S / \chi_\beta, \quad (K.4)$$

where S is the source strength in $\mu m^{-2}/min$ and the sample volume is in μm^3. Here as elsewhere in this type of discussion χ_β refers to the linear portion of the beta growth characteristic.

Defining the parameter *a*, of which the physical significance is discussed later, as

$$a = \dot{x}/(13S), \quad (K.5)$$

we obtain

$$\frac{\xi_s v}{\chi_\beta} = 13\,a\,. \quad (K.6)$$

It is to be assumed that all samples used for artificial irradiations are 'alpha-thin', that is, there is complete penetration of the sample layer, and if, as above, we are concerned with dependence on energy, it is to be assumed that the incident particles were approximately perpendicular to the plane of the layer, the emergent energy being not much less than the incident energy.

AGE EQUATION

The rate at which equivalent dose is generated in the sample by alpha particles during antiquity is given by

$$\dot{Q}_\alpha = \dot{G}_\alpha / \chi_\beta$$
$$= \left(\frac{\xi v}{\chi_\beta}\right)\left(\frac{4\dot{\alpha}_t}{A}\right) \quad (K.7)$$

from Equation K.3. Using $\xi = 0.90 \xi_s$ and Equation K.6 we obtain

$$\dot{Q}_\alpha = 0.90(13a)4\,\frac{\dot{\alpha}_t}{A}. \quad (K.8)$$

Note that it is implicit in Equations K.5 and K.6 that v is measured in μm^3 because the definition of S is based on micrometres; hence as the value of A for the standard 42-mm diameter scintillator we use $13.85 \times 10^8\,\mu m$. We also insert the measured alpha count-rate [$\dot{\alpha} = 0.835\,\dot{\alpha}_t$], the numerical factor being the electronic threshold factor discussed in Section 4.3.2, and so obtain

$$\dot{Q}_\alpha = 4.06(10^{-8}a\dot{\alpha}). \qquad (K.9)$$

If the units of $\dot{\alpha}$ are counts per kilosecond and we wish to obtain $\dot{Q}_\alpha$ in micrograys per year for insertion into the age equation, we must multiply by $10^6 \times$ (number of kiloseconds in a year), that is, $10^6 \times 31.5 \times 10^3$, thereby obtaining for the effective annual alpha dose $D'_\alpha = 1280\,a\dot{\alpha}\ \mu\text{Gy/a}$ as in Equation 4.15.

THE PHYSICAL MEANING OF a

For a 3.7-MeV alpha particle the rate of energy loss in quartz is 0.81 MeV per mg/cm² traversed, that is, 0.21 MeV/μm taking $\rho = 2.6$ g/cm³. A source of strength S delivers per minute, to an alpha-thin sample, S micrometres of track per cubic micrometre and hence $0.21S$ MeV of absorbed energy per cubic micrometre. Using the value just quoted for ρ and remembering that 1 Gy = 1 J/kg, the absorbed dose/min is

$$\frac{0.21 \times S \times 1.6 \times 10^2}{2.6} = 13\,S,$$

that is, the denominator of Equation K.5. The dose in a non-quartz sample will be $13S$ multiplied by the stopping power ratio r between sample and quartz.

The numerator of Equation K.5 represents the beta dose necessary to induce in the sample the same amount of thermoluminescence as the alpha irradiation. Hence for 3.7-MeV alpha particles the ratio between alpha effectiveness (per gray) and beta effectiveness is $r^{-1}a$. If the beta dosage was 100% efficient (which of course it is not) then the ratio would represent the fraction of the deposited alpha energy actually used, the remainder going to waste because of the traps in the core of the track being saturated.

THE RELATIONSHIP OF a-VALUE TO k-VALUE

In the k-value system (developed by Zimmerman, 1971, though he used the symbol ε in place of k) the basic definition is

$$k_{3.7} = \frac{\text{TL/Gy for 3.7-MeV alpha particles}}{\text{TL/Gy for beta irradiation}}. \qquad (K.10)$$

Hence for quartz $a = k_{3.7}$, and for other samples $a = rk_{3.7}$.

However in evaluating the effective annual dose from the actual annual alpha dose it is necessary to use k_{eff}; this refers to the spectrum of alpha particles actually received during antiquity rather than 3.7-MeV particles.

Because the spectrum received during antiquity is a degraded spectrum (it is effectively a 'thick source' situation), its average energy is less than 3.7 MeV and because k decreases rapidly with decreasing energy (see Fig. 5.12) k_{eff} is less than $k_{3.7}$. According to calculations by Zimmerman (1971a), for the thorium series $(k_{\text{eff}}/k_{3.7}) = 0.86$ and for the uranium series the ratio is 0.80, both for quartz, so that for equal activity of the two series the ratio is 0.83. In various tables, and elsewhere in this book, k signifies k_{eff}.

It is useful to note here that according to calculations by Valladas (Guérin and Valladas, 1980; Guérin, 1982a) the equal activity spectrum received by the sample during antiquity is best simulated by an artificial source emitting alpha particles of energy 3.40 MeV.

THE b-VALUE SYSTEM
(BOWMAN AND HUNTLEY, 1984)

This is based on the same assumption that the thermoluminescence per unit length of alpha track is independent of energy but the parameter b is more closely related to what is actually measured in a standard laboratory determination of alpha effectiveness. The parameter is defined as:

$$b = \frac{\text{TL per unit alpha track length per unit volume}}{\text{TL per unit absorbed beta dose}}$$

$$= \frac{\text{TL per unit track length}}{\text{TL per unit absorbed beta energy}} \times \frac{1}{\text{density}}. \quad (K.11)$$

Unlike a, b is not dimensionless and it is convenient to express it in terms of Gy μm^2, in which case:

$$b = 13a. \quad (K.12)$$

DEPENDENCE ON SAMPLE SPECTRUM

The numerical factor of 1280 relating effective alpha dose-rate to alpha count-rate incorporated $\eta = 0.90$ as the ratio between the thermoluminescence per unit length of track for the sample spectrum and that for an artificial source emitting alpha particles within the range 4–6 MeV; this was based on Bowman (1976) (see Table K.1). Whereas the values of η for the thorium and uranium series separately are only 1% apart, for the pre-gas parts of the series the average is $2\frac{1}{2}$% below that for the full series (see Table K.1). This is because the pre-gas alpha energies are lower than the later ones and there is a fall-off of thermoluminescence per unit length of track below 2 MeV (see Figure 5.11). However, the threshold factors are lower also, by $5\frac{1}{2}$%, and the net effect is to increase the numerical factor by 3% (see line 23 of Table G.4).

In any particular case of disequilibrium (such as the dating of calcite, see Wintle 1978b) the adjustment to the numerical factor can be estimated as follows. The appropriate value of η for an alpha emission having range R μg/mm² is given by

$$\eta = 1 - \frac{6.4}{R} \qquad (K.13)$$

This relation has been obtained from the definition of η given with Table K.1 using the representative energy dependence of thermoluminescence per unit track length given by Aitken and Bowman (1975). This latter is based on data reported by Aitken *et al.* (1974) and consists of a constant level for energies above 2 MeV with a linear rise between 0 and 2 MeV from 25% of that constant level. In the case of calcite, before there has been any build-in of thorium-230 or protoactinium-231 the only alpha emissions are from uranium-238, uranium-234 and uranium-235. The ranges of these, obtained from Appendix J, are 39.5, 48.5 and 42.5 μg/mm², respectively, and on substitution the values obtained for η are 0.84, 0.87 and 0.85. Making allowance for the activity of uranium-235 being only 4.4% of the total, the average value of η is 0.85. The threshold fraction is given in Appendix J as

$$1 - 10.4/R. \qquad (K.14)$$

Substitution of the ranges for the three emissions gives a weighted average for the threshold fraction of 0.76. The net effect on the numerical factor relating effective alpha dose-rate to alpha count-rate is to increase it by

$$\frac{0.85}{0.90} \times \frac{0.835}{0.76} = 1.04.$$

Except for a young sample, the adjustment is barely significant because as the series tends to equilibrium the correction factor is diluted.

COARSE GRAINS AND SLICES

All of the foregoing refers to an alpha-thin layer of fine grains. We now discuss the evaluation of a-value in two situations for which there is complete stoppage of the alpha particles in the sample. Initially the motivation for so obtaining a-values was inability to achieve adequate suppression of spurious thermoluminescence from fine-grain layers of calcite and flint. With improved anti-spurious measures this is no longer a problem but there are circumstances in which it is advantageous to use coarse grains and slices. For instance, if coarse grains are being used without etching, then the a-value is needed in order to assess the alpha particle contribution to the outer skin of the grain. Because impurities may have diffused into the grain from the clay matrix during firing it is likely that the outer skin will have a higher a-value than the interior of the grain; in this case the a-value obtained from fine-grains will not be the relevant one; also there is the possibility that the act of crushing may create additional traps or destroy existing ones. It is because of the adverse effects of crushing that slices are used for porcelain dating. Another context in which the use of coarse grains for a-value may be useful is ESR dating. To obtain sufficient sample for this technique from fine-grains it is necessary to make several dozen fine-grain discs, dose with alpha radiation, and then scrape off the fine-grains for measurement in the ESR cavity.

In using slices it is necessary to make careful allowance for optical attenuation. Whereas the thermoluminescence induced by artificial alpha irradiation comes from the top 20 μm of the slice, the beta- or gamma-induced thermoluminescence with which it is being compared comes from the whole thickness of the sample and so is liable to suffer much greater optical attenuation. Of course if the transparency is very low then the thermoluminescent collection depth may be less than the alpha range and in that case the comparison of effectiveness is independent of optical attenuation to a first approximation. A further consideration is the uniformity of the administered dose; in the case of beta irradiation build-up effects need to be taken into account (see Section 5.2). Calculations (Aitken and Wintle, 1977) give the expression for a as

$$a = \frac{\varepsilon_\beta \dot{x} d}{0.013 \eta' \varepsilon_\alpha L}, \qquad (K.15)$$

where ε_α and ε_β are optical attenuation factors appropriate to the radiations indicated, d mm is the thickness of the slice, η' is the ratio, for alpha source used, between the average thermoluminescence per unit track length for total absorption and thermoluminescence per unit track length for a thin layer of fine grains, $\dot{x}$ Gy is the beta dose equivalent to 1 min of irradiation with the alpha source, and L μm^{-1}/min is the track length delivered by the source to unit area of the surface of the slice. In measurements on two slices of flint, Wintle and Aitken (1977a) determined the optical attenuation coefficient as $\sim$12 per mm for the wavelength of thermoluminescence accepted by the detection system and obtained values for ε_α and ε_β of $\sim$0.9 and $\sim$0.25, the slice thickness being 0.4 mm. The source strength is now specified as L because we are now concerned with total absorption; L is equal to the previously defined strength, S, multiplied by the range (in micrometres) of the particles being delivered. The values obtained for a, 0.2 and 0.3, were considered to be in reasonable agreement with the fine-grain estimates.

In the case of coarse grains, calculations are straightforward as long as it is assumed that the grains are spherical. Because this is not usually true in reality, Valladas and Valladas (1982) use a vibrating pan to carry the grains while being irradiated; the jumping of the grains ensures that they

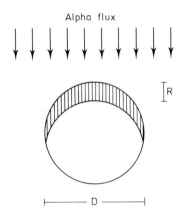

Figure K.1 A spherical grain being irradiated with a parallel flux.

receive an omnidirectional flux, thus minimizing error due to non-spherical shape. Consider a spherical grain of diameter D μm irradiated with a parallel flux of S particles/μm²/min, as depicted in Figure K.1. Then $\frac{1}{4}\pi D^2 S$ is the rate at which particles strike the grain and if R μm is the range the rate at which track length is delivered to the grain is $\frac{1}{4}\pi D^2 L$, where $L = RS$ as above. This neglects the fact that a small number of particles only 'graze' the grain and are not stopped in it; however for grains of the order of 100 μm these only amount to a few percent of the total. The thermoluminescence induced is obtained by multiplying by ξ'_s, the average thermoluminescence per unit length of track for the case of total absorption of an alpha particle.

In the case of beta or gamma irradiation of the grain, suppose, as usual, $\dot{x}'$ Gy is the dose equivalent to 1 min of alpha irradiation. Then the energy absorbed by the grain is $\dot{x}'$ times its mass and the thermoluminescence induced is

$$\tfrac{4}{3}\pi(D/2)^3\rho\psi\dot{x}' \times 10^{-18},$$

where ψ is the thermoluminescence induced per joule and ρ kg/m³ is the density. Equating this to the alpha thermoluminescence induced per minute we obtain

$$\tfrac{1}{4}L\xi'_s = \tfrac{1}{6}D\rho\psi\dot{x}' \times 10^{-18} \tag{K.16}$$

The quantities ξ'_s and ψ are related by means of the a-value. Consider a thin layer, area A and thickness d, of the same material and let $\dot{x}$ Gy be the beta or gamma dose equivalent to 1 min of irradiation with an alpha source of strength S μm/min. Equating beta and alpha thermoluminescence we obtain

$$\dot{x}Ad\rho\psi = S\xi_s Ad \times 10^{18}. \tag{K.17}$$

Appendix K: The a-Value System

Using the definition of a given in Equation K.5 we have

$$a = \frac{\dot{x}}{13S}$$

$$= \frac{\xi_s \times 10^{18}}{13\rho\psi}. \qquad (K.18)$$

Writing $\eta' = \xi'_s/\xi_s$ and substituting from Equation K.15 we obtain

$$a = \frac{2}{3}\frac{\dot{x}'D}{13\eta'L}$$

$$= \frac{2}{3}\frac{\dot{x}'D}{13\eta'RS}. \qquad (K.19)$$

If we now define a' as the apparent value of a that would be obtained if the sample irradiated was assumed to be an alpha-thin layer, then

$$a' = \frac{\dot{x}'}{13S},$$

hence

$$a = \frac{2}{3}\frac{D}{\eta'R}a'. \qquad (K.20)$$

Substituting $D = 100$ μm, and taking the incident alpha energy to be 4 MeV so that $R = 14$ μm (for quartz) and $\eta' = 0.83$ (using Equation K.13),

$$a = \frac{2}{3}\frac{100}{0.83 \times 14}a'$$

$$= 5.7a'. \qquad (K.21)$$

It must be stressed that there has not as yet been any experimental testing of this relation and it should be used with caution until that has been done. No allowance is included for optical effects such as might arise because the alpha thermoluminescence is emitted only from the surface layer of the grain; also it is assumed that the grains are spherical, and that the alpha flux is parallel.

APPENDIX L

The Portable 4-Channel Gamma Spectrometer

The use of this instrument for evaluation of the components of the annual gamma dose from the soil has been outlined in Section 4.3.4. Although the total gamma dose can be obtained by means of the scintillometer, which is simpler and quicker, in considering the degree of importance of such effects as radon escape and leaching it is advantageous to know the partition of the dose between potassium, thorium, and uranium. Effectively this means that the concentrations are determined too and in the case of sediment dating, where sample and soil are one and the same, this allows evaluation of the alpha and beta annual dosage also.

As discussed in the main text the pulse height (energy) spectrum obtained from a sodium iodide crystal consists of a photopeak (with a full width at half height of around 8%) and a Compton continuum spreading from a little below the photopeak energy down to near zero (see Figure 4.14). Although a pulse height window set astride 2.6 MeV will receive counts from thallium-208 of the thorium series only, in the case of the windows used for potassium and uranium there is a substantial contribution from the Compton continua of higher energy gamma emissions and in addition a small contribution from over-lapping photopeaks. Hence the need for 'spectrum stripping' as discussed in the first section of this Appendix. Having so obtained individual count-rates the second step is to interpret them in terms of concentrations (or parent activities) and annual gamma dose contributions; this is done in the second and third sections. Finally in the fourth section the all-important need for temperature stabil-

Appendix L: The Portable 4-Channel Gamma Spectrometer

ity is emphasized and the technique of spectrum stabilization by which it is achieved is outlined.

SPECTRUM STRIPPING

In order to determine the contributions to the count-rates in the potassium and uranium windows from the thorium series and so forth it is necessary to know the relative count-rates into each of the three windows when the detector is exposed separately to potassium, thorium, and uranium sources. Due to degradation, spectra resulting from distributed radioactivity, as in the soil, are not the same as spectra from discrete sources. Therefore it is necessary to build three quasi-infinite distributed sources; this has been done by various workers, for example, by Løvborg and Kirkegaard (1974) for 2π geometry, and by Bowman (1976; Murray, Bowman and Aitken, 1978) for 4π geometry as appropriate to archaeological application. In the latter case the 'sources' consist of 3 concrete blocks, each a 0.51-m cube, doped respectively with potassium, thorium, and uranium. Although these blocks are not quite large enough for the dose-rate at the centre to be equal to the infinite matrix dose-rate it is presumed that the spectral shape, especially at the lower energy end, will be indistinguishable from that for a true infinite matrix. Access to the centre is by means of a 3-inch diameter channel, and by placing the detector there stripping factors such as given in Table L.1 are obtained.

When used in the field the observed count-rates in each window may be written as

$$W_h = H + (uh)U + R_h, \quad \text{(L.1)}$$

$$W_u = U + (hu)H + R_u, \quad \text{(L.2)}$$

and

$$W_k = K + (hk)H + (uk)U + R_k, \quad \text{(L.3)}$$

with W representing the window count-rate with suffixes k, h, and u denoting potassium, thorium, and uranium, respectively, K representing the contribution from potassium into its own window, (similarly for H and U), R representing the contribution from cosmic rays and radioactivity in the detector itself (see note 3 of Table L.1). Hence (!), by measuring the window count-rates the 'own spectrum' contributions can be determined from

$$H = \frac{W_h - (uh)W_u + (uh)R_u - R_h}{1 - (uh)(hu)}, \quad \text{(L.4)}$$

TABLE L.1
Typical Stripping Factors for 2" Sodium Iodide Crystal[a]

	Window		
Spectrum	K	U	Th
K	1	<0.01	<0.01
U	1.53	1	0.22
	(uk)		(uh)
Th	0.76	0.72	1
	(hk)	(hu)	

[a] 1. The factors are the ratio, for a given spectrum, between the count-rate in the window specified and that in the spectrum's 'own' window, for example, $(uh) = 0.22$ indicates that for the U spectrum the ratio between the Th window count-rate and the U window count-rate is 0.22.

2. The window widths are 0.15 MeV for K and U and 0.3 MeV for Th. The windows are centred approximately, on 1.46, 1.76, and 2.61 MeV, respectively. The detector was a sodium iodide crystal of diameter 1.75 inches (44 mm) by 2 inches (50mm) high, having a resolution of 8% (full width at half height, for cesium-137.

3. The factors were obtained using three concrete blocks, each a cube of 0.51 metre, doped respectively with 5% K, 170 ppM of U, and 190 ppM of Th (with an associated impurity of 7 ppM of U); these are nominal values. The contribution from the concrete itself was measured using a fourth block, and subtracted. This subtraction also removes the barely significant contribution from cosmic rays and radioactive impurities in the detector itself. The combined contribution from these two latter, found by immersing the detector beneath 1 metre of water in a deep river, are 2.25 0.85 and 1.12 counts per minute for the K, U, and Th windows respectively.

4. The measured dose-rates quoted in the text are substantially lower than would be predicted from the nominal concentrations. This is partly because blocks are not big enough to be truly infinite and partly due to escape of radon and thoron.

$$U = \frac{W_u - (hu)W_h + (hu)R_h - R_u}{1 - (uh)(hu)}, \qquad (L.5)$$

and

$$K = \{W_k - R_k\} - \{W_h - R_h\}\left\{\frac{(hk) - (uk)(hu)}{1 - (uh)(hu)}\right\}$$

$$- \{W_u - R_u\}\left\{\frac{(uk) - (hk)(uh)}{1 - (uh)(hu)}\right\} \qquad (L.6)$$

Substituting values from the Table we obtain

$$H = 1.19W_h - 0.26W_u - 1.17, \qquad (L.7)$$

Appendix L: The Portable 4-Channel Gamma Spectrometer

$$U = 1.18 W_u - 0.86 W_h - 0.01, \quad \text{(L.8)}$$

$$K = W_k + 0.41 W_h - 1.62 W_u - 1.35. \quad \text{(L.9)}$$

These are given for illustration only; it must be stressed that numerical values of the stripping factors vary from detector to detector depending on crystal size and resolution as well as on the window width and window position set in the electronics.

CONVERSION TO ANNUAL DOSE AND CONCENTRATION

The above equations enable the individual contributions H, U, and K to be determined, but conversion factors need to be evaluated before annual dose components can be obtained. One way to do this would be to build blocks large enough for the dose-rate at the centre to be equal to the infinite matrix dose-rate; then, if the concentrations of the nuclides in the blocks are known the dose-rates can be evaluated from the tables of Appendix G. According to Murray (1981), for a density of 1.9 g/cm³ a cube of 1 metre is necessary in the case of uranium; for thorium the required size would be larger. However by using an alternative procedure it is possible to obtain the conversion factors by means of the ½-metre blocks.

This is done by using thermoluminescence dosimeters to measure the gamma dose-rate at the centre of each block and hence obtain the ratio between the observed K, H, and U for the blocks and the respective dose-rates. The latter have been measured (by C. J. Gaffney and P. A. Clark) and after subtraction of dose-rate measured in an undoped block the values obtained are 1.23, 6.84, and 12.43 mGy/a for the potassium, thorium, and uranium blocks, respectively; the dosimeters used were 0.7-mm walled copper capsules containing natural fluorite and a correction factor of 1.10, discussed shortly, has been applied to the measured dose-rates in order to obtain the unshielded dose-rate to quartz-like material. With these values the conversion factors from K, H, and U to dose-rate, when using the same detector system as for Table L.1, are 2.8, 10.6, and 12.4 μGy/a per count/minute, respectively. This deals with the individual annual dose components from potassium, thorium, and uranium; the cosmic ray component is obtained from the count-rate above an electronic threshold corresponding to 3 MeV. This is too high for any contribution from natural radioactivity, and the conversion factor is obtained by observing the above-threshold count-rate below ½ metre of soil; using the value of 185 μGy/a quoted in Appendix I the appropriate conversion factor for the detector under discussion is 6.5 μG/a/cpm.

The soil concentrations of potassium, thorium, and uranium can be obtained by using the conversion factors given on line 11 of Table G.4 and line 6 of Table G.5. For the detector under discussion the conversion factors are 72.5 cpm/% K, 4.82 cpm/ppm Th, and 9.32 cpm/ppm U. As elsewhere if there is loss of radon then the uranium concentration is liable to be underestimated, but as far as annual dose is concerned the conversion factor is not much affected because most of the dose is carried by the gamma ray being measured.

THE CAPSULE CORRECTION FACTOR

The factor of 1.10 used to convert the observed gamma dose-rate in the fluorite of the capsule to the gamma dose-rate that would be experienced by unshielded quartz-like material is necessary because of attenuation in the 0.7-mm copper wall of the capsule and because of the difference between the absorption coefficient of the fluorite powder and that of quartz. It is based on calculations by Murray (1981; Murray et al., 1978) using spectra derived for distributed sources (Jain, Evans and Close, 1979). The calculations were made for a capsule having a wall thickness of 1.5 mm and by experimental comparison it has been found that the 0.7-mm wall capsule records a dose higher than that recorded by the 1.5 mm wall capsule by 9% for typical soil radioactivity (Aitken et al., 1985).

The calculated factor (the ratio between infinite matrix dose and capsule dose) has been checked by Murray in the case of uranium using a concrete block doped with 3000 ppm of uranium. This was a cube of 1 metre, built up from smaller 20-cm cubes, and therefore sufficient in size to be considered as infinite. The concentration of uranium was measured, as well as the degree of radon escape. The experimentally derived factor was within 2% of the predicted one; however the estimated error limits in the absolute accuracy of these factors were ±5% and ±2%, respectively.

Murray also calculates the ratio between the infinite matrix dose and the dose to quartz shielded by a glass capsule of 3-mm wall thickness to be 1.13. Hence the ratio between the latter dose and the dose to fluorite in a 0.7-mm copper capsule is 0.97; this is in line with the experimental observation by Mejdahl (1978) quoted in the main text that the dose to aluminium oxide (similar to quartz in absorption characteristics) in a 2.5-mm wall quartz capsule is 0.95 times the dose to $CaSO_4$:Dy shielded by 1.5 mm of steel.

Direct calibration of a gamma spectrometer in this 'hot' block is not possible because the count-rate is so high that there is serious electronic pile-up.

SPECTRUM STABILIZATION

For a given energy of gamma ray, the size of the scintillations from the sodium iodide crystal has some dependence on temperature, falling by about 5% on going from 20°C to 0°C; also there may be changes in the gain of the photomultiplier and of the electronics. A given percentage change in overall amplification (i.e., in the pulse height for a given energy of gamma ray) causes a much larger change in the window count-rate because if the amplification changes the photopeaks will be only partially within the electronic window that has been set; a change of 5% in amplification can cause a change in window count-rate of the order of 50% (depending on the relative proportions of K, Th and U). As a result it is not practical to use the spectometer in field conditions unless automatic control of amplification is incorporated.

Such 'spectrum stabilization' is achieved by means of a weak radioisotope reference source having an X-ray or gamma emission that is low enough in energy not to make significant contribution to the count-rate in the three measurement windows. This source is incorporated in the detector head, and by means of electronic circuitry the voltage to the multiplier is adjusted so that the photopeak from the reference source is midway between two closely-spaced reference windows. If the count-rate in the upper window is below that in the lower window then the voltage is raised, and vice versa, until there is equality. With the reference photopeak in the correct position with respect to its windows the spectrum being measured will be correctly positioned with respect to the measurement windows. Effectively, by means of this automatic change in photomultiplier gain the overall amplification is held constant even though there may be changes in individual components.

Suitable reference sources are americium-241 (0.06 MeV), barium-133 (0.35 MeV), and manganese-54 (0.83 MeV). The advantage of americium-241 is that its energy is very remote from the lowest measurement window (potassium at 1.46 MeV) making it preferable for low-level work; on the other hand, there is a greater 'lever arm' involved between its energy and the measurement energies than in the case of a higher energy reference source, and any slight bending of this arm (due to component non-linearity) limits the degree of stabilization. Nevertheless excellent performance can be achieved, such as a change of only $\frac{1}{2}$% for a 50°C change in temperature, that is, 0.01% per °C.

When measuring high levels of radioactivity there is a possible side effect from spectrum stabilization that needs to be borne in mind. There are contributions to the count-rates in the reference windows from the low energy part of the spectrum being measured, and if these become appreciable the compensation mechanism may be upset; thus, if from the spectrum being measured there are more counts in the lower reference window than in the higher one the mechanism will erroneously raise the photomultipier voltage above its correct level. However by making the width of the lower window less than that of the higher one by an amount that matches the average spectrum (so that equal counts are put into each window) this 'pulling' of the voltage can be made unimportant.

As far as the scintillometer is concerned, in which all gamma rays above about 0.4 MeV are counted, only americium-241 can be used as the reference source. On the other hand, because in this mode a 1% change in overall amplification causes a change of only 1–2% in count-rate, the temperature dependence of the count-rate is comparatively weak, and depending on the precision required and the climatic conditions, it may be possible to operate without spectrum stabilization.

MISCELLANEOUS

An important consideration in archaeological application is the size of the auger hole necessary to accommodate the detector head. While it is quite practical to use a detector housing of diameter $2\frac{1}{2}''$ (64 mm), sufficient for the size of crystal discussed, a lesser disturbance to the archaeologist's site would be advantageous. Narrower housings are available but usually at the price of lower count rates and/or poorer resolution (and hence bigger stripping factors). An alternative type of crystal now becoming available is bismuth germanate (BGO); although the pulse height for a given gamma energy is lower by a factor of 10 compared to sodium iodide (which is compensated by using higher gain in the photomultiplier) and its temperature dependence is higher, the photopeak count-rates for a given level of gamma flux are higher by a factor of nearly three. Use of this material would thus allow use of a smaller crystal; also, it is more robust.

The instrument that has been used by the author is the one made for geological surveys by the NUTMAQ unit at AERE, Harwell U.K. There are a number of other manufacturers including Geometrics (U.S.A.), Scintrex (Canada) and Chemtron (South Africa). Some models incorporate a microcomputer which performs the spectrum stripping directly. An alternative to the 4-channel device is to use a portable multichannel analyser and obtain the window count-rates by setting appropriate regions of interest.

APPENDIX M

Miscellaneous Phosphor Notes

MBLE FLUORITE

Although there are many sources of this mineral (natural CaF_2) around the world, only a few yield material that combines high sensitivity and absence of fading. Fluorite from one particular source has been developed by a Belgian company (Manufacture Belge de Lamps et de Material Electronique) and this, usually referred to as MBLE Super S, is excellent for low level measurements except with respect to its radioactive impurities, which give it a self-dose of about 0.1 mGy/a. The thermoluminescent emission is in the violet and ultraviolet range spreading from 350 to 450 nm; hence the standard colour filter for dating, such as a Corning 7-59 can be used, though some advantage is gained in minimization of spurious thermoluminescence by using a filter more restricted to shorter wavelengths such as a Corning 7-51. The minimum detectable dose is on the order of 0.1 μGy, using stringent anti-spurious precautions and a heating rate of 20°C/sec. The basic dosimetric properties of MBLE fluorite have been reported by Shayes *et al.* (1967), and its use in the context of dating by Aitken (1968a, 1969, 1979). Studies of Indian fluorite have been reported by Sunta (1971).

In preparing material for dosimetry the crushed mineral is washed in dilute aqua regia for several hours at about 80°C; this removes surface discolouration and improves light output. After thorough washing in distilled water, sodium hexametaphosphate, and alcohol, the powder is sieved into grain size fractions and usually the 90–150 μm fraction is selected for dosimetry purposes. Until the light sensitivity has been attenuated and the natural dose erased, great care needs to be taken to avoid contaminating a thermoluminescence laboratory with fine particles of fluorite because in its raw state fluorite has a thermoluminescent output several million times above other material being measured. It is therefore strongly recommended that before the above preparation is started, the heating procedure given below should be carried out.

The acquisition of thermoluminescence through exposure to light is due to phototransfer from traps responsible for glow peaks at 525°C and 600°C into traps responsible for a peak on the upper shoulder of peak III, the maxima being separated by only 30°C; this new peak is termed III', see Figure M.1. There is also some transfer, although to a lesser extent, into the traps responsible for peak II. The deep traps carry a geological dose and this is not emptied by the standard 'dose-erasure' heating to around 400°C. Even a sample that has been heated to 500°C will acquire substantial thermoluminescence after only 1 min exposure to 1000 lux from an electric light bulb (1000 lux is the illumination from a 60-W bulb at about 18 inches). Its sensitivity to daylight with its higher ultraviolet content is

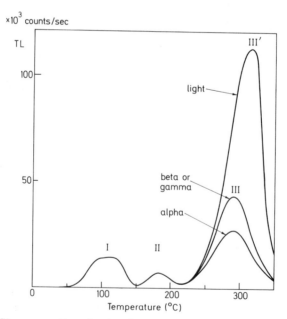

Figure M.1 Glow curves from fluorite after exposure to alpha particles, beta or gamma radiation, and light.

very much greater and unless stringent precautions are to be taken during sample handling for measurement, this sensitivity must be reduced by heating to around 600°C for about 10 min. Even then operation in subdued red light is advised, because not only does some such sensitivity remain, particularly to daylight and fluorescent light, but also there is liable to be some bleaching of peak II (see below).

The heating must be done in an atmosphere of dry oxygen-free nitrogen, otherwise the radiation sensitivity will be impaired due to diffusion into the grains of oxygen and water vapour. There will also be reduction in sensitivity unless the cooling is fairly rapid. In practice this is achieved by pulling the crucible out of the oven in a stream of nitrogen and placing it on a cold metal surface in air; it seems that while still in the stream of nitrogen the fluorite cools to a temperature at which exposure to air no longer has a deleterious effect. This treatment reduces the light-induced thermoluminescence in peak III' to the order of 1 μGy for 1000 lux-min of bulb light. Once this has been done for a batch of fluorite it does not have to be repeated unless there has been enough exposure to radiation for the deep traps to be refilled to a significant extent.

The presence of a light-induced signal shows itself not only by an apparent broadening of peak III due to the superposition of peak III' but also by a change in the height relative to peak II. Whereas for beta and gamma radiation the III/II ratio is usually around 7/1, for light-induced thermoluminescence the III'/II ratio is about 20/1.

As mentioned above the traps of peak II are susceptible to bleaching and the heat treatment does not reduce this; of course this effect shows itself only in a sample that has been irradiated. For bulb light the initial rate of loss is about 1% per 1000 lux-min and for a fluorescent light it is about 1% per min at a distance of 4 feet. The loss in peak III is negligible by comparison.

Alpha irradiation of fluorite gives a III/II peak ratio of about 5/1 instead of the 7/1 for beta or gamma irradiation. This is because the a-value for peak II is 0.12 whereas that for peak III is 0.09. The actual values vary a little from batch to batch presumably depending on impurity content and heat treatment.

If fluorite is used at a level well above that relevant to soil dose measurement, for example, for source intercalibration, then it is necessary to beware of the pre-dose effect. If a sample that has been given a dose of several gray is heated to 500°C, as in the course of a glow-curve, its sensitivity will increase by several percent. However if the glow-curve is terminated at about halfway down the high temperature side of peak III, which is enough for erasure of the thermoluminescence, the effect can be avoided to any significant degree as long as the total dosage accumulated by the sample prior to heating has not got too high.

OTHER SYNTHETIC PHOSPHORS: a-VALUE AND AVAILABILITY

Although CaF_2: Dy is almost an order of magnitude more sensitive than MBLE fluorite, it suffers from rather severe fading. However it is of interest for special applications both because of its high a-value, approximately 0.5, and because it is available in solid form, for example, as rods $1 \times 1 \times 6$ mm and as plates $3 \times 3 \times 1$ mm. Also, whereas the useful upper dose limit for MBLE fluorite is 50 Gy, that for CaF_2: Dy is 10 kGy. It is marketed by Harshaw as TLD 200.

The a-value for $CaSO_4$: Dy is approximately 0.3. This too is available commercially; for instance, in the USA from Harshaw and Teledyne, in the U.K. from Vinten, and in Japan from National Panasonic. Other commercially available phosphors are $CaSO_4$: Tm, $CaSO_4$: Mn, CaF_2: Mn, and $Li_2B_4O_7$: Mn. Recipes for preparation have been given for some. For example for $CaSO_4$: Dy and $CaSO_4$: Tm by Yamashita *et al.* (1971), for activated Al_2O_3 by Metha and Sengupta (1976), for magnesium silicate by Bhasin, Sasidharan and Sunta (1976). The latter is available commercially from Kasei Optonics (2-7-18 Hamamatsu Cho, Minato-Ku, Tokyo).

ENERGY DEPENDENCE

For gamma rays below 0.1 MeV there is enhanced absorption of energy, relative to quartz, by phosphors which have a higher effective Z

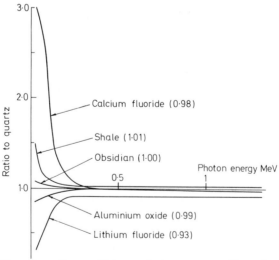

Figure M.2 Mass absorption coefficients relative to quartz. The values of the ratio at 1 MeV are given in parenthesis (from Murray, 1981, based on Storm and Israel, 1970).

Appendix M: Miscellaneous Phosphor Notes

TABLE M.1
Primary Spectra[a]

Energy (keV)	Relative Intensity	Parent Isotope	Energy (keV)	Relative Intensity	Parent Isotope
Thorium Series[b]			Uranium Series[b]		
2614.5	0.145	Tl 208	2448.0	0.008	Bi 214
1587.9	0.015	Ac 228	2204.3	0.026	Bi 214
968.9	0.073	Ac 228	2118.7	0.006	Bi 214
964.6	0.023	Ac 228			
911.1	0.121	Ac 228	1847.6	0.011	Bi 214
860.5	0.017	Tl 208	1764.6	0.081	Bi 214
794.8	0.020	Ac 228	1729.8	0.016	Bi 214
727.3	0.027	Bi 212	1661.4	0.006	Bi 214
583.1	0.125	Tl 208	1583.3	0.004	Bi 214
510.7	0.033	Tl 208	1509.3	0.011	Bi 214
463.0	0.019	Ac 228	1408.0	0.013	Bi 214
338.4	0.050	Ac 228	1401.6	0.007	Bi 214
328.0	0.014	Ac 228	1385.4	0.004	Bi 214
300.1	0.014	Pb 212	1377.7	0.021	Bi 214
270.3	0.016	Ac 228	1281.0	0.007	Bi 214
241.0	0.015	Ra 224	1238.2	0.030	Bi 214
238.6	0.179	Pb 212	1155.3	0.008	Bi 214
209.4	0.019	Ac 228	1120.4	0.073	Bi 214
129.1	0.012	Ac 228	1001.1	0.004	Pa 234m
84.4	0.064	Th 228	934.0	0.015	Bi 214
			806.2	0.005	Bi 214
			786.0	0.004	Pb 214
			768.4	0.023	Bi 214
			665.6	0.007	Bi 214
			609.4	0.208	Bi 214
			352.0	0.177	Pb 214
			295.2	0.092	Pb 214
			258.8	0.004	Pb 214
			241.9	0.037	Pb 214
			186.0	0.019	Ra 226
			185.7	0.012	U 235
			92.8	0.017	Th 234
			92.3	0.015	Th 234
			63.3	0.028	Th 234

[a] This table gives the primary spectra used by Jain et al. (1979) for calculation of the distributed source spectra shown in Figure M.3.

[b] For the thorium series the spectrum has been limited to emissions above 84 keV, and for the uranium spectrum, above 63 KeV. The emissions used represent 95% and 90%, respectively, of the total intensity; in terms of energy they represent more than 99% of the total.

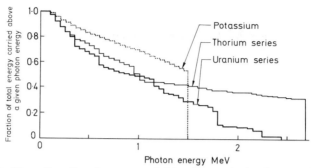

Figure M.3 Proportion of total energy carried by photons, both primary and secondary, above a given energy (from Murray, 1981, based on Jain *et al.*, 1979). The spectra have been calculated for a shale-like soil of infinite extent.

than quartz; this is because the photoelectric absorption coefficient depends on Z^4/A. The energy dependence of a number of materials is shown in Figure M.2. The dose received by a phosphor relative to that received by quartz will depend therefore on the gamma spectrum incident on the phosphor. For a distributed source, as in soil, the flux will be partly primary uncollided photons and partly secondary photons degraded in energy through Compton interactions, roughly of equal importance in as far as deposited dose is concerned; Figure M.3 shows the infinite matrix spectra to be expected. The capsule wall will modify these spectra, with relatively greater attenuation of the regions below 0.1 MeV for materials such as copper which have a higher Z value than the effective value for soil; this greater attenuation tends to compensate for the overresponse of the fluorite in this energy region. Calculated factors relating phosphor dose to quartz dose for some phosphor/wall combinations have been quoted in Section 4.3.3 and for fluorite in the section on the capsule correction factor of Appendix L; factors for some other combinations will be found in Murray (1981). Table M.1 gives the primary spectra on which the degraded spectra of Figure M.3 are based.

THERMOLUMINESCENT RESPONSE PER GRAY

There is also experimental evidence of an additional effect in some phosphors (e.g., Tochilin and Goldstein, 1968); this is that the thermoluminescence per unit absorbed dose at low gamma energies is different to the value above 0.1 MeV. For MBLE fluorite Shayes *et al.* (1963) report an overresponse, whereas for synthetic CaF_2: Dy Almond *et al.* (1968) find no variation.

REFERENCES

The abbreviation *PACT* signifies the Journal of the European Study Group on Physical, Chemical, Biological and Mathematical Techniques Applied to Archaeology. It is obtainable from T. Hackens, 28a, av. Leopold, B-1330, Rixensart, Belgium.

Aitken, M. J.
- 1968a   Evaluation of effective radioactive content by means of thermoluminescent dosimetry. In *Thermoluminescence of Geological Materials,* edited by D. J. McDougall, pp. 463–470. Academic Press, New York.
- 1968b   Low-level environmental radiation measurements using natural calcium fluoride. *Proceedings of the 2nd International Conference on Luminescent Dosimetry,* Gatlinburg, edited by J. A. Auxier, K. Becker, and E. M. Robinson, CONF-680920, pp. 281–290. U.S. National Bureau Standards, Washington, D.C.
- 1969   Thermoluminescent dosimetry of environmental radiation on archaeological sites. *Archaeometry* 11, 109–114.
- 1976   Thermoluminescent age evaluation and assessment of error limits: revised system. *Archaeometry* 18, 233–238.
- 1978   Radon loss evaluation by alpha counting. *PACT* 2, 104–117.
- 1979   Interlaboratory calibration of alpha and beta sources. *PACT* 3, 443–447.
- 1984   Non-linear growth: allowance for alpha particle contribution. *Ancient TL* 2, 2–5.

Aitken, M. J., Tite, M. S., and Reid, J.
- 1964   Thermoluminescent dating of ancient ceramics. *Nature* 202, 1032–1033.

Aitken, M. J., Fleming, S. J., Doell, R. R., and Tanguy, J. C.
- 1968a   Thermoluminescent study of lavas from Mt. Etna and other historic flows: preliminary results. In *Thermoluminescence of Geological Materials,* edited by D. J. McDougall, pp. 359–366. Academic Press, New York.

Aitken, M. J., Zimmerman, D. W., and Fleming, S. J.
- 1968b   Thermoluminescent dating of ancient pottery. *Nature* 219, 442–444.

Aitken, M. J., and Fleming, S. J.
- 1971   Preliminary application of thermoluminescent dating to the eruptions of Thera.

Acta of the 1st International Scientific Congress on the Volcano of Thera, pp. 293–302.

Aitken, M. J., Moorey, P. R. S., and Ucko, P. J.
1971   The authenticity of vessels and figurines in the Hacilar style. *Archaeometry* 13, 89–141.

Aitken, M. J., and Alldred, J. C.
1972   The assessment of error limits in thermoluminescent dating. *Archaeometry* 14, 257–267.

Aitken, M. J., and Bowman, S. G. E.
1975   Thermoluminescent dating: assessment of alpha particle contribution. *Archaeometry* 17, 132–138.

Aitken, M. J., and Huxtable, J.
1975   Thermoluminescence and Glozel: a plea for caution. *Antiquity* 49, 223–6.

Aitken, M. J., Huxtable, J., Wintle, A. G., and Bowman, S. G. E.
1975   Age determination by TL: review of progress at Oxford. *Proceedings of the 4th International Conference on Luminescent Dosimetry,* Krakow, pp. 1045–1055.

Aitken, M. J., and Wintle, A. G.
1977   Thermoluminescence dating of calcite and burnt flint: the age relation for slices. *Archaeometry* 19, 100–105.

Aitken, M. J., and Bussell, G. D.
1979   Zero-glow monitoring (ZGM). *Ancient TL* no. 6, 13–15.

Aitken, M. J., and Bussell, G. D.
1982   TL Dating of fallen stalactites. *PACT* 6, 550–554.

Aitken, M. J., Clark, P. A., Gaffney, C. G., and Løvborg, L.
1985   Beta and gamma gradients in sample and soil. *Nuclear Tracks and Radiation Measurements,* 10, in press.

Aitken, M. J., Huxtable, J., and Debenham, N. C.
1985   Thermoluminescence dating in the paleolithic. *Bulletin de l'Association français pour l'Etude du Quaternaire,* in press.

Almond, P. R., McCray, K., Espejo, D., and Watanabe, S.
1968   The energy response of lithium fluoride, calcium fluoride and lithium borate from 26 keV to 22 MeV. *Proceedings of the 2nd International Conference on Luminescent Dosimetry,* Gatlinburg, edited by J. A. Auxier, K. Becker, and E. M. Robinson, CONF-680920, pp. 410–423. National Bureau Standards, Washington, D.C.

Bailiff, I. K.
1976   Use of phototransfer for the anomalous fading of thermoluminescence. *Nature* 264, 531–533.
1979   Pre-dose dating: high S_o sherds. *PACT* 3, 345–355.
1980   A beta irradiator for use in TL dating. *Ancient TL* no. 10, 12–14.
1982   Beta TLD apparatus for small samples, *PACT* 6, 72–75.
1983   Sensitization of R-traps? *PACT* 9, 207–214.
1985   Pre-dose and inclusion dating: an attempted comparison using Iron Age pottery from north Britain. *Nuclear Tracks and Radiation Measurements* 10, in press.

Bailiff, I. K., Bowman, S. G. E., Mobbs, S. F., and Aitken, M. J.
1977   The phototransfer technique and its use in thermoluminescence dating. *Journal of Electrostatics* 3, 269–280.

Bailiff, I. K., and Aitken, M. J.
1980   Use of thermoluminescence dosimetry for evaluation of internal beta dose-rate in archaeological dating. *Nuclear Instruments and Methods* 173, 423–429.

References

Bailiff, I. K., and Haskell, E. H.
  1984 Use of the pre-dose technique for environmental dosimetry. *Radiation Protection Dosimetry* 6, 245–248.
Bangert, U., and Henning, G. J.
  1979 Effects of sample preparation and the influence of clay impurities on the TL-dating of calcite cave deposits. *PACT* 3, 281–289.
Barbetti, M. F.
  1976 Archaemomagnetic analyses of six Glozelian ceramic artifacts. *Journal of Archaeological Sciences* 3, 137–151.
Barbetti, M. F., and McElhinney, M. W.
  1972 The Lake Mungo geomagnetic excursion. *Philosophical Transactions of the Royal Society,* London A281, 515–542.
Bechtel, F., Schvoerer, M., Rouanet, J. F., and Gallois, B.
  1979 Extension à la prehistorie, à l'oceanographie et à la volcanologie de la methode de datation par thermoluminescence. *PACT* 3, 481–492.
Bell, W. T.
  1976 The assessment of the radiation dose-rate for thermoluminescence dating. *Archaeometry* 18, 107–111.
  1977 Thermoluminescence dating: revised dose-rate data. *Archaeometry* 19, 99–100.
  1978 Studies in thermoluminescence dating in Australasia. Unpublished Ph.D. thesis, Department of Physics, School of General Studies, Australian National University Canberra.
  1979a Attenuation factors for the absorbed radiation dose in quartz grains for thermoluminescence dating. *Ancient TL* no. 8, 2–13.
  1979b Thermoluminescence dating: radiation dose-rate data. *Archaeometry* 21, 243–245.
  1980 Alpha dose attenuation in quartz grains for thermoluminescence dating. *Ancient TL* no. 12, 4–8.
Bell, W. T., and Zimmerman, D. W.
  1978 The effect of HF acid etching on the morphology of quartz inclusions for thermoluminescence dating. *Archaeometry* 20, 63–65.
Bell, W. T., and Mejdahl, V.
  1981 Beta source calibration and its dependency on grain transparency. *Archaeometry* 23, 231–240.
Beltrao, M. C., Danon, J., Enriquez, C. R., Poupeau, G., and Zuleta, E.
  1982 Datations par thermoluminescence des silex brulés du site archeologique Alice Boer, Brésil, *Comptes Rendus de l'Academie de Sciences (Paris)* 295, 629–632.
Benko, L.
  1983 TL properties of individual quartz grains. *PACT* 9, 175–182.
Berger, G. W., Mulhern, P. J., and Huntley, D. H.
  1980 Isolation of silt-sized quartz from sediments. *Ancient TL* no. 11, 8–9.
Berger, G. W., Brown, T. A., Huntley, D. J., and Wintle, A. G.
  1982 5 spurious tidbits. *Ancient TL* no. 18, 7–11.
Berger, G. W., and Huntley, D. J.
  1983 Dating volcanic ash by thermoluminescence. *PACT* 9, 581–592.
Berger, G. W., Huntley, D. J., and Stipp, J. J.
  1984 Thermoluminescence studies on a C-14 dated marine core. *Canadian Journal of Earth Sciences* 21, 1145–1150.
Berger, M. J.
  1971 Distribution of absorbed dose around point sources of electrons and beta particles in water and other media. *Journal of Nuclear Medicine* (Suppl.) 5, 5–23.

1973   *Improved point kernels for electron and beta-ray dosimetry.* NBSIR 73-107, National Bureau of Standards, Washington, D.C., 20234.

Berger, M. J., and Seltzer, S. M.
1982   *Stopping powers and ranges of electrons and positrons* (2nd ed.) NBSIR 82-2550-A, National Bureau of Standards, Washington D.C., 20234.

Bhasin, B. D., Sasidharan, R., and Sunta, C. M.
1976   Preparation and thermoluminescent characteristics of terbium doped magnesium orthosilicate phosphor. *Health Physics* 30, 139–142.

Boag, J. W.
1971   Radiation Physics. In *Manual of Radiation Haematology* (Technical Report 123) pp. 31–43. International Atomic Energy Agency, Vienna.

Bonhommet, N., and Babkine, J.
1967   Sur la presence d'aimantations inversées dans la Chaîne des Puys. *Comptes Rendus de l'Academie de Sciences* (Paris) B264, 92–94.

Bothner, M. H., and Johnson, N. M.
1969   Natural thermoluminescence dosimetry in late Pleistocene pelagic sediments. *Journal of Geophysical Research* 74, 5331–5338.

Bøtter-Jensen, L., Bundgaard, J., Mejdahl, V.
1983   An HP-85 microcomputer-controlled automated reader system for TL dating. *PACT* 9, 343–350.

Bøtter-Jensen, L., and Mejdahl, V.
1985   Determination of potassium in feldspars by beta counting using a GM multicounter system. *Nuclear Tracks and Radiation Measurements* 10, in press.

Bowman, S. G. E.
1975   Dependence of supralinearity on pre-dose: some observations. *Archaeometry* 17, 129–132.
1976   Thermoluminescent dating: the evaluation of radiation dosage. Unpublished D.Phil. thesis, Faculty of Physical Sciences, Oxford University.
1978   Dose-rate dependence of natural calcium fluoride. *PACT* 6, 292–294.
1979   Phototransferred thermoluminescence in quartz and its potential use in dating. *PACT* 3, 381–400.
1982a   Alpha particle ranges in pottery. *PACT* 6, 61–66.
1982b   Thermoluminescence studies on burnt flint. *PACT* 6, 353–361.

Bowman, S. G. E., and Seeley, M. A.
1978   The British Museum flint dating project. *PACT* 2, 151–164.

Bowman, S. G. E., Loosemore, R. P. W., Sieveking, G. de G., and Bordes, F.
1982   Preliminary dates for Pech de l'Aze IV. *PACT* 6, 362–369.

Bowman, S. G. E., and Sieveking, G. de G.
1983   Thermoluminescence dating of burnt flint from Combe Grenal. *PACT* 9, 253–268.

Bowman, S. G. E., and Huntley, D. J.
1984   A new proposal for the expression of alpha efficiency in TL dating. *Ancient TL* 2, 6–11.

Boyle, R.
1664   Experiments and Considerations upon Colours with Observations on a Diamond that Shines in the Dark. Henry Herringham, London.

Braunlich, P.
1968   Thermoluminescence and thermally stimulated current: tools for determination of trapping parameters. *Thermoluminescence of Geological Materials,* edited by D. J. McDougall, pp. 61–88. Academic Press, New York.

References

Braunlich, P., Gasiot, J., and Fillard, J. P.
   1982   Laser heating in thermoluminescence dosimetry. *Journal of Applied Physics* 53, 5200–5209.

Brou, R., and Valladas, G.
   1975   Appareil pour la mesure de la thermoluminescence de petits enchantillons. *Nuclear Instruments and Methods* 127, 109–113.

Cameron, J. R., Suntharalingam, N., and Kenney, G. N.
   1968   *Thermoluminescent Dosimetry*. University of Wisconsin Press, Madison.

Carriveau, G. W.
   1974   Application of thermoluminescent dating to prehistoric metallurgy. In *Application of Science to the Dating of Works of Art*. Museum of Fine Arts, Boston.

Carriveau, G. W.
   1977   Cleaning quartz grains. *Ancient TL* no. 1, 6.

Carriveau, G. W., and Harbottle, G.
   1983   Ban Chiang pottery: thermoluminescence dating problems. *Antiquity* 57, 56–59.

Charalambous, S., and Michael, C.
   1976   A new method of dating pottery by TL. *Nuclear Instruments and Methods* 137, 565–567.

Charalambous, S., Hasan, F., Michael, C., Siona, A., and Tzamarias, S.
   1982   Dating using the shape of the glow-curve. *PACT* 6, 265–271.

Chen, R.
   1969   On the calculation of activation energies and frequency factors from glow curves. *Journal of Applied Physics* 40, 570–585.
   1976   Methods for kinetic analysis of thermally stimulated processes. *Journal of Material Science* 11, 1521–1541.
   1979   Saturation of sensitization of the 110 C TL peak in quartz and its potential application in the pre-dose technique. *PACT* 3, 325–335.

Chen, R., and Bowman, S. G. E.
   1978   Superlinear growth of TL due to competition during irradiation. *PACT* 2, 216–230.

Chen, R., and Kirsch, Y.
   1981   *Analysis of thermally stimulated processes*. Pergamon Press, London.

Chen, R. S., McKeever, S. W. S., and Durrani, S. A.
   1982   Dose rate dependence of thermoluminescence: the possibility of different behaviours of different peaks. *PACT* 6, 295–302.

Christodoulides, C., and Fremlin, J. H.
   1971   Thermoluminescence of biological materials. *Nature* 232, 257–258.

Crawford, T. D.
   1977   Les tablettes inscrites de Glozel. *Revue Archeologique du Centre* 63–64, 377–389.

Cross, W. G.
   1968   Variation of beta dose attenuation in different media. *Physics in Medicine and Biology* 13, 611–618.

Cross, W. G., Ing, H., and Freedman, N.
   1983   A short atlas of beta-ray spectra. *Physics in Medicine and Biology* 28, 1251–1260.

Dalrymple, G. B., and Doell, R. R.
   1970   Thermoluminescence of lunar samples from Apollo 11. *Proceedings of the Apollo 11 Lunar Science Conference, Geochimica et Cosmoschimica* (Acta Suppl. 1) 3, 2081–2092.

Daniel, G.
  1974   Editorial. *Antiquity* 46, 261–264.
  1977   Editorial. *Antiquity* 51, 89–91.
Daniels, F., Boyd, C. A. and Saunders, D. F.
  1953   Thermoluminescence as a research tool. *Science* 117, 343–349.
Debenham, N. C.
  1978   More hints on spurious reduction. *Ancient TL* no. 5, 3–4.
  1985   Use of UV emission in TL dating of sediments. *Nuclear Tracks and Radiation Measurements* 10, in press.
Debenham, N. C., Driver, H. S. T. and Walton, A. J.
  1982   Anomalies in the TL of young calcites. *PACT* 6, 555–562.
Debenham, N. C., and Walton, A. J.
  1983   TL properties of some wind-blown sediments. *PACT* 9, 531–538.
Debenham, N. C., and Aitken, M. J.
  1984   Thermoluminescence dating of stalagmitic calcite. *Archaeometry* 26, 155–170.
Delbecq, C. J., Toyozawa, Y., and Yuster, P. H.
  1974   Tunnelling recombination of trapped electrons and holes in KCl:AgCl and KCl:TlCl. *Physical Review* B9, 4497–4505.
Dexter, D. L.
  1954   X-ray coloration of alkali halides. *Physical Review* 93, 985–992.
Diamond, H., and Gindler, J. E.
  1963   Alpha half-lives of Po-216, At-217 and Rn-218. *Journal of Inorganic Nuclear Chemistry* 25, 143–164.
Dreimanis, A., Hutt, G., Raukas, A., and Whippey, P. W.
  1978   Dating methods of Pleistocene deposits: thermoluminescence. *Geoscience Canada* 5, 55–60.
Driver, H. S. T.
  1979   The preparation of thin slices of bone and shell for TL. *PACT* 3, 290–297.
Durrani, S. A., Groom, P. J., Khazal, K. A. R., and McKeever, S. W. S.
  1977   The dependence of the thermoluminescence sensitivity upon the temperature of irradiation in quartz. *Journal of Physics D: Applied Physics* 10, 1351–1361.
Elitzsch, C., Pernicka, E., and Wagner, G. A.
  1983   Thermoluminescence dating of archaeometallurgical slags. *PACT* 9, 271–286.
Evans, R. D.
  1955   *The Atomic Nucleus.* McGraw-Hill, New York, Toronto and London.
Ewbank, W. B.
  1979   Status of transactinium nuclear data in the evaluated nuclear structure date file. *Transactinium Isotope Nuclear Data-1979,* TECDOC-233 IAEA, Nuclear Data Section, Wagramerstrasse 5, A-1400, Vienna.
Flanagan, F. J.
  1969   U.S. Geological Survey standards II. *Geochimica et Cosmochimica Acta* 33, 81–120.
  1972   U.S. Geological Survey standards III. *Geological survey professional papers* 840, 131–183. U.S. Department of Interior.
Fleischer, R. L., Price, P. B., and Walter, R. M.
  1975   *Nuclear Tracks in Solids,* pp. 489–526. University of California Press, Berkeley and Los Angeles.
Fleming, S. J.
  1966   Study of thermoluminescence of crystalline extracts from pottery. *Archaeometry* 9, 170–173.

References

1969   *The acquisition of radioluminescence by ancient ceramics.* Unpublished D.Phil. thesis, Faculty of Physical Sciences, Oxford University.
1970   Thermoluminescent dating: refinement of the quartz inclusion method. *Archaeometry* 12, 133–147.
1973   The pre-dose technique: a new thermoluminescence dating method. *Archaeometry* 15, 13–30.
1975a   *Authenticity in art.* The Institute of Physics, London and Bristol.
1975b   Supralinearity corrections in fine grain thermoluminescence dating: a re-appraisal. *Archaeometry* 16, 91–95.
1979   *Thermoluminescence techniques in archaeology.* Clarendon Press, Oxford.

Fleming, S. J., and Thompson, J.
1970   Quartz as a heat-resistant dosimeter. *Health Physics* 18, 567–568.

Fleming, S. J., and Stoneham, D.
1973a   The subtraction technique of thermoluminescent dating. *Archaeometry* 15, 229–238.

Fleming, S. J., and Stoneham, D.
1973b   Thermoluminescent authenticity study and dating of Renaissance terracottas. *Archaeometry* 15, 239–247.

François, H., McKerrell, H., and Mejdahl, V.
1977   Thermoluminescence dating of ceramics from Glozel, *Proceedings of the 5th International Conference on Luminescence Dosimetry,* pp. 469–479. Sao Paulo, Brazil.

Fremlin, J. H., and Srirath, S.
1964   Thermoluminescent dating: examples of non-uniformity of luminescence. *Archaeometry* 7, 58–62.

Gallois, B., Nguyen, Ph. Hao, Bechtel, Fr., and Schvoerer, M.
1979   Datation par TL de coraux fossiles des Caraibes: observation et interpretation de mecanismes inhabituels de TL. *PACT* 3, 493–505.

Ganguly, S., and Kaul, I. K.
1984   Analysis of thermoluminescence glow peaks from natural calcium fluoride. *Modern Geology* 8, 155–162.

Garlick, G. F. J., and Gibson, A. F.
1948   The electron trap mechanism of luminescence in sulphide and silicate phosphors. *Proceedings of the Physical Society* 60, 574–590.

Garlick, G. F. J., Lamb, W. E., Steigmann, G. A., and Geake, J. E.
1971   Thermoluminescence of lunar samples and terrestrial plagioclases. *Proceedings of the 2nd Lunar Science Conference, Geochimica et Cosmochimica Acta* (Suppl. 2) 3, 2277–2284.

Garlick, G. F. J., and Robinson, I.
1972   The thermoluminescence of lunar samples. In *The Moon,* edited by S. K. Runcorn and H. Urey, pp. 324–329. International Astronomers Union, Dordrecht.

Garrison, E. G., Rowlett, R. M., Cowan, D. L., and Holroyd, L. V.
1981   ESR dating of ancient flints. *Nature* 290, 44–45.

Gillot, P. Y., Valladas, G., and Reyss, J. L.
1978   Dating of lava flow using a granitic enclave: application to the Laschamp magnetic event. *PACT* 2, 165–173.

Gillot, P. Y., Labeyrie, J., Laj, C., Valladas, G., Guérin, G., Poupeau, G., and Delibrias, G.
1979   Age of the Laschamp paleomagnetic excursion revisited. *Earth Planetary Science Letters* 42, 444–450.

Goedicke, C.
1984 Microscopic investigations of the quartz etching technique for TL dating. *Nuclear Tracks* 9, 87–93.

Göksu, H. Y., and Fremlin, J. H.
1972 Thermoluminescence from unirradiated flints: regeneration thermoluminescence. *Archaeometry* 14, 127–132.

Göksu, H. Y., Fremlin, J. H., Irwin, H. T., and Fryxell, R.
1974 Age determination of burned flint by a thermoluminescent method. *Science* 183, 651–654.

Göksu, H. Y., and Turetken, N.
1979 Source identification of obsidian tools by TL. *PACT* 3, 356–358.

Gorbics, S. G., Nash, A. E., and Attix, F. H.
1968 Thermal quenching of luminescence in six thermoluminescent dosimetry phosphors. *Proceedings of the 2nd International Conference on Luminescent Dosimetry*, Gatlinburg, edited by J. A. Auxier, K. Becker, and E. M. Robinson, CONF-680920, pp. 587–605. National Bureau Standards, Washington, D.C.

Greening, J. R.
1981 *Fundamentals of Radiation Dosimetry*. Adam Hilger, Bristol.

Grogler, N., Houtermans, F. G., and Stauffer, H.
1960 Ueber die Datierung von Keramik und Ziegel durch Thermolumineszenz. *Helvetica Physica Acta* 33, 595–596.

Groom, P. J., Durrani, S. A., Khazal, K. A. R., and McKeever, S. W. S.
1978 The dose-rate dependence of the thermoluminescence response in quartz. *PACT* 2, 200–210.

Guérin, G.
1982a Croissance de la thermoluminescence en fonction de la dose des feldspaths d'origine volcanique. *PACT* 6, 417–425.
1982b L'evenment Laschamp dans les laves de la Chaine des Puys. *Modern Geology* 8, 121–126.
1983 La thermoluminescence des plagioclases, methode de datation du volcanisme. Thèse de doctorat d'etat, l'université Pierre de Marie Curie, Paris.

Guérin, G., and Valladas, G.
1980 Thermoluminescence dating of volcanic plagioclases. *Nature* 286, 697–699.

Halperin, A., and Braner, A. A.
1960 Evaluation of thermal activation energies from glow curves. *Physical Review* 117, 408–415.

Haskell, E. H.
1983 Beta dose-rate determination: preliminary results from an interlaboratory comparison of techniques. *PACT* 9, 77–86.

Haskell, E. H., and Bailiff, I. K.
1985 Diagnostic and corrective procedures for TL analysis using the pre-dose technique. *Nuclear Tracks and Radiation Measurements* 10, in press.

Haskell, E. H., Kaipa, P. L., and Wrenn, M. E.
1984 The use of thermoluminescence analysis for atomic bomb dosimetry: estimating and minimizing total error. US–Japan Workshop for Reassessment of Atomic Bomb Dosimetry in Hiroshima and Nagasaki, November 1983, published by the Radiation Effects Research Foundation, 5-2 Hijiyama Park, Minami-ku, Hiroshima, 730, Japan, pp. 32–44.
1985 Accident dosimetry using the pre-dose TL technique. *Nuclear Tracks and Radiation Measurements* 10, in press.

References

Hedges, R. E. M., and McLellan, M.
1976   On the cation exchange capacity of fired clays and its effect on the chemical and radiometric analysis of pottery. *Archaeometry* 18, 203–207.

Hennig, G. J., and Grun, R.
1984   ESR dating in Quaternary geology. *Quaternary Sciences Reviews* 2, 157–238.

Hennig, G. J., Herr, W., Weber, E., and Xirotiris, N. I.
1981   ESR dating of the fossil hominid cranium from Petralona Cave. *Nature* 292, 533–536.

Hoogenstraaten, W.
1958   Electron traps in zinc sulphide phosphors. *Philips Research Reports* 13, 515–562.

Howarth, J. L.
1965   Calculation of the alpha-ray absorbed dose to soft tissue. *British Journal of Radiology* 38, 51–56.

Hoyt, H. P., Kardos, J. L., Miyajima, M., Seitz, M. G., Sun, S. S., Walker, R. M., and Wittels, M. C.
1970   Thermoluminescence, X-ray and stored energy measurements of Apollo 11 samples. *Proceedings of the Apollo 11 Lunar Science Conference, Geochimica et Cosmochimica Acta* (Suppl. 1) 3, 2269–2288.

Hoyt, H. P., Walker, R. M., Zimmerman, D. W., and Zimmerman, J.
1972   Thermoluminescence of individual grains and bulk samples of lunar fines. *Proceedings of the 3rd Lunar Scientific Conference* 3, 2997–3008.

Hubbell, J. H.
1977   Photon mass attenuation and mass-absorption coefficients for H, C, N, O, Ar and seven mixtures from 0.1 keV to 20 MeV. *Radiation Research* 70, 58–81.

Huntley, D. J.
1985   On the zeroing of the thermoluminescence of sediments. *Physics and Chemistry of Minerals*, in press.

Huntley, D. J., and Johnson, H. P.
1976   Thermoluminescence as a potential means of dating siliceous ocean sediments. *Canadian Journal of Earth Sciences* 13, 593–596.

Huntley, D. J., and Bailey, D. C.
1978   Obsidian source identification by thermoluminescence. *Archaeometry* 20, 159–170.

Huntley, D. J., and Wintle, A. G.
1981   The use of alpha scintillation counting for measuring Th-230 and Pa-231 contents of ocean sediments. *Canadian Journal of Earth Science* 18, 419–432.

Huntley, D. J., Godfrey-Smith, D. I., and Thewalt, M. L. W.
1985   Optical dating of sediments. *Nature* 313, 105–107.

Hutt, G., and Smirnov, A.
1982   Thermoluminescence dating in the Soviet Union, *PACT* 7, 97–103.

Hutt, G., and Smirnov, A.
1983   Thermoluminescence dating of sediments by means of the quartz and feldspar inclusion methods. *PACT* 9, 463–472.

Huxtable, J.
1978   Fine grain dating. *PACT* 2, 7–11.
1981   Light bleaching of archaeological flint samples: a warning. *Ancient TL* no. 16, 2–4.
1982   Fine grain thermoluminescence techniques applied to flint dating. *PACT* 6, 346–352.

Huxtable, J., Aitken, M. J., and Weber, J. C.
1972   Thermoluminescent dating of baked clay balls of the Poverty Point culture. *Archaeometry* 14, 269–275.
Huxtable, J., Aitken, M. J., Hedges, J. W., and Renfrew, A. C.
1976   Dating a settlement pattern by thermoluminescence: the burnt mounds of Orkney. *Archaeometry* 18, 5–17.
Huxtable, J., and Aitken, M. J.
1977   Thermoluminescent dating of Lake Mungo geomagnetic polarity excursion. *Nature* 265, 40–41.
Huxtable, J., and Aitken, M. J.
1978   Thermoluminescence dating from Sham Wan, Lamma Island. *Journal Monograph Hong Kong Archaeological Society* III, 116–124.
Huxtable, J., Aitken, M. J., and Bonhommet, N.
1978   Thermoluminescence dating of sediment baked by lava flows of the Chaine des Puys. *Nature* 275, 207–209.
Huxtable, J., and Jacobi, R. M.
1982   Thermoluminescence dating of burned flints from a British Mesolithic site: Longmoor Inclosure, East Hampshire. *Archaeometry* 24, 164–169.
Huxtable, J., and Barton, R. N. E.
1983   New dates for Hengistbury Head, Dorset. *Antiquity* 52, 133–135.
Huxtable, J., and Aitken, M. J.
1985   Flint dating by TL: European results from the Mesolithic to the Lower Palaeolithic. *Proceedings of the 24th Archaeometry Symposium,* Washington, D.C. *Smithsonian Institute* Washington, D.C., in press.
Iacconi, P.
1980   Origin of the thermoluminescence in zircon. *Nuclear Instruments and Methods* 175, 222–223.
Ichikawa, Y.
1965   Dating of ancient ceramics by thermoluminescence. *Bulletin of the Institute of Chemical Research, Kyoto University* 43, 1–6.
Ichikawa, Y., and Nagatomo, T.
1978   Thermoluminescence dating of burnt sandstone from Senpukuji Cave. *PACT* 2, 174–178.
1983   Thermoluminescent dating and its application to gamma ray dosimetry. US–Japan joint Workshop for Reassessment of Atomic Bomb Dosimetry in Hiroshima and Nagasaki, February 1983, published by the Radiation Effects Research Foundation, 5-2 Hijiyama Park, Minami-ku, Hiroshima, 730, Japan, pp. 104–144.
Ikeya, M.
1978   Electron spin resonance as a method of dating. *Archaeometry* 20, 147–158.
1980   Electron spin resonance dating of animal and human bones. *Science* 207, 977–979.
Jain, M., Evans, M. L., and Close, D. A.
1979   Nondestructive assay technology for uranium resource evaluation infinite medium calculations. *Los Alamos Report* LA-7713-MS/UC-51. Los Alamos Scientific Laboratory, Los Alamos, NM.
Jain, V. K.
1978   Thermoluminescence glow curve and spectrum of zircon. *Bulletin of Mineralogy* 101, 358–362.
Jasinka, M., and Niewiadomski, T.
1970   Thermoluminescence of biological materials. *Nature* 227, 1159–1160.

References

Jensen, H. E., and Prescott, J. R.
- 1983   The thick-source alpha particle counting technique: comparison with other techniques and solutions to the problem of overcounting. *PACT* 9, 25–36.

Jerlov, N. G.
- 1970   Light: general introduction. In *Marine Ecology* edited by O. Kinne, pp. 95–102. Wiley-Interscience, New York.

Johnson, N. M.
- 1960   Thermoluminescence in biogenic calcium carbonate. *Journal of Sedimentary Petrology* 30, 305–313.
- 1963   Thermoluminescence in contact metamorphosed limestone. *Journal of Geology* 71, 596–616.

Johnson, N. M., and Blanchard, R. L.
- 1967   Radiation dosimetry from the natural thermoluminescence of fossil shell. *American Mineralogy* 52, 1297–1310.

Johnson, T. L.
- 1974   Quantitative analysis of the growth and decay of the TL peaks in LiF. *Proceedings of the 4th International Conference on Luminescence Dosimetry*, Krakow, pp. 197–218.

Kaipa, P. L., and Haskell, E. H.
- 1985   Use of the sensitized 210 C peak of quartz for in situ dosimetry of intact ceramics. *Nuclear Tracks and Radiation Measurements*, 10, in press.

Kalefezra, J., and Horowitz, Y. S.
- 1979   Electron backscattering corrections for beta dose-rate estimations in archaeological objects. *PACT* 3, 428–438.

Kennedy, G. C., and Knopff, L.
- 1960   Dating by thermoluminescence. *Archaeology* 13, 147–148.

King, K. J., and Johnson, T. L.
- 1983   Dependence of thermoluminescence output on temperature during irradiation for several thermoluminescence phosphors. *Naval Research Laboratory Report 8761*. Naval Research Laboratory, Washington, D.C.

Kirkegaard, P., and Løvborg, L.
- 1980   Transport of terrestrial gamma radiation in plane semi-infinite geometry. *Journal of Computational Physics* 36, 20–34.

Kristianpoller, N.
- 1983   Effects of ultraviolet irradiations on quartz. *PACT* 9, 153–162.

Kronborg, C.
- 1983   Preliminary results of age determination by TL of interglacial and interstadial sediments. *PACT* 9, 595–606.

Lang, A. R., and Miuscov, V. F.
- 1967   Dislocations and fault surfaces in synthetic quartz. *Journal of Applied Physics* 38, 2477.

Langouet, L., Roman, A., Deza, A., Brito, P., Concha, G., and Asenjo de Roman, C.
- 1979   Datation relative par thermoluminescence, méthode DATE. *Revue d'Archeometrie* 3, 57–67.

Langouet, L., Roman, A., and Gonzales, R.
- 1980   Datation de poteries anciennes par la méthode DATE. *Proceedings of the 16th International Symposium of Archaeometry, National Museum of Antiquities of Scotland*, Edinburg, pp. 312–320.

Lederer, C. M., and Shirley, V. S., (Eds.)
- 1978   *Table of Isotopes* (7th ed.). John Wiley, New York.

Leeper, G. W.
1964   *Introduction to Soil Science.* Melborne University Press, Melborne.

Levy, P. W.
1979   TL studies having applications to geology and archaeometry. *PACT* 3, 466–480.
1982   Thermoluminescence and optical bleaching in minerals exhibiting second order kinetics and other retrapping characteristics. *PACT* 6, 224–242.
1983   Characteristics of thermoluminescence glow curves for materials exhibiting more than one glow peak. *PACT* 9, 109–122.

Liritzis, Y., and Galloway, R. B.
1982   A new approach to the beta dosimetry of ceramics for thermoluminescence dating. *Nuclear Instruments and Methods* 201, 503–506.

Lorenz, A., (Ed.)
1983   *Proposed Recommended List of Heavy Element Radionuclide Decay Data,* INDC (NDS)-149/NE. IAEA Nuclear Data Section, Wagramerstrasse 5, A-1400 Vienna.

Løvborg, L., and Kirkegaard, P.
1974   Response of 3″ × 3″ NaI(Tl) detector to terrestrial gamma radiation. *Nuclear Instruments and Methods* 121, 239–251.

McKeever, S.
1979   A note on the plateau test as used in TL dating. *Ancient TL* no. 6, 13–16.
1983   Dating of meteorite falls using thermoluminescence. *PACT* 9, 321–334.
1985   Thermoluminescence of solids. Cambridge University Press, London.

McKerrell, H., Mejdahl, V., François, H., and Portal, G.
1974   Thermoluminescence and Glozel. *Antiquity* 48, 265–272.
1975   Thermoluminescence and Glozel: a plea for patience. *Antiquity* 49, 267–272.

McKerrell, H. and Medjahl, V.
1980   Progress and problems with automated TL dating. *Proceedings of the 16th International Symposium of Archaeometry, National Museum of Antiquities of Scotland,* Edinburgh.

McKinlay, A. F.
1981   *Thermoluminescence Dosimetry.* Adam Hilger, Bristol.

Mangini, A., Pernicka, E., and Wagner, G. A.
1983   Dose-rate determination by radiochemical analysis. *PACT* 9, 49–56.

Martini, M., Piccinini, G., and Spinolo, G.
1983a   A new dosimetric system for x-ray dose-rate determination in soils. *PACT* 9, 87–92.

Martini, M., Spinolo, G., and Dominici, G.
1983b   Alpha spectrometry as a tool for annual dose-rate determination in pottery and soil. *PACT* 9, 9–18.

Maruyama, T., Kumamoto, Y., Noda, Y., Tamada, H., Okamoto, Y., Fujita, S., and Hashizume, T.
1983   Reassessment of gamma ray dose estimates from thermoluminescent yields in Hiroshima and Nagasaki. US–Japan Joint Workshop for Reassessment of Atomic Bomb Dosimetry in Hiroshima and Nagasaki, February 1983, published by the Radiation Effects Research Foundation, 5-2 Hijiyama Park, Minami-ku, Hiroshima, 730, Japan, pp. 122–137.

May, R. J.
1977   Thermoluminescence dating of Hawaiian alkalic basalts. *Journal of Geophysical Research* 82, 3023–3029.

References

Mazess, R. B., and Zimmerman, D. W.
- 1966   Pottery dating from thermoluminescence. *Science* 152, 347–348.

Meakins, R. L., Dickson, B. L., and Kelly, J. C.
- 1978   The effect of thermoluminescent dating of disequilibrium in the uranium decay chain. *PACT* 2, 97–103.
- 1979   Gamma ray analysis of K, U and Th for dose-rate estimation in thermoluminescent dating. *Archaeometry* 21, 79–86.

Mejdahl, V.
- 1969   Thermoluminescence dating of ancient Danish ceramics. *Archaeometry* 11, 99–104.
- 1970   Measurement of environmental radiation at archaeological excavation sites. *Archaeometry* 12, 147–160.
- 1978   Measurement of environmental radiation at archaeological sites by means of TL dosimeters. *PACT* 2, 70–83.
- 1979   Thermoluminescence dating: beta-dose attenuation in quartz grains. *Archaeometry* 21, 61–73.
- 1983   Feldspar inclusion dating of ceramics and burnt stones. *PACT* 9, 351–364.
- 1985a   TL dating based on feldspars. *Nuclear Tracks and Radiation Measurements* 10, in press.
- 1985b   Thermoluminescence dating of partially bleached sediments. *Nuclear Tracks and Radiation Measurements* 10, in press.

Mejdahl, V., and Winther-Nielsen, M.
- 1982   TL dating based on feldspar inclusions. *PACT* 6, 426–437.

Mejdahl, V., and Wintle, A. G.
- 1984   Applications to archaeological and geological dating. In *Thermoluminescence and Thermoluminescent Dosimetry*, (Vol. 3), edited by Y. S. Horowitz, pp. 133–190. Boca Raton, Florida.

Melcher, C. L.
- 1981a   Thermoluminescence of meteorites and their terrestrial ages. *Geochimica et Cosmochimica Acta* 45, 615–626.
- 1981b   Thermoluminescence of meteorites and their orbits. *Earth and Planetary Science Letters* 52, 39–54.

Melcher, C. L., and Zimmerman, D. W.
- 1977   Thermoluminescent determination of prehistoric heat treatment of chert artefacts. *Science* 197, 1359–1362.

Metha, S. K., and Sengupta, S.
- 1976   Gamma dosimetry with aluminum oxide thermoluminescence phosphor. *Physics in Medicine and Biology* 21, 955–964.

Miallier, D., Fain, J., Sanzelle, S., Daugas, J. P., and Raynal J. P.
- 1983   Dating of the Butte de Clermont basaltic maar by means of the quartz inclusion method. *PACT* 9, 487–498.

Miallier, D., Fain, J., Sanzelle, S., Raynal, J. P., Daugas, J. P., and Paquereau, M. M.
- 1985   Single quartz grain thermoluminescence dating. *Nuclear Tracks and Radiation Measurements* 10, in press.

Mobbs, S. F.
- 1978   Low temperature optical re-excitation in thermoluminescence dating. Unpublished M.Sc. thesis, Faculty of Physical Sciences, Oxford University.
- 1979   Phototransfer at low temperatures. *PACT* 3, 407–413.

Morozov, G. V.
1968 The relative dating of Quaternary Ukrainian sediments by the thermoluminescence method. *8th International Quaternary Association Congress,* Paris, p. 167. U.S. Geological Survey Library, Washington, D.C., Cat. No. 208M8280.

Morozov, G. V., and Shelkoplyas, V. N.
1980 TL of quartz from glacial deposits and determination of age of glacial and limnoglacial formations. *Abstracts of a Seminar on Field and Laboratory Methods of Research into Glacial Sediments,* Tallinn, pp. 101–102.

Murray, A. S.
1981 *Environmental radioactivity studies relevant to thermoluminescence dating.* Unpublished D.Phil. thesis, Faculty of Physical Sciences, Oxford University.
1982 Studies of the stability of radioisotope concentrations and their dependence on grain size. *PACT* 6, 216–223.

Murray, A. S., Bowman, S. G. E., and Aitken, M. J.
1978a Evaluation of the gamma dose-rate contribution. *PACT* 2, 84–96.
1978b Environmental radiation studies relevant to thermoluminescence dating. Proceedings of NRE III Conference, Houston.

Murray, A. S., and Wintle, A. G.
1979 Beta source calibration. *PACT* 3, 419–427.

Murray, A. S., and Aitken, M. J.
1982 The measurement and importance of radioactive disequilibria in TL samples. *PACT* 6, 155–169.

Murray, A. S., and Heaton, B.
1983 Alpha spectrometry without chemistry. *PACT* 9, 37–48.

Nambi, K. S. V.
1984 Alpha-radioactivity-related upper age limit for thermoluminescence dating?, Proceedings of the Indian Academy of Science. *Earth and Planetary Science Letters* 93, 47–56.

National Council on Radiation Protection (NCRP)
1976 Environmental radiation measurements. *National Council on Radiation Protection, Report 50.* NCRP Publications, Bethesda, MD.

Nichols, A. L., and James, M. F.
1981 *Radioactive Heavy Element Decay Data for Reactor Calculations,* UKAEA Report AEEW-R 1407. Obtainable from IAEA Nuclear Data Section, Wagramerstrasse 5, A-1400 Vienna.

Northcliffe, L. C., and Schilling, R. F.
1970 Range and stopping-power tables for heavy ions. *Nuclear Data Tables A* 7, 233–463.

Oberhofer, M., and Scharmann, A. (Eds.)
1981 *Applied Thermoluminescence Dosimetry.* Adam Hilger, Bristol.

Pages, L., Bertel, E., Joffre, H., and Sklavenitis, L.
1972 Energy loss, range and bremsstrahlung yield for 10 keV to 100 MeV electrons. *Atomic Data* 4, 1–127.

Parker, C. A.
1968 *Photoluminescence of Solutons.* Elsevier, London.

Pernicka, E., and Wagner, G. A.
1982 Radioactive equilibrium and dose-rate determination in TL dating. *PACT* 6, 132–144.

Plachy, A. L., and Sutton, S. R.
1982 Determination of the dose-rate to quartz in granite. *PACT* 6, 188–194.

References

Portal, G.
  1981   Preparation and properties of principal TL products. In *Applied Thermoluminescence Dosimetry*, edited by M. Oberhofer and A. Scharmann, pp. 97–122. Adam Hilger, Bristol.

Poupeau, G., Sutton, S., Walker, R. M., and Zimmerman, D. W.
  1976   Thermoluminescent dating of fired rocks: application to the site of Pincevent, France. *Proceedings of the 9th Congress of the Union International of Prehistoric Sciences*, Nice.

Prescott, J. R.
  1983   Thermoluminescence dating of sand dunes at Roonka, South Australia. *PACT* 9, 505–512.

Prescott, J. R., and Stephan, L. G.
  1982   Contribution of cosmic radiation to environmental dose. *PACT* 6, 17–25.

Prescott, J. R., Polach, H. A., Pretty, G. L., and Smith, B. W.
  1983   Comparison of C-14 and thermoluminescence dates from Roonka South Australia. *PACT* 8, 205–213.

Pye, K.
  1984   Loess. *Progress in Physical Geography* 8, 176–217.

Ralph, E. K., and Han, M. C.
  1966   Dating of pottery by thermoluminescence. *Nature* 210, 245–247.

Randall, J. T., and Wilkins, M. H. F.
  1945   Phosphorescence and electron traps. *Proceedings of the Royal Society of London* A 184, 366–407.

Rendell, H. M.
  1985   The precision of water content estimates in the thermoluminescence dating of loess from Northern Pakistan. *Nuclear Tracks and Radiation Measurements* 10, in press.

Renfrew, A. C.
  1975   Glozel and the two cultures. *Antiquity* 49, 219–222.

Riehl, N.
  1970   Tunnel luminescence and infrared stimulation. *Journal of Luminescence* 1, 1–16.

Robins, G. V., Seely, N. J., Symons, M. C. R., and MacNeil, D. A. C.
  1981   Manganese (II) as an indicator of ancient heat treatment in flint. *Archaeometry* 23, 103–108.

Sanderson, D. C. W., Warren, S. E., and Hunter, J. R.
  1983   The TL properties of archaeological glass. *PACT* 9, 287–298.

Sankaran, A. V., Nambi, K. S. V., and Sunta, C. M.
  1982   *Current Status of Thermoluminescence Studies on Minerals and Rocks*. Bhaba Atomic Research Centre, Bombay, India.

Sankaran, A. V., Nambi, K. S. V., and Sunta, C. M.
  1983   Progress of thermoluminescence research on geological materials. *Proceedings of the Indian National Science Academy* 49, 18–112.

Sanzelle, S., Fain, J., and Miallier, D.
  1983   Thermoluminescence dating: alpha dosimetry using solid state track detectors. *PACT* 9, 59–68.

Sasidharan, R., Sunta, C. M., and Nambi, K. S. V.
  1979   Phototransfer method of determining archaeological dose of pottery sherds. *PACT* 3, 401–406.

Schvoerer, M., Lamarque, P., and Rouanet, J. F.
1974   Datation absolute par thermoluminescence: gres brulés provenant de niveaux magdelenians IV et VI. *(Paris) B Comptes Rendus de l'Academie de Sciences* 279, 191–194.

Schvoerer, M., Bechtel, F., Deshouilles, J. M., Dautant, A., Gallois, B.
1982   Datation par gamma-thermoluminescence: recherches sur une nouvelle méthode. *PACT* 6, 86–101.

Schwarcz, H. P.
1980   Absolute age determination of archaeological sites by uranium series dating of travertines. *Archaeometry* 22, 3–24.

Sears, D. W.
1975   Thermoluminescence studies and the preatmospheric shape and mass of the Estacado meteorite. *Earth and Planetary Science Letters* 26, 97–104.

Sears, D. W., and Mills, A. A.
1973   Temperature gradients and atmospheric ablation rates for the Barwell meteorite. *Nature* 242, 25–26.
1974   Thermoluminescence and the terrestrial age of meteorites. *Meteorites* 9, 47–67.

Sears, D. W., and Durrani, S. A.
1980   Thermoluminescence and the terrestrial age of meteorites: some recent results. *Earth and Planetary Science Letters* 46, 159–166.

Shalgaonkar, C. S., and Narlikar. A. V.
1972   Review: recent methods for determining trap depth from glow curves. *Journal of Material Science* 7, 1465–1471.

Shaw, J.
1979   Rapid changes in the magnitude of the archaeomagnetic field. *Geophysical Journal of the Royal Astronomical Society* 58, 107–116.

Shayes, R., Lorthoir, M., and Lheureux, M.
1963   Thermoluminescent dosimetry IV. *Revue mble* 6, 1–55.

Shayes, R., Brooke, C., Kozlowitz, I., and Lheureux, M.
1967   Thermoluminescent properties of natural calcium fluoride. In *Luminescence Dosimetry*, pp. 138–157. Available as CONF-650637 from U.S. National Bureau Standards, Washington, D.C.

Shelkoplyas, V. N.
1971   Dating of the Quaternary deposits by means of thermoluminescence. In *Chronology of the Glacial Age*, pp. 155–160. Geographic Society of the USSR, Pleistocene Commission, Leningrad.

Shenker, D., and Chen, R.
1971   Numerical curve fitting of general order kinetics glow peaks. *Journal of Physics D: Applied Physics* 4, 287–291.

Sieveking, G. de G., Bush, P., Ferguson, J., Craddock, P. T., Hughes, M. J., and Cowell, M. R.
1972   Prehistoric flint mines and their identification as sources of raw material. *Archaeometry* 14, 151–176.

Singhvi, A. K., and Aitken, M. J.
1978   Americium-241 for alpha irradiations. *Ancient TL* no. 3, 2–9.

Singhvi, A. K., Sharma, Y. P., and Agrawal, D. P.
1982   Thermoluminescence dating of sand dunes in Rajasthan, India. *Nature* 295, 313–315.

References

Skinner, A. F.
    1983   Overestimate of stalgmitic calcite ESR dates due to laboratory heating. *Nature* 304, 152–154.

Smalley, I. J.
    1975   Loess lithology and genesis. *Benchmark Papers in Geology* (no. 26). John Wiley & Sons, New York.

Smith, B. W.
    1983   *New applications of thermoluminescence dating.* Unpublished Ph.D. thesis, Department of Physics, Adelaide University.

Stoneham, D.
    1983   Porcelain dating. *PACT* 9, 227–239.

Storm, E., and Israel, H. I.
    1970   Photon cross sections from 1 keV to 100 MeV for elements 1 through 100. *Nuclear Data Tables* A 7, 565–681.

Strickertsson, K.
    1985   The thermoluminescence of potassium feldspars—glow curve characteristics and initial rise measurements. *Nuclear Tracks and Radiation Measurements* 10, in press.

Suhr, N. H., and Ingamells, C. O.
    1966   Solution technique for analysis of silicates. *Analytical Chemistry* 38, 730–734.

Sunta, C. M.
    1971   TL of natural calcium fluoride and its applications. *Proceedings of the 3rd International Conference of Luminescence Dosimetry*, Risø, Denmark, pp. 392–409.

Sunta, C. M., and David, M.
    1983   Firing temperature of pottery from pre-dose sensitization. *PACT* 6, 460–467.

Sutton, S. R.
    1985a   TL measurements on shock-metamorphosed sandstone and dolomite from Meteor Crater, Arizona: Pt. 1 shock dependence of TL properties. *Journal of Geophysical Research* 9, 3683–3689.
    1985b   TL measurements on shock-metamorphosed sandstone and dolomite from Meteor Crater, Arizona: Pt. 2 TL age of Meteor Crater. *Journal of Geophysical Research* 9, 3690–3700.

Sutton, S. R., and Zimmerman, D. W.
    1976   Thermoluminescent dating using zircon grains from archaeological ceramics. *Archaeometry* 18, 125–134.
    1977   Hints on spurious reduction. *Ancient TL* no. 1, 7–9.
    1978a   Attempts to circumvent anomalous fading. *Ancient TL* no. 3, 10–12.
    1978b   Thermoluminescence dating: radioactivity in quartz. *Archaeometry* 20, 67–69.
    1979   The zircon natural method: initial results and low level TL measurement. *PACT* 3, 465.

Tanner, A. B.
    1964   Radon migration in the ground. In *The natural radiation environment*, edited by J. A. S. Adams and W. M. Lowder, pp. 161–190. University of Chicago Press.

Templer, R. H.
    1985a   The dating of zircons by autoregenerated TL at low temperatures. *Nuclear Tracks and Radiation Measurements* 10, in press.
    1985b   The removal of anomalous fading in zircons. *Nuclear Tracks and Radiation Measurements* 10, in press.

Templer, R. H., and Walton, A. J.
1983 Image intensifier studies of TL in zircons. *PACT* 9, 300–308.
Templer, R. H., Amin, T., Walton, A., and Watt, S.
1985 Zoning in zircons. *Nuclear Tracks and Radiation Measurements* 10, in press.
Thompson, J.
1970 The influence of previous irradiation on thermoluminescent sensitivity. Unpublished D.Phil. thesis, Faculty of Physical Sciences, Oxford University.
Thompson, R., and Berglund, B.
1976 Late Weichselian geomagnetic "reversal" as a possible example of the reinforcement syndrome. *Nature* 263, 490–491.
Tite, M. S.
1966 Thermoluminescent dating of ancient ceramics: a reassessment. *Archaeometry* 9, 155–169.
Tochilin, E., and Goldstein, N.
1968 The quality and LET dependence of three thermoluminescent dosimeters. *Proceedings of the 2nd International Conference on Luminescence Dosimetry*, Gatlinburg, pp. 424–437. Available as CONF-680920 from U.S. National Bureau Standards, Washington, D.C.
Turner, R. C., Radley, J. M., and Mayneord, W. V.
1958 The alpha-ray activity of human tissues. *British Journal of Radiology* 31, 397–402.
Valladas, G.
1982 Measure de la dose gamma annuelle de l'environnement d'un site par un dosimetre TL. *PACT* 6, 77–85.
Valladas, G., and Lalou, C.
1973 Etude de la thermoluminescence de la meteorite Saint Severin. *Earth and Planetary Science Letters* 18, 168–171.
Valladas, G., and Gillot, P. Y.
1978 Dating of the Olby lava flow using heated quartz pebbles: some problems. *PACT* 2, 141–150.
Valladas, G., Gillot, P. Y., and Guérin, G.
1979 Dating plagioclases. *PACT* 3, 251–256.
Valladas, G., and Ferreira, J.
1980 On the dose-rate dependence of thermoluminescence response of quartz. *Nuclear Instruments and Methods* 175, 216–218.
Valladas, G., and Valladas, H.
1982 Influence du debit de dose sur la thermoluminescence du quartz. *PACT* 6, 281–291.
1983 A variant of the thermoluminescence technique for beta-ray dosimetry. *PACT* 9, 73–76.
Valladas, H.
1978 Thermoluminescence dating of burnt stones from a prehistoric site, *PACT* 2, 180–183.
1979 La datation des foyers prehistoriques. *La Recherche* 98, 297–299.
1983 Estimation de la temperature de chauffe de silex prehistoriques par leur thermoluminescence. *Paris Comptes Rendus de l'Academie des Sciences* 296, 993–996.
Valladas, H., and Valladas, G.
1982 Effet de l'irradiation alpha sur des grains de quartz. *PACT* 6, 171–178.

References

Visocekas, R.
1979 Miscellaneous aspects of artificial TL of calcite: emission spectra, athermal detrapping and anomalous fading. *PACT* 3, 258–265.

Visocekas, R.
1985 Tunnelling radiative recombination associated with anomalous fading of thermoluminescence in labradorite. *Nuclear Tracks and Radiation Measurements* 10, in press.

Visocekas, R., Ceva, T., Marti, C., Lefaucheux, F., and Robert M. C.
1976 Tunnelling processes in afterglow of calcite. *Physica Status Solidi* A 35, 315–327.

Visocekas, R., Ouchene, M., and Gallois, B.
1983 Tunnelling afterglow and anomalous fading in dosimetry with calcium sulphate. *Nuclear Instruments and Methods in Physics Research* 214, 553–555.

Walton, A. J.
1982 An image intensifier spectrograph for thermoluminescence studies. *PACT* 6, 524–532.

Walton, A. J., and Debenham, N. C.
1982 Dating of paleolithic calcite by TL: observation of spatial inhomogeneity. *PACT* 6, 202–208.

Walton, A. J., Templer, R., and Reynolds, G. T.
1984 Depth of field measurements relevant to single photon detection using image-intensifier microscopy. *Journal of Physics E: Scientific Instruments*, in press.

Wang, W.
1983 Ultrathin TLD measurement of alpha dose-rate and comparison with alpha counting, *Ancient TL* 1, 2–4.

Wang, W., and Zhou, Z.
1983 Thermoluminescence dating of Chinese pottery. *Archaeometry* 25, 99–106.

Warren, S. E.
1978 Thermoluminescence dating of pottery: an assessment of the dose-rate from rubidium. *Archaeometry* 20, 69–70.

Watson, I. A., and Aitken, M. J.
1985 Firing temperature analysis using the 110 C TL peak of quartz. *Nuclear Tracks and Radiation Measurements* 10, in press.

Whittle, E. H.
1975 Thermoluminescent dating of Egyptian pre-dynastic pottery from Hemamieh and Qurna-Tarif. *Archaeometry* 17, 119–122.

Whittle, E. H., and Arnaud, J. M.
1975 Thermoluminescent dating of Neolithic and Chalcolithic pottery from sites in Central Portugal. *Archaeometry* 17, 5–24.

Williams, F.
1968 Donor-acceptor pairs in semiconductors. *Physica Status Solidi* (a) 25, 493–512.

Williamson, C. F., Boujot, J. P., and Picard, J.
1966 Tables of range and stopping power of chemical elements for charged particles of energy 0.05 to 500 MeV. *Centre d'Etudes Nucleaires de Saclay,* Saclay, Essonne, France, Rapport *CEA-R3042.*

Wintle, A. G.
1973 Anomalous fading of thermoluminescence in mineral samples. *Nature* 245, 143–144.

1974 Factors determining the thermoluminescence of chronologically significant materials. Unpublished D.Phil. thesis, Faculty of Physical Sciences, Oxford University.
1975a Effects of sample preparation on the thermoluminescence characteristics of calcite. *Modern Geology* 5, 165–167.
1975b Thermal quenching of thermoluminescence in quartz. *Geophysical Journal of the Royal Astronomical Society* 41, 107–113.
1977a Detailed study of a thermoluminescent mineral exhibiting anomalous fading. *Journal of Luminescence* 15, 385–393.
1977b Thermoluminescence dating of minerals: traps for the unwary. *Journal of Electrostatics* 3, 281–288.
1978a Anomalous fading. *PACT* 2, 240–243.
1978b A thermoluminescence dating study of some Quaternary calcite: potential and problems. *Canadian Journal of Earth Sciences* 15, 1977–1986.
1983 Thermoluminescence. In *Geomagnetism of Baked Clays and Recent Sediments,* edited by K. M. Creer, P. Tucholka, and C. E. Barton, pp. 63–71. Elsevier, London.
1985 Stability of the TL signal in fine grains from loess. *Nuclear Tracks and Radiation Measurements* 10, in press.

Wintle, A. G., Aitken, M. J., and Huxtable, J.
1971 Abnormal thermoluminescence fading characteristics. *Proceedings of the 3rd International Conference on Luminescence Dosimetry,* Riso, Denmark, pp. 105–131.

Wintle, A. G., and Oakley, K. P.
1972 Thermoluminescent dating of fired rock-crystal from Bellan Bandi Palassa, Ceylon. *Archaeometry* 14, 277–279.

Wintle, A. G., and Aitken, M. J.
1977a Absorbed dose from a beta source as shown by thermoluminescence dosimetry. *International Journal of Applied Radioisotopes* 28, 625–627.
1977b Thermoluminescence dating of burnt flint: application to a Lower Palaeolithic site, Terra Amata. *Archaeometry* 19, 111–130.

Wintle, A. G., and Murray, A. S.
1977 Thermoluminescence dating: reassessment of the fine grain dose-rate. *Archaeometry* 19, 95–98.

Wintle, A. G., and Huntley, D. J.
1979 Thermoluminescence dating of a deep-sea ocean core. *Nature* 279, 710–712.
1980 Thermoluminescence dating of ocean sediments. *Canadian Journal of Earth Sciences* 17, 348–360.
1982 Thermoluminescence dating of sediments. *Quaternary Science Reviews* 1, 31–53.

Wintle, A. G., Shackleton, N. J., and Lautridou, J. P.
1984 Thermoluminescence dating of periods of loess deposition and soil formation in Normandy. *Nature* 310, 491–493.

Yamashita, Y., Nada, N., Anichi, H., and Kitamura, S.
1971 Calcium sulfate activated by thulium or dysprosium for thermoluminescence dosimetry. *Health Physics* 21, 295–300.

Yokoyama, Y., Quaegebeur, J. P., Bibron, R., and Leger, C.
1983 ESR dating of Palaeolithic calcite: thermal annealing experiment and trapped electron lifetime *PACT* 9, 371–380.

Zeller, E. J., Wray, J. L., and Daniels, F.
1955   Thermoluminescence induced by pressure and by crystallization. *Journal of Chemical Physics* 23, 2187–2189.

Zimmerman, D. W.
1967   Thermoluminescence from fine grains from ancient pottery. *Archaeometry* 10, 26–28.
1970   The dependence of thermoluminescence on energy and type of ionizing radiation and its significance for archaeologic age determination. Unpublished D.Phil. thesis, Faculty of Physical Sciences, Oxford University.
1971   Thermoluminescent dating using fine grains from pottery. *Archaeometry* 13, 29–52.
1972   Relative thermoluminescence effects of alpha- and beta-radiation. *Radiation Effects* 14, 81–92.
1978   Introduction to basic procedures for sample preparation and TL measurement of ceramics. *PACT* 2, 1–6.
1979a   Radioactive recombination and anomalous fading. *PACT* 3, 257.
1979b   A study of phototransfer in zircon. *Ancient TL* no. 7, 2–9.

Zimmerman, D. W., Yuhas, M. P., and Meyers, P.
1974   Thermoluminescence authenticity measurements on core material from the Bronze Horse of the New York Metropolitan Museum of Art. *Archaeometry* 16, 19–30.

Zimmerman, J.
1971   The radiation-induced increase of thermoluminescence sensitivity of fired quartz. *Journal of Physics C: Solid State Physics* 4, 3277–3291.

INDEX

A

a-value, 11, 25, 85, 133, 308, 327, 328
a.d., 31
A.D., 31
Abri Vaufrey, 196
Absorbed dose, 15
Absorption coefficient, 66, 114, 117, 148
Acceptance criteria, 266
Acceptor centre, 237
Acceptor trap, 168
Accrued dose, 9
Accumulated dose, 9
Accuracy, 30
Additive dose, 19, 126, 162, 167, 198, 222
Aeolian, 219
Afterglow, 45, 276
Age equation, 2, 11, 18, 29, 239, 311
Age plateau, 10
Albite, 181, 189
Alice Boer, 196
Alpha attenuation, 150
Alpha counting, 26, 82, 244, 287, 299, 309
   thin sample, 39, 94
Alpha effectiveness, 11, 25, 129, 133, 165, 195, 199, 205, 212, 215, 217, 228, 237, 239, 246, 263
Alpha irradiation, 128, 244
Alpha ranges, 203
Alpha source, 129, 149
   strength, 311
Alpha spectrometry, 94, 111
Alpha spectrum, 312, 313
Aluminium oxide, 98, 114, 306, 322, 328

Ambient activation, 160
Americium-241, 129, 146, 149, 323
Analbite, 189
Ancient TL, 14
Andesine, 189
Annual dose, 10, 12, 21, 23, 66, 82, 183, 185, 193, 236, 239, 243, 246, 252, 282, 321
Anomalous fading, 9, 28, 54, 60, 96, 169, 171, 172, 175, 178, 189, 193, 197, 209, 237, 274
Anorthite, 181, 189
Anti-correlation, 12, 172, 175, 187, 207
Antiperthite, 189
Apatite, 69, 275
Aragonite, 203, 211
Archaeodose, 9
Archaeometry, 14
Arrhenius plot, 270
Art ceramics, 32
Artificial irradiation, 113
Athermal fading, 55
Atomic absorption, 109
Atomic number, effective, 306
Atomic weight, 63
   effective, 306
Attenuation
   alpha, 13, 102, 112, 253
   beta, 18, 99, 112, 120, 216, 244, 249, 258, 295
   gamma, 98, 104, 111, 116, 245, 249, 289, 322
   optical, 125, 215, 234, 256, 315
Attenuation coefficient, 148, 151

B

b-value, 86, 313
b.c., 31
B.C., 31
b.p., 31
B.P., 31, 32
Backscattering, 116, 122
Barium-133, 323
Becquerel (Bq), 64
Before Present, 32
Benign peak, 20
Beta counting, 183
Beta irradiation, 118
Beta source, 118, 144, 148
Beta spectrum, 120, 283
Beta thermoluminescence dosimetry, 99
Biache, 196
Bimolecular decay, 51
Bioluminescence, 46
Bismuth germanate (BGO), 324
Black-body, 1
Bleaching, 18, 45, 96, 139, 189, 194, 198, 220, 221, 265, 267, 327
Bone, 210, 212
Boyle, R., 3
Bragg-Kleeman rule, 303
Bremsstrahlung, 121, 143, 145
Bronze, 33
Brunhes, 213
Build-up
  beta dose, 48, 122
  gamma dose, 116
Build-up factor, 289
Burnt flint, 191, 264, 296, see also Flint
Burnt stone, 196
Bytownite, 189

Auger electron, 258, 283
Auger hole, 23, 102, 266, 324
Authenticity testing, 32, 82, 172
Autoradiography, 108
Autoregeneration, 177
Averaging, 250

C

Caesium-137, 114, 115
Calcite, 58, 65, 71, 74, 81, 114, 188, 202, 203, 212, 237, 281, 296, 306, 313, 314

Calcium sulphate, 21, 95, 98, 114, 132, 137, 150, 306, 310, 322, 328
Calibrating dose, 154
Calibration standards, 304
Capsule, 22, 97, 245, 266, 321, 322
Cariguela Cave, 196
Cathodoluminescence, 45, 175, 262
Caune de l'Arago, 208, 210
Chain disintegration, 66
Chaîne des Puys, 198, 200, 214, 263
Chalcedony, 191
Characteristic temperature, 47
Charged particle equilibrium, 117
Chemiluminescence, 46
Chemithermoluminescence, 46
Chert, 192
Clandestine irradiation, 33, 172
Closed system, 211
Coarse-grain technique, 254, 314, see also Quartz inclusion technique
Cobalt-60, 114, 115, 117
Cold light, 3
Colour, of sample, 89
Colour filter, 6, 96, 189, 198, 205, 216, 228, 238, 325
Combe Grenal, 196, 216
Commercial manufacturers, 14
Competition, 136, 150
Compton process, 103, 116, 289, 318
Conduction band, 44
Contact-baked sediment, 200
Contamination, 10, 33
Coral, 211
Cosmic radiation, 68, 74, 186, 246, 297, 321
Curie (Ci), 64
Curium source, 149, 150
Curve-fitting techniques, 270

D

Date citation, 31, 250
Daybreak equipment, 91, 302
Decay chain, 62
Defect creation, 14
Defect diffusion, 281
Desert loess, 220
Dolomite, 203
Donor trap, 168, 238
Dordogne, 208, 216
Dose-rate, 11

Dose-rate effect, 140
Dosimetry, by thermoluminescence, 21, 95
Dunite, 306

E

Effective dose, 11, 12, 85
Electron spin resonance (ESR), 4, 192, 204, 211, 216, 315
Electron-volt, 59
Element independence, 38
Energy-level diagram, 43, 156
Equivalent dose, 9, 19, 23, 26, 28, 239, 244, 246
Error limits, 24, 30, 241
Escape probability, 47
Etching, 18, 126, 187, 255, 259, 261
Etiolle, 196
Exposoure, 117

F

Feldspar, 55, 56, 60, 69, 185, 187, 196, 197, 199, 215, 221, 232, 235, 238, 274, 306, 310
Feldspar dating, 180
Fine-grain technique, 13, 24, 68, 227, 247, 253
Fireplace, 214
Firing temperature, 165
First-order kinetics, 51, 269, 270, 273
Fission track, 108, 195, 200, 216, 262, 263
Flame photometry, 82, 109
Flint, 71, 74, 191, 215, see also Burnt flint
Flowstone, 202
Fluorapatite, 55, 171
Fluorite, 21, 49, 52, 95, 97, 114, 132, 137, 141, 150, 245, 270, 306, 310, 321, 322, 325, 329
Forgeries, 33
Frequency factor, 49, see also Trap depth
Frosty grains, 126, 256

G

Gamma irradiation, 115
Gamma spectrometer, 102, 266, 318
Gamma spectrometry, high resolution, 106
Gamma spectrum, 329
Gamma-thermoluminescence, 186

Gas cell, 93
Geochemical mobility, 81
Geomagnetic excursions, 213
Geomagnetic intensity, 37
Germanium crystal, 107
Glacial grinding, 220
Glass, 197, 202
Glozel controversy, 35
Gothenberg, 214
Granite, 200, 215
Gray (Gy), 15
GSP sand, 306

H

Hacilar ware, 34, 155
Half-life, 59
Hawaii, 217
Health hazard, 142
Hectorite, 306
Helsingor conference, 237
Hengistbury Head, 196
Hydrofluoric acid, 38, 182, 217, 238, 256
Hydrofluosilicic acid, 238

I

Illite, 306
Image intensifier, 176, 195, 206, 216
Incandescence, 1, 2
Infinite matrix, 66, 70, 282
Indian desert, 223
Initial rise method, 53, 270
Internal conversion electron (IC), 258, 283
Internal radioactivity, 259, 263
Intrinsic dose, 263
Ionium, 203
Iron oxide coating, 217
Iron oxide component, 256
Irradiation, at elevated temperature, 141, 180
Isothermal decay method, 270

K

k-value, 11, 25, 85, 186, 243, 308, 312
Kappa value, 165
Killer centre, 44
Kinetics, 51, 178, 269
Knapping, 192

L

La Cotte, 196
Labradorite, 56, 60, 189, 274
Lake Mungo, 213, 256
Lake sediment, 220
Laschamp, 214
Laser, 6, 227
Lava, 197, 198, 214, 263, 275
Leaching, 80
Lead-210, 79, 91, 94
Lifetime, 43, 50, 59, 179, 193, 236, 269
Limestone, 55, 204
Linear mean, 251
Lithium fluoride, 306, 328
Loess, 219, 237
Logarithmic dependence, 237, 276, 278
Longmoor, 196
Luminescence centre, 42, 54, 138, 156, 275, 280
Lunar samples, 60

M

Magnesium silicate, 95, 328
Magnetic reversal, 213
Magnetic separation, 38, 256
Magnetism, remanent, 37
Malign peak, 20
Manganese-54, 323
MBLE, 325
Mercury lamp spectrum, 233
Mesvin, 196
Meteor craters, 201
Meteorites, 200
Microcline, 189, 310
Mineral separation, 173, 181, 228, 262
Moisture content, 22, 69, 74, 77, 195
Monalbite, 189,
Monazite, 147, 306
Monitor capsule, 98, 266
Monomolecular decay, 51
Multiple activation, 158, 166

N

Neutrino, 148, 283, 288
Neutron activation, 109, 216
Non-radiation induced (NRI) emission, 46, see also Spurious thermoluminescence
Normalization, 19, 38, 126, 163, 215, 216

O

Obsidian, 202, 328
Obsidian Creek, 202
Ocean sediment, 220, 236, 237
Oligoclase, 189
On-site measurements, 266
Open system, 211
Optical dating, 4, 227
Orkney Islands, 196
Orthoclase, 189, 310
Overcounting, 92
Overresponse, 98

P

PACT, 14, 331
Pairs, 90, 301
Paleodose, 7, 9, 11, 18, 19, 23, 28, 29, 155, 184, 188, 193, 205, 216, 217, 221, 224, 239
Paleolithic, 191, 219
Parameter uncertainties, 249
Partial bleach method, 225
Pastiche, 33
Peak shift, 48, 270
Peak temperature, 48, 270
Pech de l'Aze, 196
Perthite, 189
Petralona Cave, 213
Phallic symbols, 36
Phonon-assistance, 234
Phosphorescence, 270, 275, 276
Photocathode, 4, 5, 6, 87, 173, 176, 189, 216
Photoelectric effect, 116
Photoeviction, 45, 168, 221, 237
  dependence on temperature, 234
Photoluminescence, 45
Photomultiplier, 4, 26, 87, 173, 323
Photopeak, 103, 318
Photostimulated luminescence (PSL), 4, 227
Photo-thermoluminescence, 45
Phototransfer spectrum, 232
Phototransferred thermoluminescence (PTTL), 168, 226, 275, 326
Piezoluminescence, 45
Piezo-thermoluminescence, 45
Pitchblende, 147, 306
Plagioclase, 185, 263
Plateau, 7, 8, 9, 10, 52, 55, 199, 210

Plateau test, 280
Polonium-210, 79, 91, 94, 109
Polonium-215, 63, 90
Polonium-216, 63, 90
Pontnewydd, 196
Porcelain, 32, 155
Porosity, 75
Pot-boilers, 196
Predicted error, 251
Pre-dose dating, 153
Pre-dose effect, 19, 128, 137, 139, 174, 327
Principle of superposition, 245, 291
Prompt luminescence, 45, 171, 175
Protactinium-231, 217, 236, 286, 314
Pseudo-fading, 281
Pumice, 276

Q

Quantum efficiency, 5
Quartz, 55, 58, 69, 98, 114, 117, 126, 132, 137, 141, 150, 154, 168, 170, 178, 182, 185, 214, 221, 232, 235, 253, 259, 306, 310, 322
Quartz inclusion technique, 13, 17, 68, 179, *see also* Coarse-grain technique
Quenching correction, 157

R

R-Γ method, 225
Rad, 15, 150
Radiation bleaching, 136
Radiative recombination, 276
Radioactive disequilibrium, 65, 79, 80, 109, 110, 195, 203, 209, 218, 236, 250, 313
Radioactive equilibrium, 64, 66
Radioactive inclusion dating, 13
Radioactive inclusions, 69
Radioactive series, 62
Radiobiological equivalence, 150
Radiocarbon dating, 32, 213
Radiological protection, 143
Radioluminescence, 45
Radon, 63, 65, 67, 76, 91, 112, 195, 250, 259, 285, 287, 304, 313, 322
Random error, 242, 246, 250
Range, alpha, 130
Range, of alpha particles, 302
Recoil signal, 213
Reconstitution, 33, 37

Red-hot glow, 1, 2, 3
Reflectivity, 89
Regeneration, 223
Regeneration thermoluminescence, 192
Relative dating, 13
Rem, 15
Replica, of sample, 216
Reservoir trap, 156
Retrapping, 51
Roentgen, 15, 117, 150
Root-mean-square deviation, 251
Royat flow, 215

S

Sample collection, 264
Sample preparation, 18, 25
Sand dunes, 220
Sanidine, 60, 189
Santorini Island, 197
Saturation, 135, 139, 165, 193, 229, 235
Scintillation screen, 26, 86, 93, 246
Scintillometer, 105, 324
Sclayn, 196
Seattle region, 197
Seclin, 196
Secondary flux, 289
Secondary photons, 329
Second-glow growth, 20, 38, 128, 139
Second-order kinetics, 51
Sediment dating, 219, 267
Self-dose, 70, 245, 258
Sensitivity, for thermoluminescence, 11, 188
Sensitization, 137, 174, 224
Settling times, 25
Shell, 211, 212
Shiny grains, 126, 256
Sievert (Sv), 15, 150
Siliceous material, 237
Silicone oil, 124, 205, 217
Silicone resin, 21
Single-grain studies, 275
Slag, 201
Slice technique, 192, 204, 215, 314
Sodium hexametaphosphate, 325
Sodium iodide crystal, 102, 318, 320, 324
Solar resetting, 219
Solar simulator, 231, 233
Solar simulator spectrum, 234
Source calibration, 247

Specific activity, 66
Specivex filter, 217
Spectrum, of thermoluminescence, 205, 207, 229, 275
Spectrum stabilization, 323
Spectrum stripping, 104, 319
Spike technique, 163
Spring deposit, 203
Spurious thermoluminescence, 6, 10, 38, 45, 193, 197, 204, 217, 327
Stability, 7, 47, 193, 209, 235, 238
Stalactite, 202
Stalagmite, 65, 81, 202, 212
Standard deviation, 246
Standard error, 251
Stopping power, 112, 114, 119, 132, 148
Storage tests, 278
Stripping factors, 320
Strontium-90, 118, 128, 145, 148
  decay of, 128
Sublinearity, 136
Subtraction dating, 29, 184
Sunlamp, 233
Sunlight spectrum, 231, 234
Supralinearity, 9, 19, 135, 139, 205, 239
Supralinearity intercept, 28
Systematic error, 242, 247, 250

T

Tautavel, 210
Temperature dependence
  of irradiation, 141
  of photoeviction, 234
Temperature-sensitive paint, 111
Tephra, 197
Terra Amata, 196
Test-dose, 154
Test programme, 241
Thera, 197
Thermal activation, 154
Thermal activation characteristic (TAC), 158
Thermal de-activation, 159
Thermal equilibrium, 139, 179
Thermal lag, 270
Thermal quenching, 53, 169, 270
Thermal radiation, 1
Thermal signal, 189
Thermally-stimulated electron emission (TSEE), 4

Thermoluminescence dosimetry, 21, 95
Thorium-230, 63, 66, 81, 203, 209, 236, 285, 314
Thorium/uranium ratio, 27, 216, 249
Threshold, 88, 300, 312
Tissue, equivalent, 95
Total dose, 9
Transparency, 125, 315
Trap depth, 180, 209, 217, 269
Travel monitor, 98, 266
Travertine, 217
Triboluminescence, 45
Tribo-thermoluminescence, 45
Tunnelling, 55, 275, 276

U

Unbleachable component, 221, 227
Uncollided flux, 289
Underfiring, 9, 37
Uranium/thorium ratio, 27, 216, 249
UV reversal, 157, 187

V

Vacuum, 7, 130
Valence band, 44
Vice traps, 281
Vicing, 18, 25, 92, 192, 204
Volcanic eruptions, 197, 214

W

Waddon Hill, 23
Wakefield compound, 228
Water content, *see* Wetness
Wetness, 22, 69, 74, 77, 195, 236, 243, 244, 245, 250, 265, 267
Wipe tests, 145, 146
Wörms conference, 237
Wratten filters, 232

X

X-ray fluorescence, 109
Xenon lamp, 233

Y

Yttrium-90, 118

Z

Zero-glow monitoring (ZGM), 38, 127, 204, 228
Zero setting, 9, 97, 155, 164, 171, 193, 195, 203, 205, 219, 221
Zinc sulphide, 26, 86, 93, 110
Zircon, 55, 69, 169, 171, 172, 187, 189, 262, 275, 306
Zoning, 172, 175, 206